Nanomaterials in Water Applications

Nanomaterials in Water Applications

Editors

Lei Huang
Hongguo Zhang
Junye Cheng
Zhenxing Wang

MDPI

Basel • Beijing • Wuhan • Barcelona • Belgrade • Novi Sad • Cluj • Manchester

Editors

Lei Huang
Guangzhou University
Guangzhou
China

Hongguo Zhang
Guangzhou University
Guangzhou
China

Junye Cheng
Shenzhen MSU-BIT
University
Shenzhen
China

Zhenxing Wang
Ministry of Ecology and
Environment of the People's
Republic of China
Guangzhou
China

Editorial Office
MDPI AG
Grosspeteranlage 5
4052 Basel, Switzerland

This is a reprint of articles from the Special Issue published online in the open access journal *Nanomaterials* (ISSN 2079-4991) (available at: https://www.mdpi.com/journal/nanomaterials/special_issues/nano_water_app).

For citation purposes, cite each article independently as indicated on the article page online and as indicated below:

Lastname, A.A.; Lastname, B.B. Article Title. *Journal Name* **Year**, *Volume Number*, Page Range.

ISBN 978-3-7258-2219-5 (Hbk)
ISBN 978-3-7258-2220-1 (PDF)
doi.org/10.3390/books978-3-7258-2220-1

Cover image courtesy of Lei Huang

Contents

nanomaterials

Editorial

Nanomaterials in Water Applications: Adsorbing Materials for Fluoride Removal

Lei Huang [1], Kuilin Wan [1], Jia Yan [1], Lei Wang [1], Qian Li [1], Huabin Chen [1], Hongguo Zhang [1,2,*] and Tangfu Xiao [1,3]

1 School of Environmental Science and Engineering, Guangzhou University, Guangzhou 510006, China; huanglei@gzhu.edu.cn (L.H.); klwan1996@163.com (K.W.); jiayan@gzhu.edu.cn (J.Y.); dreamboatrn@163.com (L.W.); qianli@gzhu.edu.cn (Q.L.); xun1597820419@163.com (H.C.); tfxiao@gzhu.edu.cn (T.X.)
2 Guangzhou University-Linköping University Research Center on Urban Sustainable Development, Guangzhou University, Guangzhou 510006, China
3 State Key Laboratory of Geohazard Prevention and Geoenvironment Protection, Chengdu University of Technology, Chengdu 610059, China
* Correspondence: hgzhang@gzhu.edu.cn

Citation: Huang, L.; Wan, K.; Yan, J.; Wang, L.; Li, Q.; Chen, H.; Zhang, H.; Xiao, T. Nanomaterials in Water Applications: Adsorbing Materials for Fluoride Removal. *Nanomaterials* **2021**, *11*, 1866. https://doi.org/10.3390/nano11071866

Received: 15 June 2021
Accepted: 14 July 2021
Published: 20 July 2021

Fluoride is an important pollutant in many countries, such as China, India, Australia, the United States, Ethiopia, etc. Too low concentrations of fluoride cause osteoporosis and saprodontia, leading to the use of toothpaste with fluoride. However, more regions have higher concentrations of fluoride than is needed due to dry weather and geological conditions, especially in industries containing fluoride pollution. The drinking water standard of fluoride is ruled by World Health Organization as 1.5 ppm, and the regulated standard in China is 1.0 ppm [1]. Taking in too much fluoride during a long period brings about Ayers' disease, fluorosis of bone, dental fluorosis, kidney stone, intestinal and liver disorders, etc. Therefore, different treatment technologies are investigated to deal with excessive fluoride.

C. S. Boruff reported the treatment of wastewater containing fluoride by using sufficient calcium hydroxide to precipitate in January 1934 [2]. The precipitation method opens the floodgates to remove fluoride. Adsorption, ion exchange, electric flocculation, membrane technology, solvent extraction, and electro-adsorption were applied to remove fluoride from nature, life, and industries [3–10]. Ion exchange needs ion-exchange resin, which makes it easy to exchange fluoride in an ion exchange column. However, the ion-exchange resin is easy to reach saturation and often needs to regenerate. Electric flocculation makes use of electrical energy that could change metal to metal ions. The metal ions could combine fluoride to bring about flocculation. It brings metallic contamination and power consumption. Fluoride can be obstructed by the pore diameter of a membrane. Membrane fouling is an important risk to the technology. Solvent extraction requires extraction and reverse extraction. The redundant process limits the application.

Adsorption and electro-adsorption use the bonding ability of materials with fluoride. Adsorption is also an important way of dealing with water pollution [11]. Electro-adsorption is the development of adsorption that applies an electric field to enhance the binding capacity of materials for removing fluoride. Adsorbing materials are the dominant factor for improving the adsorption capacity, adsorption rate, high selectivity, range of pH, price, and recycling property. In this paper, we will discuss the adsorbing material of fluoride due to it being most researched in this field This paper covers: (1) the past of adsorbing material from 1930 to 2000: the initial preparation, application for removing fluoride; (2) the present of adsorbing material from 2001 to 2021: modified, mechanism about fluoride removal; (3) the future for developing adsorbents: design, screen for capturing fluoride. This provides the timeline of the development of adsorbing material for dealing with wastewater containing fluoride.

The past of adsorbing material: Ralph H. McKee and William S. Johnston reported that four different carbons were applied to remove fluoride, and some kinds of carbons were more efficient and promising [12]. C. S. Boruff also carried out similar work by using contact beds with the activated alumina or aluminum compounds [2]. This research provided a good attempt and guidance for the possibility of adsorbing fluoride. G. J. FINK and F. K. LINDSAY used Al_2O_3 to adsorb fluoride, and the adsorption capacity could decline with the increase in pH. Others research proved that the addition of magnesium may improve the ability to remove fluoride [13]. Then, many researchers investigated alumina, until now. At present, aluminium oxide is recommended as the optimizing and commercial sorbent by the World Health Organization. H. Farrah and W. F. Pickering reported hydrous iron oxides [14], Toshishige M. SUZUKI applied hydrous zirconium oxide and La/chelating resin [15,16], Chong Mou Wang and Thomas E. Mallouk investigated titanium dioxide [17], and J. Nomura, H. Imai and T. Miyake used hydrous cerium oxide [18] to remove fluoride from wastewater. In this stage, most adsorbing materials were immediately purchased or simply prepared to find the possibility of fluoride removal. This period offered the fundaments and inspiration for the subsequent research in adsorbing fluoride.

The present of adsorbing material: with the development of industry and society, high concentrations of, and complex wastewater containing fluoride need to be disposed. Hence, biomass, thallus, nano-metal oxide, graphene, metal organic frameworks, carbon nanotubes, polymers, layered double hydroxides, and perovskite are applied to improve the performances of removing fluoride [19–27]. There are also other different adsorbing materials for fluoride removal. The developments of adsorptive directions are summarized as: (A) finding new kinds of materials: Sn(II)-TMA metal organic framework (MOF) demonstrated efficient adsorption efficiencies at a wide pH in simulated wastewater [28]; (B) modifying adsorbing materials: $Fe_3O_4/\gamma\text{-}MnO_2$ meso-porous nanocomposite furnished adsorption sites O-Mn-OH to exchange -OH with -F. The study offered an effective modified approach [29]; (C) density functional theory: intrinsic, B-doped, and Al-doped graphene were revealed to affect the adsorption changes of F^- ions and HF molecules. Among them, the Al-doped graphene was more easily combined with F^- ions. The molecule of HF could be only chemisorbed on Al-doped graphene [30]; (D) complicated wastewater: Acinetobacter H12 was investigated to deal with wastewater containing calcium, fluoride, and nitrate [31]. Layered double oxides were studied for remediating the industrial wastewater containing manganese and fluoride [32]; (E) technological upgrade: $Ti(OH)_4$/activated carbon was used as the electrode to remove fluoride that can reach 115.2 mg/g when a voltage of +1.2 V was investigated and regenerated quickly at −1.6 V [33].

The future for developing adsorbents: There were more efforts to understand the process of fluoride clearly. (1): Mechanisms and models: the mechanism of fluoride can be revealed more clearly with the development of materials' characterization and theoretical calculation, such as synchrotron radiation source, cryoelectron microscopy, etc. The in-depth mechanism can generate adsorption models for removing fluoride; (2): electrosorption: this newly developed technology will have caused more interest. Many problems need to be further investigated; (3): complicated wastewater: wastewater contains too many constituents for modern society. The high selectivity and specificity of fluoride are one goal for complicated wastewater. What is more, the simultaneous adsorption of contaminants is also another goal; (4): machine learning and artificial intelligence will be important tools to design the adsorbing materials. They offer an efficient and goal-oriented way to sieve adsorbing materials and modified crystal structures; (5): advanced materials: the advanced materials will open a new possibility to exhibit new series adsorbing materials, for example, MXene, graphdiyne, etc. These design ideas will be applied to modify advanced materials to improve the capability of fluoride removal.

Humans and nature are finding the equilibrium between fluoride and health. It requires the discovery of new technology and material. Urbanization keeps the urban drinking water lower than the fluoride drinking standard of 1 ppm. However, rural drinking water is still confronted with a great challenge from industrial urban development. It should utilize the resource of fluoride. This Special Issue is built to promote the researchers who deliver these thoughts, ideas, and discoveries of nanomaterials in water applications. Thank you to everyone who wants to, or can contribute to this Special Issue.

Author Contributions: Writing—original draft preparation, L.H.; reading and collecting, K.W., J.Y.; discussion, L.W., Q.L., H.C., T.X.; design, H.Z. All authors have read and agreed to the published version of the manuscript.

Funding: The work was supported by the National Natural Science Foundation of China (nos. 51208122, 51778156, 51708142), Pearl River S & T Nova Program of Guangzhou (201806010191), Talent Cultivation Program of Guangzhou University (RP2021014, YJ2021005).

Institutional Review Board Statement: Not applicable.

Informed Consent Statement: Not applicable.

Data Availability Statement: Data sharing not applicable.

Conflicts of Interest: The authors declare no conflict of interest.

References

1. Wen, T.; Hong, Z.; Xiang, B.P.; Peng, W.; Hai, Y.W.; Lei, H. Removal of fluoride from wastewater solution using Ce-AlOOH with oxalic acid as modification. *J. Hazard. Mater.* **2020**, *384*, 121373.
2. Boruff, C.S. Removal of Fluorides from Drinking Waters. *Ind. Eng. Chem.* **1934**, *26*, 69–71. [CrossRef]
3. Aliaskari, M.; Schäfer, A.I. Nitrate, arsenic and fluoride removal by electrodialysis from brackish groundwater. *Water Res.* **2021**, *190*, 116683. [CrossRef]
4. Castañeda, L.F.; Rodríguez, J.F.; Nava, J.L. Electrocoagulation as an affordable technology for decontamination of drinking water containing fluoride: A critical review. *Chem. Eng. J.* **2021**, *413*, 127529. [CrossRef]
5. Grzegorzek, M.; Majewska-Nowak, K.; Ahmed, A.E. Removal of fluoride from multicomponent water solutions with the use of monovalent selective ion-exchange membranes. *Sci. Total Environ.* **2020**, *722*, 137681. [CrossRef] [PubMed]
6. Park, G.; Hong, S.P.; Lee, C.; Lee, J.; Yoon, J. Selective fluoride removal in capacitive deionization by reduced graphene ox-ide/hydroxyapatite composite electrode. *J. Colloid Interf. Sci.* **2021**, *581*, 396–402. [CrossRef] [PubMed]
7. Guo, Y.; Li, C.; Gong, Z.; Guo, Y.; Wang, X.; Gao, B.; Qin, W.; Wang, G. Photocatalytic decontamination of tetracycline and Cr(VI) by a novel α-FeOOH/FeS_2 photocatalyst: One-pot hydrothermal synthesis and Z-scheme reaction mechanism insight. *J. Hazard. Mater.* **2020**, *397*, 122580. [CrossRef] [PubMed]
8. Raghav, S.; Nair, M.; Kumar, D. Tetragonal prism shaped Ni-Al bimetallic adsorbent for study of adsorptive removal of fluoride and role of ion-exchange. *Appl. Surf. Sci.* **2019**, *498*, 143785. [CrossRef]
9. Sandoval, M.A.; Fuentes, R.; Thiam, A.; Salazar, R. Arsenic and fluoride removal by electrocoagulation process: A general review. *Sci. Total Environ.* **2021**, *753*, 142108. [CrossRef]
10. He, Y.; Huang, L.; Song, B.; Wu, B.; Yan, L.; Deng, H.; Yang, Z.; Yang, W.; Wang, H.; Liang, Z.; et al. Defluorination by ion exchange of SO_4^{2-} on alumina surface: Adsorption mechanism and kinetics. *Chemosphere* **2021**, *273*, 129678. [CrossRef]
11. Zhao, Y.; Xia, K.; Zhang, Z.; Zhu, Z.; Guo, Y.; Qu, Z. Facile Synthesis of Polypyrrole-Functionalized $CoFe_2O_4$@SiO_2 for Removal for Hg(II). *Nanomaterials* **2019**, *9*, 455. [CrossRef]
12. Ralph, H.M.; William, S.J. Removal of Fluorides from Drinking Water. *Ind. Eng. Chem.* **1934**, *26*, 849–851.
13. Fink, G.J.; Lindsay, F.K. Activated alumina for removing fluorides from drinking water. *Ind. Eng. Chem.* **1936**, *28*, 947–948. [CrossRef]
14. Farrah, H.; Pickering, W. Interaction of dilute fluoride solutions with hydrous iron oxides. *Aust. J. Soil Res.* **1986**, *24*, 201–208. [CrossRef]
15. Suzuki, T.M.; Chida, C.; Kanesato, M.; Yokoyama, T. Removal of Fluoride Ion by a Porous Spherical Resin Loaded with Hydrous Zirconium Oxide. *Chem. Lett.* **1989**, *18*, 1155–1158. [CrossRef]
16. Kanesato, M.; Yokoyama, T.; Suzuki, T.M. Selective Adsorption of Fluoride Ion by La(III)-loaded Chelating Resin Having Phosphonomethylamino Groups. *Chem. Lett.* **1988**, *17*, 207–210. [CrossRef]
17. Chong, M.W.; Thomas, E.M. Wide-Range Tuning of the Titanium Dioxide Flat-Band Potential by Adsorption of Fluoride and Hydrofluoric Acid. *J. Phys. Chem.* **1990**, *94*, 4276–4280.
18. Nomura, J.; Imai, H.; Miyake, T. Removal of Fluoride Ion from Wastewater by a Hydrous Cerium Oxide Adsorbent. *ACS Symp. Ser.* **1990**, *422*, 157–172. [CrossRef]
19. Bessaies, H.; Iftekhar, S.; Asif, M.B.; Kheriji, J.; Necibi, C.; Sillanpää, M.; Hamrouni, B. Characterization and physicochemical aspects of novel cellulose-based layered double hydroxide nanocomposite for removal of antimony and fluoride from aqueous solution. *J. Environ. Sci.* **2021**, *102*, 301–315. [CrossRef] [PubMed]
20. Huang, L.; Yang, Z.; Lei, D.; Liu, F.; He, Y.; Wang, H.; Luo, J. Experimental and modeling studies for adsorbing different species of fluoride using lanthanum-aluminum perovskite. *Chemosphere* **2021**, *263*, 128089. [CrossRef] [PubMed]

21. Kuang, L.; Liu, Y.; Fu, D.; Zhao, Y. FeOOH-graphene oxide nanocomposites for fluoride removal from water: Acetate mediated nano FeOOH growth and adsorption mechanism. *J. Colloid Interface Sci.* **2017**, *490*, 259–269. [CrossRef]
22. Ruan, Z.; Tian, Y.; Ruan, J.; Cui, G.; Iqbal, K.; Iqbal, A.; Ye, H.; Yang, Z.; Yan, S. Synthesis of hydroxyapatite/multi-walled carbon nanotubes for the removal of fluoride ions from solution. *Appl. Surf. Sci.* **2017**, *412*, 578–590. [CrossRef]
23. Su, R.; Xie, C.; Alhassan, S.I.; Huang, S.; Chen, R.; Xiang, S.; Wang, Z.; Huang, L. Oxygen Reduction Reaction in the Field of Water Environment for Application of Nanomaterials. *Nanomaterials* **2020**, *10*, 1719. [CrossRef]
24. Talebi, S.S.; Javid, A.B.; Roudbari, A.A.; Yousefi, N.; Ghadiri, S.K.; Shams, M.; Khaneghah, A.M. Defluoridationof drinking water by metal impregnated multi-layer green graphene fabricated from trees pruning waste. *Environ. Sci. Pollut. Res.* **2021**, *28*, 18201–18215. [CrossRef]
25. Wan, K.; Huang, L.; Yan, J.; Ma, B.; Huang, X.; Luo, Z.; Zhang, H.; Xiao, T. Removal of fluoride from industrial wastewater by using different adsorbents: A review. *Sci. Total Environ.* **2021**, *773*, 145535. [CrossRef]
26. Zhu, X.-H.; Yang, C.-X.; Yan, X.-P. Metal-organic framework-801 for efficient removal of fluoride from water. *Microporous Mesoporous Mater.* **2018**, *259*, 163–170. [CrossRef]
27. Wang, Z.; Ali, A.; Su, J.; Hu, X.; Zhang, R.; Yang, W.; Wu, Z. Batch fluidized bed reactor based modified biosynthetic crystals: Optimization of adsorptive properties and application in fluoride removal from groundwater. *Chemosphere* **2021**, *281*, 130841. [CrossRef]
28. Arnab, G.; Gopal, D. Green synthesis of a novel water-stable Sn(II)-TMA metal-organic framework (MOF): An efficient adsor-bent for fluoride in aqueous medium in a wide pH range. *New J. Chem.* **2020**, *44*, 1354.
29. Zhao, Z.; Geng, C.; Yang, C.; Cui, F.; Liang, Z. A novel flake-ball-like magnetic Fe_3O_4/γ-MnO_2 meso-porous nano-composite: Adsorption of fluorinion and effect of water chemistry. *Chemosphere* **2018**, *209*, 173–181. [CrossRef] [PubMed]
30. Chen, T.; An, L.; Jia, X. A DFT-based analysis of adsorption properties of fluoride anion on intrinsic, B-doped, and Al-doped graphene. *J. Mol. Model.* **2021**, *27*, 56. [CrossRef] [PubMed]
31. Ali, A.; Wu, Z.; Li, M.; Su, J. Carbon to nitrogen ratios influence the removal performance of calcium, fluoride, and nitrate by Acinetobacter H12 in a quartz sand-filled biofilm reactor. *Bioresour. Technol.* **2021**, *333*, 125154. [CrossRef] [PubMed]
32. Teixeira, M.A.; Mageste, A.B.; Dias, A.; Virtuoso, L.S.; Siqueira, K.P. Layered double hydroxides for remediation of industrial wastewater containing manganese and fluoride. *J. Clean. Prod.* **2018**, *171*, 275–284. [CrossRef]
33. Li, Y.; Zhang, C.; Jiang, Y.; Wang, T. Electrically enhanced adsorption and green regeneration for fluoride removal using $Ti(OH)_4$-loaded activated carbon electrodes. *Chemosphere* **2018**, *200*, 554–560. [CrossRef] [PubMed]

nanomaterials

MDPI

Article

Studies on Kinetics, Isotherms, Thermodynamics and Adsorption Mechanism of Methylene Blue by N and S Co-Doped Porous Carbon Spheres

Yongpeng Ren [1,2], Feng Chen [3], Kunming Pan [1,2], Yang Zhao [2], Lulu Ma [3] and Shizhong Wei [1,2,*]

[1] School of Materials Science and Engineering, Henan University of Science and Technology, Luoyang 471003, China; REN_YP123@163.com (Y.R.); pankunming2008@haust.edu.cn (K.P.)
[2] Henan Key Laboratory of High-Temperature Structural and Functional Materials, National Joint Engineering Research Center for Abrasion Control and Molding of Metal Materials, Henan University of Science and Technology, Luoyang 471003, China; gloryfire@126.com
[3] School of Environmental and Biological Engineering, Henan University of Engineering, Zhengzhou 451191, China; chenfeng871588@163.com (F.C.); malulu1001@163.com (L.M.)
* Correspondence: hnwsz@126.com

Abstract: Heteroatom-doped carbon is widely used in the fields of adsorbents, electrode materials and catalysts due to its excellent physicochemical properties. N and S co-doped porous carbon spheres (N,S-PCSs) were synthesized using glucose and L-cysteine as carbon and heteroatom sources using a combined hydrothermal and KOH activation process. The physicochemical structures and single-factor methylene blue (MB) adsorption properties of the N,S-PCSs were then studied. The optimized N,S-PCSs-1 possessed a perfect spherical morphology with a 2–8-μm diameter and a large specific area of 1769.41 m^2 g^{-1}, in which the N and S contents were 2.97 at% and 0.88 at%, respectively. In the single-factor adsorption experiment for MB, the MB adsorption rate increased with an increase in carbon dosage and MB initial concentration, and the adsorption reached equilibrium within 2–3 h. The pseudo-second-order kinetic model could excellently fit the experimental data with a high R^2 (0.9999). The Langmuir isothermal adsorption equation fitted well with the experimental results with an R^2 value of 0.9618, and the MB maximum adsorption quantity was 909.10 mg g^{-1}. The adsorption of MB by N,S-PCSs-1 was a spontaneous, endothermic, and random process based on the thermodynamics analyses. The adsorption mechanism mainly involved Van der Waals force adsorption, π-π stacking, hydrogen bonds and Lewis acid–base interactions.

Keywords: methylene blue; porous carbon spheres; heteroatom-doping; adsorption; mechanism

Citation: Ren, Y.; Chen, F.; Pan, K.; Zhao, Y.; Ma, L.; Wei, S. Studies on Kinetics, Isotherms, Thermodynamics and Adsorption Mechanism of Methylene Blue by N and S Co-Doped Porous Carbon Spheres. *Nanomaterials* **2021**, *11*, 1819. https://doi.org/10.3390/nano11071819

Academic Editor: George Z. Kyzas

Received: 5 June 2021
Accepted: 6 July 2021
Published: 13 July 2021

1. Introduction

Dyes are widely used in textile, plastics, and the paper and pulp industries [1,2]. The characteristics of printing and dyeing wastewater, such as complex water quality, high COD value, poor biodegradability, low light permeability, and carcinogenesis and mutagenesis, lead to allergic dermatitis, skin allergy, cancer and gene mutations in the human body [3,4]. New printing and dyeing technology also aggravate wastewater treatment. Therefore, it is urgent to solve the problem of printing and dyeing wastewater treatment, as well as to provide emission standards. To date, adsorption, reverse osmosis, precipitation, biological treatment, and other decoloring technologies have been widely used to deal with printing and dyeing wastewater [4,5]. The adsorption method possesses the advantages of a high efficiency, environmental protection, simple operation, and low cost, which draws wide interest from researchers all over the world [3,6]. A new type adsorbent with a large specific surface area, high adsorption capacity, fast adsorption rate, and special surface reactivity is essential to enhance the performance of wastewater treatment.

Porous carbon has attracted broad interests because of its advantages, such as its excellent chemical and thermal stability, good mechanical stability, controllable pore structure, and high specific surface area [7,8]. In recent years, surface functionalized heteroatomic carbon materials have already been widely used as catalysts, adsorbents and as energy storage materials due to their unique physical and chemical properties. For example, Zhu's team synthesized highly nitrogen-doped hollow carbon nanoparticles using a facile one-pot method, which showed excellent electrocatalytic activity for triiodide reduction in dye-sensitized solar cells, which was better than conventional platinum catalysts [9]. Zhang's group demonstrated that *Pennisetum alopecuroides*-derived urea-modified activated carbon could improve the adsorption of Ni (II) from aqueous solutions because of its abundant surface nitrogen-containing functional groups [10]. Wang et al. reported that nitrogen-doped porous carbon nanosheets could lead to the effective immobilization of polysulfides and a simultaneous improvement in the reaction kinetics of sulfur species in lithium–sulfur batteries [11].

Apart from single-atom-doped carbon materials, doping with two or more kinds of heteroatoms in a carbon framework is considered to be more conducive to improving the physical and chemical properties of carbon materials, thus further broadening their application fields. Some recent studies have demonstrated this viewpoint, such as nitrogen and sulfur co-doped micro-mesoporous carbon sheets for improving Cr(VI) adsorption through synergistic effects [12], the efficient removal of methylene blue with nitrogen and oxygen co-doped three-dimensional honeycomb porous carbons [13], enhanced CO_2 uptake and outstanding methylene blue adsorption capacities by heteroatom nitrogen and oxygen-doped porous carbon materials [14], cobalt and nitrogen co-doped porous carbon materials for enhanced capacitive deionization [15], and sulfur/nitrogen/oxygen tri-doped hierarchical porous carbon as high quality sulfur hosts in lithium–sulfur batteries [16]. Therefore, inspired by the aforementioned works, this study aimed to synthesize N and S co-doped porous carbon spheres (N,S-PCSs) derived from biomass for the treatment of printing and dyeing wastewater.

In this work, N and S co-doped porous carbon spheres (N,S-PCSs) were synthesized using glucose and L-cysteine as the carbon source and doping agent using a facile hydrothermal-KOH activation two-step method. The structure and physicochemical properties of N,S-PCSs were tested, and the adsorption properties of organic dyes in aqueous solution by N,S-PCSs were investigated in detail using methylene blue (MB) as a model dye. In addition, the adsorption kinetics, isotherms, thermodynamics, and adsorption mechanism of MB by N,S-PCSs were also discussed. This work provides an insight into sustainable biomass resources utilization and dye treatments, which could be simultaneously beneficial for emission reduction and wastewater reduction.

2. Materials and Methods

2.1. Synthesis of N,S-PCSs

Typically, for N,S-PCSs-1 preparation, 7.2 g glucose and 0.72 g L-cysteine were dissolved in 80 mL distilled water and stirred for 30 min. Then the solution was injected into the Teflon linings (100 mL) with a stainless-steel reactor. The reactor was heated at 180 °C for 12 h to finish the hydrothermal reaction, afterward, it was filtered and washed with distilled water and dried at 80 °C, the hydrothermal carbon was obtained. The hydrothermal carbon was heated in a Ni crucible under a N_2 atmosphere at 400 °C for 1 h with a heating rate of 5 °C min^{-1} to obtain the carbonized samples. The carbonized samples and KOH were mixed at a mass ratio of 1:4, and then activated at 800 °C for 1 h under a N_2 atmosphere with a heating rate of 5 °C min^{-1}. When cooled to room temperature, the activated samples were washed with hydrochloric acid, filtered and washed with distilled water until a pH = 7 was reached, and then dried at 80 °C for 24 h. The resulting sample was marked as N,S-PCSs-1. As a contrast, the PCSs, N,S-PCSs-2 and N,S-PCSs-3 samples were also synthesized using the same method described above except with changing the L-cysteine masses to 0 g, 1.44 g and 2.88 g, respectively. Detailed information on the reagents and material characterization are shown in the Supplementary Materials (SM).

2.2. MB Adsorption Experiments

A total of 1 g of MB was dissolved into 1 L of distilled water to obtain the MB reserve solution (1 g L^{-1}), and different concentrations of MB adsorption solutions were prepared by diluting the MB reserve solution. As with typical MB adsorption experiments, a certain mass of adsorbent was added to 50 mL of the MB solutions in conical bottles and oscillated on a thermostatic oscillator at 120 r min^{-1}. After oscillation, the supernatant liquid was collected, and the concentration of residual MB solution was measured by using a UV-Vis spectrophotometer at a wavelength of 664 nm [17,18]. The standard calibration curves of the MB solutions were plotted with a high related coefficient R^2 of 0.9993.

To evaluate the adsorption properties of different adsorbents, 4 mg of PCSs, N,S-PCSs-1, N,S-PCSs-2 and N,S-PCSs-3 were added into 5 mg L^{-1} of MB solution and oscillated at 298 K for 16 h. To confirm the optimum mass of adsorbent for removal of MB, different weights (5–15 mg) of N,S-PCSs-1 were added to 80 mg L^{-1} of MB solution and oscillated at 298 K for 16 h.

The experiments on the effect of time were carried out, ranging from 1 min to 24 h, at 298 K, containing MB with initial concentrations of 80 mg L^{-1} and 8 mg of N,S-PCSs-1, respectively. In order to explore the mechanism of the MB adsorption process, the pseudo-second-order and the internal diffusion kinetic models were used to fit the adsorption dynamics data, and the relevant equations were shown in the SM.

The initial MB concentration and temperature experiments were conducted with initial MB concentrations from 10 to 400 mg L^{-1} at 298, 308, and 318 K, respectively. The quality of N,S-PCSs-1 and contact time were set to 8 mg and 16 h, respectively. The Langmuir and Freundlich isotherm models were used to simulate the MB adsorption, and the thermodynamic parameters (ΔG^{θ}, ΔH^{θ} and ΔS^{θ}) were also calculated based on the isotherm data. The isotherms and thermodynamic equations are detailed in the SM.

The above-mentioned MB batch adsorption experiments were carried out as triplicate independent samplings, and all the experimental data were displayed with average values with standard deviations (less than 5%) as error bars. The distinctions between the experiment groups and the control group were tested for significance using a t-test at a significant level of 0.05.

The adsorption rate (η, %) was calculated using Equation (1) (where C_0 is the MB concentration before adsorption, mg L^{-1}, C_i is the MB concentration after adsorption, mg L^{-1}).

$$\eta = \frac{C_0 - C_i}{C_0} \times 100\% \quad (1)$$

The adsorption quantity (Q, mg g^{-1}) was calculated using Equation (2) (where V was the volume of the MB solution, L, and m is the adsorbent quality, mg).

$$Q = \frac{V(C_0 - C_i)}{m} \times 1000 \quad (2)$$

3. Results and Discussion

The SEM images of PCSs, N,S-PCSs-1, N,S-PCSs-2 and N,S-PCSs-3 are shown in Figure 1. The PCS sample synthesized without L-cysteine was composed of spheres with diameters of about 300–400 nm (Figure 1a). As shown in Figure 1b, N,S-PCSs-1 was composed of smooth microspheres with diameters of about 2–8 μm, which was larger than PCS due to the addition of L-cysteine. As shown in Figure 1c, microspheres with lamellar structures on their surfaces and blocks could be discovered in the N,S-PCSs-2 sample. As a contrast, N,S-PCSs-3 with excess L-cysteine addition was composed of blocks with inverse opal structures (Figure 1d). The TEM image of N,S-PCSs-1 (Figure S1) confirmed its smooth surface and spherical morphology, corresponding to its SEM image (Figure 1b). It was demonstrated that L-cysteine addition could gradually change the morphology of the carbon material, from spheres to blocks.

Figure 1. SEM images of (**a**) PCSs, (**b**) N,S-PCSs-1, (**c**) N,S-PCSs-2 and (**d**) N,S-PCSs-3.

Table 1 summarizes the elemental contents of N and S in N,S-PCSs-1, N,S-PCSs-2 and N,S-PCSs-3 samples according to the results of EDS analyses. Based on the elemental analyses, the N atomic percentages of the N,S-PCSs-1, N,S-PCSs-2, and N,S-PCSs-3 samples were 0.34 at%, 0.92 at%, and 1.57 at%, respectively. The S atomic percentages of the three samples were 0.4 at%, 2.75 at%, and 3.3 at%, respectively. It was demonstrated that the N and S elemental contents increased with the increasing quality of L-cysteine doping agent.

Table 1. Elemental contents of N,S-PCSs-1, N,S-PCSs-2 and N,S-PCSs-3.

Samples	Type	Elemental Contents	
		N	S
N,S-PCSs-1	Weight percentage (Atomic percentage)	0.39 −0.34	1.05 −0.4
N,S-PCSs-2	Weight percentage (Atomic percentage)	0.98 −0.92	6.7 −2.75
N,S-PCSs-3	Weight percentage (Atomic percentage)	1.71 −1.57	8.24 −3.3

XRD patterns of the four samples are shown in Figure 2a. The PCSs sample without N and S only showed broad peaks, corresponding to the amorphous carbon [19]. N,S-PCSs-1 with proper N and S dopant exhibited a carbon phase (JCPDS#43-1104 and #50-1083). In addition, the N,S-PCSs-2 and N,S-PCSs-3 with excess dopant exhibited additional carbon phases (JCPDS#50-0927).

Based on the Raman spectra in Figure 2b, the PCSs, N,S-PCSs-1, N,S-PCSs-2, and N,S-PCSs-3 samples exhibited two peaks at ~1580 cm^{-1} and ~1340 cm^{-1}, which corresponded to the characteristic sp^2 peak (G peak) and a disordered peak (D peak), respectively. The I_D/I_G values (R value) were calculated to evaluate their graphitization degree. The R values of PCSs, N,S-PCSs-1, N,S-PCSs-2, and N,S-PCSs-3 were 1.12, 1.13, 1.15, and 1.28, respectively. The R value increased with the increase in the quality of L-cysteine (as well as increasing N/S contents), indicating the N and S dopant contributed to the decrease in the graphite structure and the increase in defects [20,21].

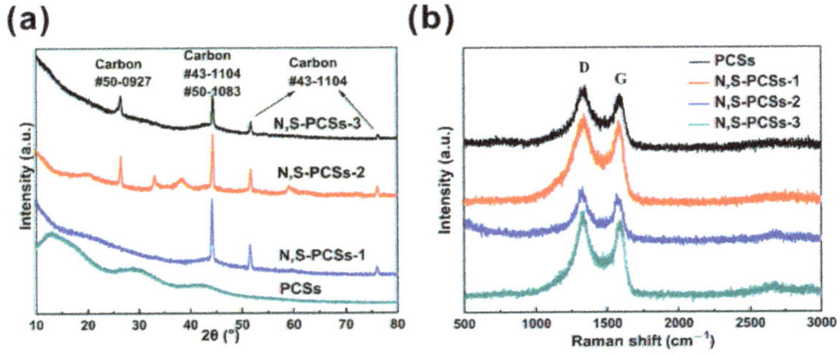

Figure 2. (**a**) XRD and (**b**) Raman spectra of PCSs, N,S-PCSs-1, N,S-PCSs-2 and N,S-PCSs-3.

According to the infrared spectroscopy of N,S-PCSs-1 (Figure S2), the peak at 1105 cm^{-1} referred to S=C vibration. The peak at 1383 cm^{-1} corresponded to -COOH and C-C stretching. The peak at 1627 cm^{-1} referred to -N=N, -O-NO_2 and -C=C stretching. The peak at 3436 cm^{-1} referred to -NH and -OH stretching [22,23]. It was demonstrated that N,S-PCSs-1 possessed the oxygen/sulfur/nitrogen-containing functional groups.

The N_2 adsorption/desorption curves and pore size distribution curve of N,S-PCSs-1 are shown in Figure 3. The adsorption/desorption curves exhibited a strong adsorption capacity under a low relative pressure, indicating the adsorption mechanism of N,S-PCSs-1 micropores and monolayers (Figure 3a). The hysteresis loop demonstrated the mesopores in N,S-PCSs-1 [24]. The pore size distribution of N,S-PCSs-1 indicated its pores were mainly distributed in micropores (diameter less than 2 nm) and mesopores between 2–4 nm (Figure 3b). As a contrast, the pores of N,S-PCSs-3 were mainly distributed in micropores (Figure S3). In addition, the specific surface area and pore volume of N,S-PCSs-1 were 1769.41 $m^2\ g^{-1}$ and 1.23 $cm^3\ g^{-1}$, respectively. In contrast, N,S-PCSs-3 possessed a specific surface area of ~1100 $m^2\ g^{-1}$, which was much lower than N,S-PCSs-1. The excess L-cysteine probably led to the reduction in the specific area. Such a hierarchical micro/mesoporous structure of N,S-PCSs-1 could provide a more favorable micro-environment for transporting paths for MB, capturing more contaminants via Van der Waals forces and trapping them by physical/chemical barriers, which might endow its good adsorption properties and could be a potential candidate for contaminant remediation [25–27].

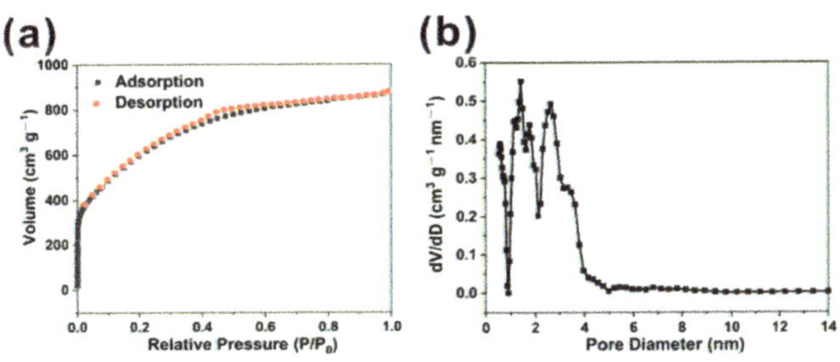

Figure 3. (**a**) N_2 adsorption/desorption curves and (**b**) pore size distribution curve of N,S-PCSs-1.

Factors such as different adsorbents, carbon amount, contact time, the MB initial concentration, and temperature were used to explore the influence on MB adsorption properties. The results are shown in Figure 4. The adsorption rates of the different carbon materials were tested under the same adsorption conditions (Figure 4a). The PCSs, N,S-PCSs-1, N,S-PCSs-2, and N,S-PCSs-3 exhibited adsorption rates of 97.24%, 99.64%, 96.16%, and 92.94%, respectively. Obviously, N,S-PCSs-1 was the optimized carbon material. According to the above characterization analysis, it could be concluded that N,S-PCSs-1 had moderate N and S contents (compared to PCSs), as well as higher specific area and pore structure (compared to N,S-PCSs-2 and N,S-PCSs-3). The adsorption performance of PCSs was restricted by its low N and S. The adsorption performances of N,S-PCSs with higher N and S contents were restricted by their low specific areas and unoptimized pore structures, which might be why it has good MB adsorption performance. Hence, the following research in this work is focused on N,S-PCSs-1.

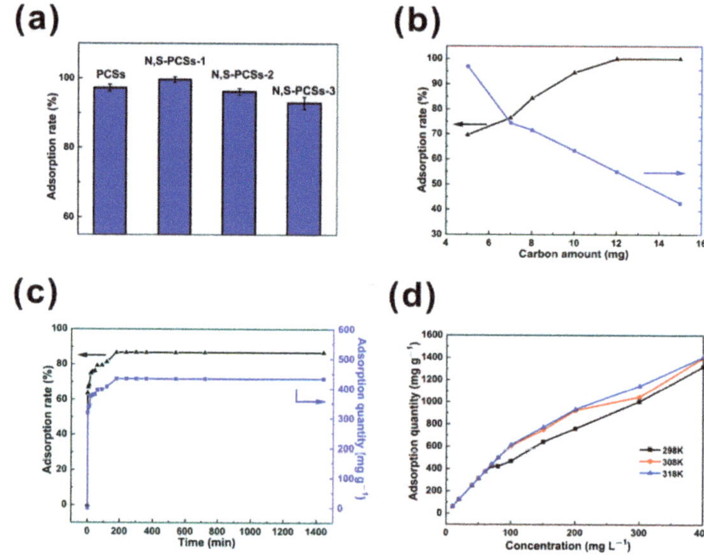

Figure 4. (**a**) Adsorption rates of PCSs, N,S-PCSs-1, N,S-PCSs-2, and N,S-PCSs-3: influences of (**b**) carbon amount, (**c**) contact time on adsorption rate and quantity of MB by N,S-PCSs-1, (**d**) influences of MB initial concentration and temperature on adsorption quantity of MB by N,S-PCSs-1.

The adsorption rate and quantity with different amounts of N,S-PCSs-1 were investigated, as shown in Figure 4b. The adsorption rate increased from 69.60% to 84.27% with a carbon amount increased from 5 to 8 mg. A total of 10 mg of N,S-PCSs-1 exhibited an adsorption rate of 94.47%. The adsorption rate showed a minor growth from 99.96% to 100% when the carbon amount increased from 12 to 15 mg. With increasing N,S-PCSs-1 amounts, the adsorption rate increased by ~30%; on the other hand, adsorption quantity was reduced by ~300 mg g^{-1}.

The effect of contact time on the adsorption property of MB by N,S-PCSs-1 was investigated (Figure 4c). With the adsorption time increasing from 1 to 120 min, the adsorption rate and quantity exhibited a drastic increase. The adsorption rate increased from 63.58% to 81.27%, and the adsorption quantity increased from 317.88 to 406.33 mg g^{-1}. With the adsorption time increasing from 120 to 180 min, the adsorption rate and quantity exhibited a minor increase. The adsorption rate increased from 81.27% to 86.61%, and the adsorption quantity increased from 406.33 to 433.03 mg g^{-1}. With the adsorption time exceeding 180 min, the adsorption rate was maintained at ~86%, and the adsorption quantity was maintained at 433 mg g^{-1}; hence, the adsorption equilibrium time of MB by N,S-PCSs-1 could be 180 min.

The influences of MB initial concentration and temperature on the adsorption quantity of MB by N,S-PCSs-1 were investigated, as shown in Figure 4d. At a temperature of 298 K, with the initial MB concentration being less than 60 mg L^{-1}, the adsorption quantity increased from 62.5 to 375 mg g^{-1}. With the initial MB concentration increasing from 70 to 400 mg L^{-1}, the adsorption quantity increased from 420.9 to 1318.5 mg g^{-1}. With the initial MB concentration increasing to 400 mg L^{-1}, the adsorption quantity approach its highest value. This phenomenon might be attributed to the hierarchical porous structures, because the MB adsorption of the micro-pores was faster than that of the meso-pores. From Figure 4d, we can also see that the adsorption quantity increased with an increase in temperature. The adsorption quantity exhibited a significant increase at 298 to 308 K, nevertheless, there was a minor increase at 308 to 318 K, demonstrating that MB absorption by N,S-PCSs-1 was an endothermic process [28,29].

Based on time-dependent adsorption data (see Figure 4c), the pseudo-second-order and internal diffusion kinetics models were used to simulate MB adsorption by N,S-PCSs-1, the results are shown in Figure 5a,b and Table 2. Based on the pseudo-second-order model (Figure 5a), the calculated q_e was 434.78 mg g^{-1}, which was well in accordance with the experimental value q_{exp} (433.01 mg g^{-1}). The calculated K_2 was 0.00067, and the R^2 was 0.9999. This manifested in the adsorption kinetics of MB by N,S-PCSs-1 being well described by the pseudo-second-order model, and the adsorption process was chemisorptions with all kinds of interactions, such as H-bond formations, electrostatic attractions, Van der Waals forces, etc. [27,30]. Based on the internal diffusion model, the adsorption process was divided into three stages (Figure 5b). Stage 1, 2 and 3 corresponded to the fast physisorption by micro-pores, the chemisorption by the functional groups on the surfaces of the N,S-PCSs-1 and the slow adsorption near equilibrium, respectively [12,31]. For all the tested MB concentrations, the lines of the three stages did not go through the origin of coordinates, suggesting that internal diffusion was not the sole rate-controlling step [31].

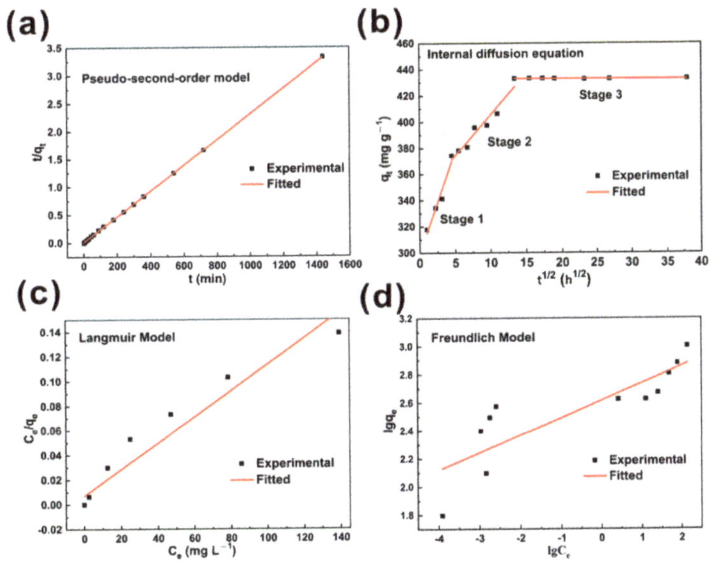

Figure 5. Experimental and fitted data of adsorption kinetics based on (**a**) pseudo-second-order model and (**b**) internal diffusion model; experimental and fitted data of adsorption isotherms based on the (**c**) Langmuir model and (**d**) the Freundlich model.

Table 2. Fitted data of adsorption kinetics and isotherms.

	(Model) Equation	Fitted Parameter		Fitted Equation
Adsorption kinetics	(Pseudo-second-order) $\frac{t}{q_t}=1/k_2q_e^2+t/q_e$ $x=t, y=t/q_t$	k_2	0.00067	$y = 0.0023x + 0.0079$
		q_e (mg g^{-1})	434.78	
		q_{exp} (mg g^{-1})	433.01	
		R^2	0.9999	
	(Internal diffusion) $q_t=k_pt^{\frac{1}{2}}+C$ $x=t^{1/2}, y=q_t$	(Stage 1) k_p	15.76	$y = 15.756x + 299.18$
		(Stage 1) R^2	0.9469	
		(Stage 2) k_p	6.25	$y = 6.2491x + 343.16$
		(Stage 2) R^2	0.9446	
		(Stage 3) k_p	−0.011	$y = -0.0106x + 433.24$
		(Stage 3) R^2	0.9786	
Adsorption isotherms	(Langmuir model) $\frac{C_e}{q_e}=\frac{1}{bq_m}+\frac{C_e}{q_m}$ $x=C_e$ $y=C_e/q_e$	qm (mg g^{-1})	909.10	$y = 0.0011x + 0.0073$
		b	0.15	
		R^2	0.9410	
	(Freundlich model) $\lg q_e=\lg k+\frac{1}{n}\lg C_e$ $x=\lg C_e, y=\lg q_e$	k	411.81	$y = 0.1237x + 2.6147$
		$1/n$	0.12	
		R^2	0.7720	

The Langmuir and Freundlich isotherm models were used to simulate the MB adsorption by N,S-PCSs-1, and the results are shown in Figure 5c,d and Table 2. The R^2 of the Langmuir model was 0.9410, which was larger than that of the Freundlich model (0.7720), indicating that the MB adsorption by N,S-PCSs-1 could be well described by the Langmuir model [28,32]. This also demonstrated that MB molecules formed a homogenous monolayer coverage on the surface of N,S-PCSs-1 [33]. Based on the Langmuir model, the maximum adsorption quantity of MB by N,S-PCSs-1 was 909.10 mg g^{-1}. According to the Freundlich model, $1/n$ was 0.12, indicating a good affinity between MB and N,S-PCSs-1 [31].

Based on adsorption experiments at different temperatures (Figure 4d), ΔG^θ and K_d were obtained. A fitted linear relation between $\ln K_d$ and $1/T$ was devoted to the calculation of ΔH^θ and ΔS^θ (Figure S4). The results are shown in Table 3. The values of ΔG^θ at 298, 308 and 318 K were less than 0, indicating that MB adsorption was a spontaneous process [34]. The absolute values of ΔG^θ increased with increasing temperature, indicating that the increase in temperature was a benefit to MB adsorption. The value of ΔH^θ was higher than 0, indicating the MB adsorption was an endothermal process [35]. The value of ΔS^θ was larger than 0, suggesting the MB adsorption by N,S-PCSs-1 was a randomness process [36].

Table 3. ΔG^θ, ΔH^θ and ΔS^θ data of MB adsorption by N,S-PCSs-1.

ΔH^θ (kJ mol^{-1})	ΔS^θ (kJ mol^{-1} k^{-1})	ΔG^θ (kJ mol^{-1})		
		298 K	308 K	318 K
5.57	19.76	−0.27	−0.61	−0.66

N,S-PCSs-1 after MB adsorption (N,S-PCSs-1/MB) showed a minor morphology change, indicating its structural stability, which also showed that the adsorption of pores via Van der Waals forces played an important role in MB adsorption (Figure S5) [25]. The infrared spectroscopy and Raman spectroscopy of N,S-PCSs-1 after MB adsorption are shown in Figure 6. The infrared spectroscopy indicated that all of the peaks have blue shift, such as the S=C and C-C peak shifted from 1105 cm^{-1} and 1112 cm^{-1} to 1383 cm^{-1} and

1385 cm^{-1}, respectively (Figure 6a). The blue shift could be attributed to the stress change caused by MB adsorption. In addition, compared to other peaks, the intensity of C-C peak exhibited an extra increase, which was attributed to the extra C-C bond introduced by the adsorbed MB molecules. Moreover, the peaks of -NH and -OH for N,S-PCSs-1 also migrated after MB adsorption, which suggested that the hydrogen bond could be involved in MB adsorption [31]. Infrared spectrum analysis showed that the surface functional groups of N,S-PCSs-1 participated in the adsorption of MB, which was consistent with the results of kinetics [28,31]. The Raman spectrum indicated that the I_D/I_G of N,S-PCSs-1/MB was 1.09, which was lower than that of N,S-PCSs-1 (1.13). This phenomenon could be attributed to the increased sp^3 carbon atom introduced by MB (Figure 6b).

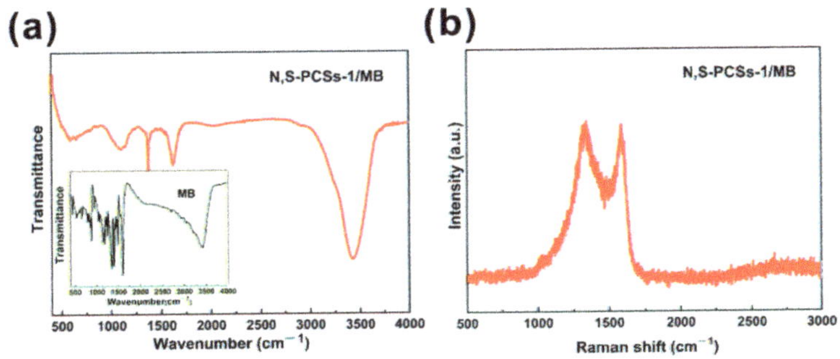

Figure 6. Infrared spectroscopy of (**a**) N,S-PCSs-1/MB and (**a**) inset MB and (**b**) Raman spectrum of N,S-PCSs-1/MB.

In order to further probe the adsorption mechanism of MB by N,S-PCSs-1, XPS tests were carried out. The XPS spectra of N,S-PCSs-1 were shown in Figure 7. According to XPS, the C, O, N and S contents of N,S-PCSs-1 were 89.45 at%, 6.7 at%, 2.97 at%, and 0.88 at%, respectively. The C_{1s} spectrum with a binding energy of ~284 eV was deconvoluted into 4 peaks centered at 283.9 eV, 285.3 eV, 286.8 eV, and 288.4 eV, corresponding to C=C/C–C, C–O/C–N/C–S, C=O, and O–C=O, respectively [31,37]. The N_{1s} spectrum at a binding energy of 398–400 eV was deconvoluted into 3 peaks centered at 397.9 eV, 399.8 eV, and 401.4 eV, corresponding to pyridinic N, pyrrolic N, and graphitic N, respectively [24,37]. The S_{2p} spectrum at a binding energy of ~163 eV was deconvoluted into 3 peaks centered at 163.2 eV, 164.4 eV, and 166.7~168.0 eV, corresponding to C–S, C=S, and -SO_x, respectively [12,38]. The O_{1s} spectrum at a binding energy of ~532 eV was deconvoluted into 3 peaks centered at 531.2 eV, 532.5 eV, and 533.5 eV, corresponding to C=O, C–OH, and C–O–C, respectively [13,39]. According to the above XPS analysis, we can see that N,S-PCSs-1 had a π-π structure in its skeleton and the oxygen/sulfur/nitrogen-containing functional groups on the surface, simultaneously, MB possessed aromatic rings in their structures. Therefore, a π-π stacking interaction should be formed between N,S-PCSs-1 and MB [13]. In addition, the MB structure contained organic ammonium ions, and the N^+ position had the capacity to receive electrons (acting as a Lewis acid), while the N, S and O atoms of N,S-PCSs-1 contained lone pair electrons, which could act as a Lewis base to donate electrons. Thus, the Lewis acid–base interactions between N,S-PCSs-1 and MB could also be beneficial to MB adsorption [12,13,40].

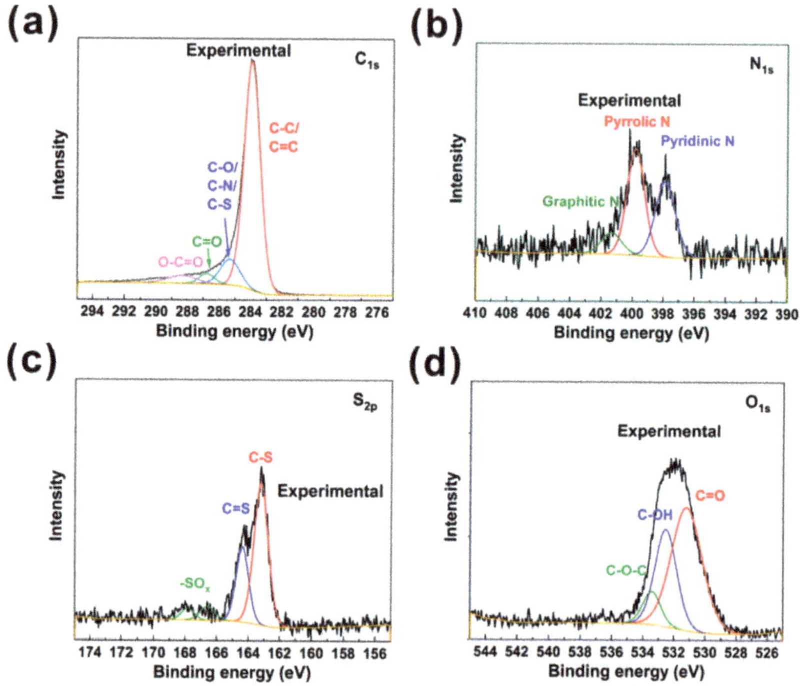

Figure 7. N,S-PCSs-1 XPS spectrums of (**a**) C_{1s} (**b**) N_{1s}, (**c**) S_{2p} and (**d**) O_{1s}.

Therefore, according to the results of the above analysis, the MB adsorption mechanism by N,S-PCSs-1 were proposed to the Van der Waals force adsorption, π-π stacking, hydrogen bond and Lewis acid–base interaction, as shown in Figure 8.

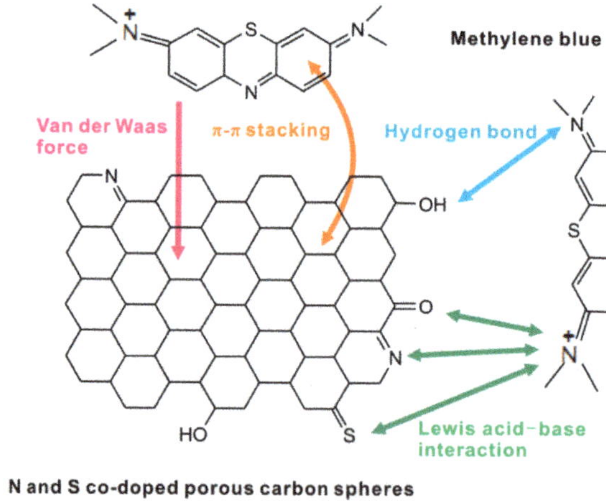

Figure 8. Schematic illustration of the MB adsorption mechanism by N,S-PCSs-1.

4. Conclusions

In conclusion, a facile method was utilized to prepared N and S co-doped porous carbon spheres (N,S-PCSs) using glucose and L-cysteine as the carbon precursor and heteroatom source. The as-prepared carbon material possessed good physical and chemical properties and was able to exhibit good properties for MB adsorption. The MB adsorption onto the optimized N,S-PCSs-1 material was fitted to follow the pseudo-second-order kinetic model and Langmuir isotherm model. The MB adsorption by N,S-PCSs-1 was a monolayer chemisorption, and the maximum adsorption quantity of MB was as high as 909.10 mg g^{-1}. Thermodynamics calculation showed that the MB adsorption was a spontaneous, endothermic, and random process. The mechanism for MB adsorption onto N,S-PCSs-1 was proposed as the Van der Waals force adsorption, π-π stacking, hydrogen bond and Lewis acid–base interactions. We believe that N,S-PCSs-1 is a green and sustainable adsorbent with good removal efficiency for the treatment of printing and dyeing wastewater.

Supplementary Materials: The following are available online at https://www.mdpi.com/article/10.3390/nano11071819/s1, Figure S1: TEM image of N,S-PCSS-1, Figure S2: Infrared spectroscopy of N,S-PCSs-1, Figure S3: N2 adsorption/desorption curves and pore size distribution curve of N,S-PCSs-3, Figure S4: Adsorption thermodynamics of MB by N,S-PCSs-1, Figure S5: SEM image of N,S-PCSs-1/MB, Experimental Details.

Author Contributions: Methodology, F.C. and L.M.; software, Y.R.; validation, S.W.; formal analysis, Y.R., F.C. and Y.Z.; investigation, Y.R.; resources, S.W.; writing—original draft preparation, Y.R.; writing—review and editing, Y.R. and F.C.; visualization, Y.R.; project administration, K.P.; funding acquisition, K.P. All authors have read and agreed to the published version of the manuscript.

Funding: This research was funded by Opening Project of State Key Laboratory for Advanced Metals and Materials, Beijing University of Science and Technology, Grants No. 2020-Z14.

Data Availability Statement: Not applicable.

Acknowledgments: This work would like to extend sincere gratitude to Liu and An, for their instructive advice to the work.

Conflicts of Interest: The authors declare no conflict of interest.

References

1. Silva, M.A.; Hilliou, L.; de Amorim, M.T.P. Fabrication of pristine-multiwalled carbon nanotubes/cellulose acetate composites for removal of methylene blue. *Polym. Bull.* **2020**, *77*, 623–653. [CrossRef]
2. Tran, T.T.; Nguyen, D.T.C.; Le, H.T.N.; Duong, C.D.; Bach, L.G.; Nguyen, H.T.T.; Nguyen, T.D. Facile synthesis of manganese oxide-embedded mesoporous carbons and their adsorbability towards methylene blue. *Chemosphere* **2019**, *227*, 455–461. [CrossRef]
3. Tang, Z.; Hu, X.; Ding, H.; Li, Z.; Liang, R.; Sun, G. Villi-like poly(acrylic acid) based hydrogel adsorbent with fast highly efficient methylene blue removing ability. *J. Colloid Interf. Sci.* **2021**, *594*, 54–63. [CrossRef]
4. Xiao, M.; Yue, H.D.; Feng, X.J.; Wang, Y.T.; He, M.T.; Chen, Q.; Zhang, Z.H. A double-layered neutral cadmium-organic framework for selective adsorption of cationic organic dyes through electrostatic affinity. *J. Solid State Chem.* **2020**, *288*, 121376. [CrossRef]
5. Zheng, S.; Huang, M.; Sun, S.; Zhao, H.; Meng, L.; Mu, T.; Song, J.; Jiang, N. Synergistic effect of MIL-88A/g-C_3N_4 and MoS_2 to construct a self-cleaning multifunctional electrospun membrane. *Chem. Eng. J.* **2021**, *3*, 129621. [CrossRef]
6. Udrea, M.L.; Paduretu, C.C.; Suica-Bunghez, I.R.; Doncea, S.M.; Ion, R.M. Methylene blue removal from residual water using fir wood sawdust as adsorbent. *J. Sci. Arts* **2021**, *1*, 261–272. [CrossRef]
7. Min, H.S. Removal of Dyes from Wastewater by Adsorption onto Activated Carbon: Mini Review. *J. Geosci. Environ. Prot.* **2020**, *8*, 120–131.
8. Zhou, X.; Chen, F.; Yang, J.; Ma, L.; Bai, T.; Long, B.; Liao, Q.; Liu, C. Dual protection of sulfur by interconnected porous carbon nanorods graphene sheets for lithium–sulfur batteries. *J. Electroanal. Chem.* **2015**, *747*, 59–67. [CrossRef]
9. Jia, R.; Chen, J.; Zhao, J.; Zheng, J.; Song, C.; Li, L.; Zhu, Z. Synthesis of highly nitrogen-doped hollow carbon nanoparticles their excellent electrocatalytic properties in dye-sensitized solar cells. *J. Mater. Chem.* **2010**, *20*, 10829–10834. [CrossRef]
10. Kang, Y.; Guo, Z.; Zhang, J.; Xie, H.; Liu, H.; Zhang, C. Enhancement of Ni(II) removal by urea-modified activated carbon derived from Pennisetum alopecuroides with phosphoric acid activation. *J. Taiwan Inst. Chem. E* **2016**, *60*, 335–341. [CrossRef]
11. Guo, C.; Xu, J.; Lv, L.; Chen, S.; Sun, W.; Wang, Y. Two-dimensional imine-based covalent-organic-framework derived nitrogen-doped porous carbon nanosheets for high-performance lithium-sulfur batteries. *New J. Chem.* **2021**, *45*, 8683. [CrossRef]
12. Chen, F.; Zhang, M.; Ma, L.; Ren, J.; Ma, P.; Li, B.; Wu, N.; Song, Z.; Huang, L. Nitrogen and sulfur codoped micro-mesoporous carbon sheets derived from natural biomass for synergistic removal of chromium (VI): Adsorption behavior and computing mechanism. *Sci. Total Environ.* **2020**, *730*, 138930. [CrossRef]

13. Zhuang, Q.Q.; Cao, J.P.; Wu, Y.; Zhao, M.; Zhao, X.Y.; Zhao, Y.P.; Bai, H.C. Heteroatom nitrogen and oxygen co-doped three-dimensional honeycomb porous carbons for methylene blue efficient removal. *Appl. Surf. Sci.* **2021**, *546*, 149139. [CrossRef]
14. Chen, B.; Yang, Z.; Ma, G.; Kong, D.; Xiong, W.; Wang, J.; Zhu, Y.; Xia, Y. Heteroatom-doped porous carbons with enhanced carbon dioxide uptake and excellent methylene blue adsorption capacities. *Microporous Mesoporous Mater.* **2018**, *257*, 1–8. [CrossRef]
15. Hu, X.; Min, X.; Wang, H.; Li, X.; He, Y.; Yang, W. Enhanced capacitive deionization boosted by Co and N co-doping in carbon materials. *Sep. Purif. Technol.* **2021**, *266*, 118590.
16. Zhang, H.; Zhou, W.; Huang, D.; Ou, L.; Lan, Z.; Liang, X.; Huang, H.; Huang, D.; Guo, J. Functionalized hierarchical porous carbon with sulfur/nitrogen/oxygen tri-doped as high quality sulfur hosts for lithium-sulfur batteries. *J. Alloys Compd.* **2020**, *858*, 157647. [CrossRef]
17. Liu, Y.; Lin, D.; Yang, W.; An, X.; Sun, A.; Fan, X.; Pan, Q. In situ modification of ZIF-67 with multi-sulfonated dyes for great enhanced methylene blue adsorption via synergistic effect. *Micropor. Mesopor. Mat.* **2020**, *303*, 110304. [CrossRef]
18. Jeng, H.B.; Teck, H.L.; Jin, H.L.; Lai, J.C. Cellulose nanofibril-based aerogel derived from sago pith waste its application on methylene blue removal. *Int. J. Biol. Macromol.* **2020**, *160*, 836–845.
19. Zhou, X.; Wang, P.; Zhang, Y.; Wang, L.; Zhang, L.; Zhang, L.; Xu, L.; Liu, L. Biomass based nitrogen-doped structure-tunable versatile porous carbon materials. *J. Mater. Chem. A* **2017**, *5*, 12958–12968. [CrossRef]
20. Niu, S.; Lv, W.; Zhou, G.; He, Y.; Li, B.; Yang, Q.H.; Kang, F. N and S co-doped porous carbon spheres prepared using L-cysteine as a dual functional agent for high-performance lithium-sulfur batteries. *Chem. Commun.* **2015**, *51*, 17720–17723. [CrossRef]
21. Chen, M.; Jiang, S.; Huang, C.; Wang, X.; Cai, S.; Xiang, K.; Zhang, Y.; Xue, J. Honeycomb-like Nitrogen and Sulfur Dual-doped Hierarchical Porous Biomass Carbon for High-energy-density Lithium-sulfur Batteries. *ChemSusChem* **2017**, *10*, 1803–1812. [CrossRef]
22. Zhang, H.; Huang, Y.; Hu, Z.; Tong, C.; Zhang, Z.; Hu, S. Carbon dots codoped with nitrogen and sulfur are viable fluorescent probes for chromium(VI). *Microchim. Acta* **2017**, *184*, 1547–1553. [CrossRef]
23. Sun, D.; Ban, R.; Zhang, P.H.; Wu, G.H.; Zhang, J.R.; Zhu, J.J. Hair fiber as a precursor for synthesizing of sulfur- and nitrogen-co-doped carbon dots with tunable luminescence properties. *Carbon* **2013**, *64*, 424–434. [CrossRef]
24. Gong, J.; Lin, H.; Antonietti, M.; Yuan, J. Nitrogen-doped porous carbon nanosheets derived from poly(ionic liquid): Hierarchical pore structures for efficient CO_2 capture dye removal. *J. Mater. Chem. A* **2016**, *4*, 7313. [CrossRef]
25. Li, C.; Zhang, L.; Gao, Y.; Li, A. Facile synthesis of nano ZnO/ZnS modified biochar by directly pyrolyzing of zinc contaminated corn stover for Pb(II), Cu(II) and Cr(VI) removals. *Waste Manag.* **2018**, *79*, 625–637. [CrossRef]
26. Sriramoju, S.K.; Dash, P.S.; Majumdar, S. Meso-porous activated carbon from lignite waste and its application in methylene Blue adsorption and coke plant effluent treatment. *J. Environ. Chem. Eng.* **2021**, *9*, 104784. [CrossRef]
27. Thta, H.T.; Le, H.H.; Pham, T.H.; Nguyen, D.T.; La, D.D.; Chang, S.W.; Lee, S.M.; Chung, W.J.; Nguyen, D.D. Comparative study on methylene blue adsorption behavior of coffee husk-derived activated carbon materials prepared using hydrothermal soaking methods. *J. Environ. Chem. Eng.* **2021**, *9*, 105362.
28. Jiang, L.; Wen, Y.; Zhu, Z.; Liu, X.; Shao, W. A Double cross-linked strategy to construct graphene aerogels with highly efficient methylene blue adsorption performance. *Chemosphere* **2021**, *265*, 129169. [CrossRef] [PubMed]
29. Wang, Y.; Pan, J.; Li, Y.; Zhang, P.; Li, M.; Zheng, H.; Zhang, X.; Li, H.; DU, Q. Methylene blue adsorption by activated carbon, nickel alginate/activated carbon aerogel, and nickel alginate/graphene oxide aerogel: A comparison study. *J. Mater. Res. Technol.* **2020**, *9*, 12443–12460. [CrossRef]
30. Thi, H.T.; Hoang, A.L.; Huu, T.P.; Nguyen, D.T.; Chang, S.W.; Chung, W.J.; Nguyen, D.D. Adsorption isotherms and kinetic modeling of methylene blue dye onto a carbonaceous hydrochar adsorbent derived from coffee husk waste. *Sci. Total Environ.* **2020**, *725*, 138325. [CrossRef] [PubMed]
31. Wu, Z.; Hua, Z.; Yuan, X.; Wang, H.; Wang, L.; Chen, X.; Zeng, G.; WU, Y. Adsorptive removal of methylene blue by rhamnolipid-functionalized graphene oxide from wastewater. *Water Res.* **2014**, *67*, 330–344. [CrossRef] [PubMed]
32. Do, T.H.; Nguyen, V.T.; Dung, N.Q.; Chu, M.N.; Kiet, D.V.; Ngan, T.H.K.; Tan, L.V. Study on methylene blue adsorption of activated carbon made from Moringa oleifera leaf. *Mater. Today Proc.* **2021**, *38*, 3405–3413. [CrossRef]
33. Asuha, S.; Fei, F.; Wurendaodi, W.; Zhuang, X. Activation of kaolinite by a low-temperature chemical method and its effect on methylene blue adsorption. *Powder Technol.* **2020**, *361*, 624–632. [CrossRef]
34. Wang, Z.; Zhang, G.; Li, Y. Preparation of Chitosan/Polyacrylamide/Graphene Oxide Composite Membranes and Study of Their Methylene Blue Adsorption Properties. *Materials* **2020**, *13*, 4407. [CrossRef] [PubMed]
35. Cheng, S.; Zhang, L.; Xia, H.; Peng, J.; Shu, J.; Li, C.; Jiang, X.; Zhang, Q. Adsorption behavior of methylene blue onto waste-derived adsorbent and exhaust gases recycling. *RSC Adv.* **2017**, *7*, 27331–27341. [CrossRef]
36. Alver, E.; Metin, A.; Brouers, F. Methylene blue adsorption on magnetic alginate/rice husk bio-composite. *Int. J. Biol. Macromol.* **2020**, *154*, 104–113. [CrossRef]
37. Chen, F.; Ma, L.; Ren, J.; Luo, X.; Liu, B.; Zhou, X. Sandwich-Type Nitrogen and Sulfur Codoped Graphene-Backboned Porous Carbon Coated Separator for High Performance Lithium-Sulfur Batteries. *Nanomaterials* **2018**, *8*, 191. [CrossRef]
38. Li, M.; Li, Y.; Zhang, X.; Zheng, H.; Zhang, A.; Chen, T.; Liu, W.; Yu, Y.; Liu, J.; Du, Q.; et al. One-step generation of S and N co-doped reduced graphene oxide for high-efficiency adsorption towards methylene blue. *RSC Adv.* **2020**, *10*, 37757–37765. [CrossRef]

39. Chen, F.; Ma, L.; Ren, J.; Luo, X.; Bing, L.; Song, Z.; Zhou, X. Wheat Straw-Derived N-, O-, and S-Tri-doped Porous Carbon with Ultrahigh Specific Surface Area for Lithium-Sulfur Batteries. *Materials* **2018**, *11*, 989. [CrossRef]
40. Cheng, Z.Q.; Wang, Z.W.; Wu, P.C.; Wang, Y.; Fu, J. Mass fabrication of oxygen and nitrogen co-doped 3D hierarchical porous carbon nanosheets by an explosionassisted strategy for supercapacitor and dye adsorption application. *Appl. Surf. Sci.* **2020**, *529*, 147079. [CrossRef]

nanomaterials

Article

Facile Preparation of Oxygen-Vacancy-Engineered MoO_x Nanostructures for Photoreversible Switching Systems

Hao Xu [1,2,†], Liangjing Zhang [2,†], Aiwu Wang [2,*], Juan Hou [1,*] and Xuhong Guo [3,*]

1 Key Laboratory of Ecophysics, Department of Physics, College of Science, Shihezi University, Shihezi 832003, China; xu_hao88@sina.com
2 Center for Advanced Material Diagnostic Technology, Shenzhen Technology University, Shenzhen 518118, China; zhangliangjing@sztu.edu.cn
3 School of Chemical Engineering, East China University of Science and Technology, 130 Meilong Road, Shanghai 200237, China
* Correspondence: wangaiwu@sztu.edu.cn (A.W.); hjuan05@sina.com (J.H.); guoxuhong@ecust.edu.cn (X.G.)
† These authors contributed to the work equally.

Abstract: Photochromic materials have attracted increasing attention. Here, we report a novel photo-reversible color switching system based on oxygen-vacancy-engineered MoO_x nanostructures with water/N-methyl-2-pyrrolidone (NMP) as solvents. In this work, the system rapidly changed from colorless to blue under UV irradiation (360–400 nm) and slowly recovered its colorless state under visible light irradiation. The obtained oxygen vacancy-engineered MoO_x nanostructures exhibited good repeatability, chemical stability, and cycling stability. Upon UV light irradiation, H^+ was intercalated into layered MoO_x nanostructures and the Mo^{6+} concentration in the H_xMoO_x decreased, while the Mo^{5+} concentration increased and increased oxygen vacancies changed the color to blue. Then, it recovered its original color slowly without UV light irradiation. What is more, the system was highly sensitive to UV light even on cloudy days. Compared with other reported photochromic materials, the system in this study has the advantage of facile preparation and provides new insights for the development of photochromic materials without dyes.

Keywords: photochromic; MoO_x nanostructures; UV light; color switching

Citation: Xu, H.; Zhang, L.; Wang, A.; Hou, J.; Guo, X. Facile Preparation of Oxygen-Vacancy-Engineered MoO_x Nanostructures for Photoreversible Switching Systems. *Nanomaterials* **2021**, *11*, 3192. https://doi.org/10.3390/nano11123192

Academic Editor: Nikolai V. Tkachenko

Received: 16 September 2021
Accepted: 13 November 2021
Published: 25 November 2021

1. Introduction

Materials that exhibit a change in their optical properties under the influence of either an electric field or light are known as discoloration materials. These materials are of interest in the realm of electrochromism and photochromism [1–4] and are of particular significance in novel technologies, such as sensors, pregnancy tests, electrochromic and photochromic displays, mirrors, and smart windows [5–7].

As an important semiconductor material, MoO_3 shows strong localized surface plasmon resonance (LSPR) absorption, and the introduction of oxygen vacancies, aliovalent ions, and hydrogen ions induces a notable color change, thus making it an ideal photochromic material [8–10].

MoO_x ($2 \leq x \leq 3$) exists in different crystalline phases from MoO_2 to MoO_3. Pure MoO_3 can involve different polymorphs, namely a thermodynamically stable orthorhombic phase (α-MoO_3), hexagonal phase (h-MoO_3), and metastable monoclinic crystal phase β-MoO_3 (similar to ReO_3 type structure). In fact, all reported MoO_x are basically constructed in different ways based on the construction of MoO_6 octahedra, which exhibit layered structures, each layer connected by van der Waals force so that they can serve as hosts for dopant intercalants [11,12]. For example, MoO_3 exhibiting a low oxygen vacancy concentration is white in color, whereas MoO_3 with a high oxygen vacancy concentration is often blue or dark blue in color [13]. This phenomenon can be explained through the crystal rearrangements induced in the Mo−O units at the surface of the material during the reduction process, which results in a disordered surface structure containing many oxygen vacancies, thus imparting enhanced visible light absorption. The chromogenic response demonstrated by MoO_3 demonstrates a stronger and more uniform absorption of visible light in its colored state and displays an increased open-circuit memory in comparison to the other transition-metal oxides [14]. MoO_3 nanostructures of varying dimensions have been successfully prepared, including one-dimensional nanobelts and nanowires, alongside two-dimensional nanosheets [15–19]. However, only a limited number of studies have reported on the preparation of zero-dimension MoO_3 quantum dots (QDs), especially with respect to their photochromic properties [20–22].

Top-down fabrication of layered nanostructures does not require harsh reaction conditions, thus low-cost high-yield preparation of materials is available. Furthermore, nanostructures prepared in this method do not develop an insulating ligand coating, which is detrimental to the electrical conductivity [23]. The ligand coating also exhibits a high defect concentration, which introduces many active edge sites for reactions [24]. The extent of the photochromism and electrochromism expressed by MoO_3 is dominated by hydrogen radical intercalation or by electrochemical intercalation of lithium ions into the inter-layer gaps formed as a result of the van der Waal bonding. To date, ion intercalation has been proved to be the most effective method for increasing the free charge carrier concentrations in 2D materials [25,26].

Successful liquid phase exfoliation (LPE) of a layered material relies on overcoming the van der Waals forces between adjacent layers in the material [27–29]. This requires either an initial increase in the layer–layer spacing, which is induced through intercalation or by mixing with a solvent displaying equal surface tension properties [28,30]. During the exfoliation process, the efficiency of the exfoliation is strongly influenced by the extent of the similarity between the surface energy and Hansen solubility parameters (HSP) of the solvent and the target material [28,31,32]. As a result, using an appropriate exfoliation solvent can impart a significant change in the discoloration properties of MoO_3. For example, a change of the solvent can affect the extent of the water photo-oxidation process and H^+ ion intercalation in molybdenum oxide (MoO_x) nanoflakes upon solar light illumination.

Herein, we propose a top-down synthesis method for the fabrication of layered nanostructures, using NMP/water as the exfoliation solvent, which is assisted by the hydrothermal method. Initially, bulk MoO_3 is exfoliated to form 2D MoO_x nanoflakes in the NMP/water system, followed by high-temperature degradation into smaller particles. The photochromism of MoO_3 can be briefly described as: in a MoO_{3-x} nanostructures sample of NMP/H_2O mixed solvent, when ultraviolet light is irradiated on MoO_3, the MoO_3 on the surface is excited to separate electrons and holes, during which NMP acts as a good hole acceptor to help the separation of electrons and holes in this process, the separated holes react with surrounding water molecules, and H^+ is produced in that way. The intercalation of H^+ into MoO_3 changed its crystal structure and reduced Mo^{+6}, which resulted in the appearance of plasmon absorption peaks and the change of color. The H_xMoO_3 produced in the above process is generally considered to be an unstable metastable state. In the presence of oxygen, it tends to spontaneously dehydrate to form the corresponding oxide, thereby recovering the color. The as-obtained MoO_{3-x} nanostructures were characterized, and their optical and photochromic properties were evaluated. Particular samples displayed excellent photochromic performance and cycle stability. The influence of reaction time on nanostructures performance was also investigated in an attempt to validate their tunability through a slight adjustment in experimental parameters.

2. Materials and Methods

MoO_3 powder (99%), ethanol (99.5%), and N-methyl-2-pyrrolidone (NMP, 98%) were obtained from Aladdin (Shanghai, China). Deionized water (18 MΩ) was used across all experimental procedures, and all reagents were used as received, without further purification.

Synthetic procedure:

MoO_3 powder (0.03 g) was added into 10 mL of solvent containing deionized water and NMP in a 1:1 volume ratio (another sample used pure NMP as solvent). The mixture was then treated using sonication for 10 min, before being transferred into a 20 mL Teflon-lined stainless steel autoclave at 120 °C for 1, 3, 6, 12, 24, and 48 h to obtain samples exhibiting varying properties. Subsequently, each sample was placed in a centrifugal chamber at 12,000 rpm for 15 min and the supernatant was obtained from the samples. Finally, each sample was irradiated under a solar simulator for varying periods of time to obtain the MoO_{3-x} nanostructures. Residual powder from the samples could be collected and dried in a vacuum drying oven for reuse.

Characterization:

Transmission electron microscopy (TEM) images were taken using a JEOL JEM-2100 microscope at 200 kV (Tokyo, Japan). X-ray diffraction (XRD) patterns were recorded on a D/MAX2500 X-ray diffractometer (Tokyo, Japan) using Cu Kα radiation (λ = 1.5418 Å). The Fourier transform infrared-red (FTIR) spectroscopy was conducted on a Nicolet IS10 FT-IR spectrometer (Madison, WI, USA). X-ray photoelectron spectroscopy (XPS) was conducted on a Thermo Scientific Escalab 250 Xi X-ray photoelectron spectrometer (Waltham, MA, USA). The ultraviolet-visible absorption spectra were measured using a Thermo Evolution 220 UV-vis spectrometer (Waltham, MA, USA). The fluorescence spectra were measured using a HITACHI F-2700 fluorescence spectrophotometer (Tokyo, Japan). The source of the artificial UV light was MICROSOLAR300 Xenon lamp of Perfectlight (Beijing, China). The wavelength range was 320~780 nm. The power was 50 W, 19.6 W in visible region, 2.6 W in ultraviolet region. Dynamic light scattering (DLS) (ZEN3600, Malvern Instruments, Shanghai, China) was used to determine the size of nanoparticles. Raman spectra (Renishawplc, inVia) were recorded at 532 nm (Gloucestershire, England). Light transmittance was recorded by a FX2000 fiber spectrometer (Ideaoptics, Shanghai, China).

3. Results and Discussion

It is widely established that expanding the inter-layer spacing of layered materials through intercalation will weaken the van der Waals interactions between the layers. Previously published reports have revealed ion intercalation to be the most effective method for increasing the free charge carrier concentrations in 2D materials to induce plasmon resonance in the visible and near-infrared (NIR) regions [19]. Free electron concentration in MoO_x can be significantly increased by the introduction of oxygen vacancies through H^+ ion intercalation into the inter-layers inherent to this material. Herein, intercalated H^+ ions were used to induce an expansion of the inter-layer space, thereby increasing the free electron concentration through the introduction of oxygen vacancies [27].

N-methyl-2-pyrrolidone (NMP) was introduced as an exfoliation solvent. During the exfoliation process, the extent of the similarity between the HSP and the surface energy of the solvent and the target material strongly influences the efficiency of the exfoliation [19]. As a result, the choice of exfoliation solvent plays a considerable role in determining the plasmonic properties of MoO_x nanoparticles. Furthermore, a change of the solvent can affect the degree of the photo-oxidation of water and H^+ ion intercalation in MoO_x nanoparticles upon solar light illumination. To characterize photochromic properties of MoO_x nanostructures, the samples were exposed to UV light for varying periods of time and their absorption response to the irradiation was recorded. The gathered visible absorption spectra underwent various changes, the results of which are provided in Figure 1.

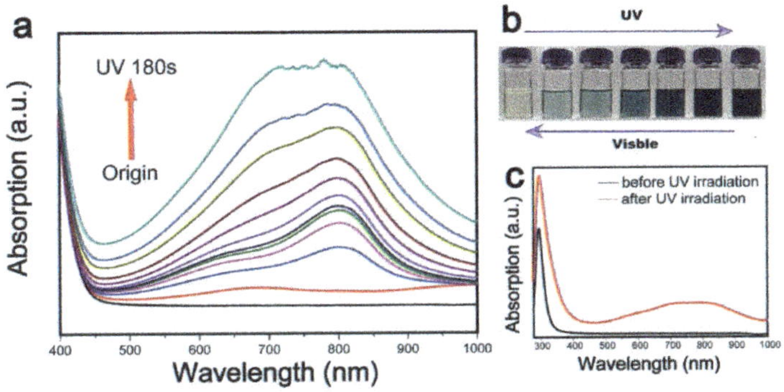

Figure 1. (**a**) The absorption spectra and (**b**) photos of MoO_{3-x} nanostructures irradiated by ultraviolet light for 180 s. (**c**) The absorption spectra of MoO_{3-x} nanostructures in the range of UV-Vis light.

In the case of the sample that was not exposed to UV light, the spectrum indicated that it exhibited a light yellow coloration, while there was no absorption peak observed in the visible light wavelength range. Upon exposure to UV light, the sample demonstrated a gradual darkening in color, the color change is reflected in its UV-visible absorption spectrum (Figure 1a), changing from light blue to dark blue to black (Figure 1b). As the exposure time increased, 600 nm broad absorption peaks gradually appeared in the <1000 nm range, while two further peaks were observed at approximately 700 and 800 nm. The appearance of the absorption peaks suggests that the MoO_{3-x} nanostructures possessed strong LSPR performance. MoO_{3-x} nanostructures had a strong absorption peak in UV range shown in Figure 1c.

The results of this study suggest that photo-excitation induces competing solvent and water oxidation processes, resulting in a reduction in H^+ intercalation. NMP forms an aprotic solvent, in which the molecules of the solvent are not hydrogen bonded to one another, meanwhile the photo-oxidation of NMP is weak. Therefore, in the NMP/H_2O system, water is photo-oxidized under ultraviolet light instead of NMP and H^+ produced in this process is intercalated into MoO_3. On the other hand, NMP is provided with the highest proton affinity ($\sim$920 kJ mol^{-1}) [20] and lowest proton dissociation rate, which induces a higher degree of H^+ intercalation in NMP in comparison to other solvents [27].

The mechanism of ultraviolet (UV) light irradiation is summarized below [33], * represent the excited state:

$$MoO_3 + hv \rightarrow MoO_3^* + h^+ + e^- \quad (1)$$

$$2h^+ + H_2O \rightarrow 2H^+ + 1/2O_2 \quad (2)$$

$$MoO_3 + XH^+ + Xe^- \rightarrow H_XMoO_{3-x} \quad (3)$$

Throughout the photo-oxidation process, H_2O provides a source of hydrogen ions, while NMP performs the role of a hole trapping acceptor, which can promote the separation of electrons and holes, thus speeding up the reaction process. This promotion induces an increase of migration of hydrogen ions in the system.

Small particles, including nanostructures, were detached from the rough MoO_3 particles, as shown in Figure 2. To further explore the structure characteristics, the XRD (Figure 2a) was carried out to obtain the crystalline phase of MoO_x nanostructures. Sharp diffraction peaks located in the range of 20°–30°, representing α-MoO_3, became wide in MoO_x nanostructures, indicating that an amorphous structure was formed. In Figure 2b, IR bands occurring in the range of 962 cm^{-1} to 835 cm^{-1} were Mo=O bond stretching vibration and Mo-O-Mo bond stretching vibration, which prove that the obtained nanostructure was molybdenum oxide [20]. DLS in Figure 2c shows the nanostructures had an average size of 30 nm, which contradicted most of our TEM images—the increase of statistical size could be attributed to the aggregation of small particles. In Figure 2d,e, the core level spectra of Mo3d in H_xMoO_3 nanostructures were introduced. The double peaks at 235.6 eV and 232.2 eV corresponded to the binding energy of 3d5/2 and 3d3/2 orbital electrons of Mo^{6+}, while 234.3 eV and the double peaks at 231.2 eV represented the binding energy of the 3d5/2 and 3d3/2 orbital electrons of Mo^{5+}. It is obvious that after ultraviolet light irradiation, the Mo^{6+} content in the H_xMoO_3 nanostructures decreased and the Mo^{5+} content increased, which indicates that the nanostructures underwent surface structure reconstruction, and the concentration of oxygen vacancies increased, accompanied by a decrease in the valence state of molybdenum.

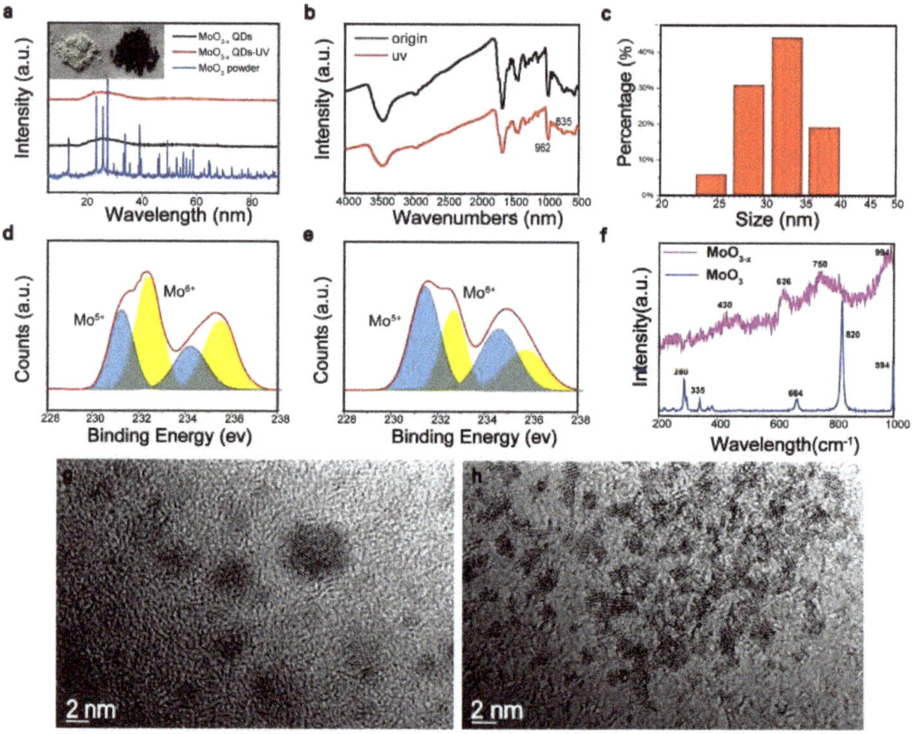

Figure 2. (**a**) XRD pattern of MoO_{3-x} nanostructures and photographs of α-MoO_3 powder and MoO_{3-x} nanostructures powder (inset image). (**b**) FTIR spectra of MoO_{3-x} nanostructures with and without UV. (**c**) DLS spectrum of MoO_{3-x} nanostructures. (**d**) Mo 3d core level spectra of MoO_{3-x} nanostructures without UV excitation and (**e**) Mo 3d core level spectra of MoO_{3-x} nanostructures without UV excitation. (**f**) Raman spectra of MoO_3 and MoO_{3-x} nanostructure. (**g**) TEM images of MoO_{3-x} nanostructures without ultraviolet light irradiation and (**h**) irradiated by ultraviolet light.

The Raman spectra of the MoO_{3-x} structure were also investigated (Figure 2f) before and after illumination, the 280 cm^{-1} represented the wagging mode of the double bond O=Mo=O, the peaks at 664 cm^{-1} and 820 cm^{-1} could be assigned to the stretching modes of the triple coordinated oxygen (Mo_3-O), and the doubly coordinated oxygen (Mo_2-O), respectively, the 335 cm^{-1} was assigned to the Mo_3-O bending mode. For the samples of MoO_{3-x} structure after illumination, traces of MoO_{3-x} were also identified by two additional peaks 430 cm^{-1} and 750 cm^{-1} due to the modification of original Mo_2-O bond [33]. From Raman analysis, the existence of some substoichiometry in the MoO_{3-x} structure can be possibly due to the intercalation of H^+ from the solvent and increase of oxygen vacancy in the presence of significant heat and force generated during the solvothermal process [19,20,33,34]. After centrifugation, the resultant MoO_{3-x} nanostructures were observed using transmission electron microscopy (TEM). The TEM images of the MoO_{3-x} nanostructures are illustrated in Figure 2g,h, revealing an average diameter of approximately several nanometers.

Concerning samples using pure NMP as solvent, these exhibited quite different optical behavior (Figure 3a). The nanostructures had fluorescent peaks at 450 nm under 360 nm excitation light. In addition, the absorption peaks in the range of 700–900 nm had no response to irradiation, while at the same time, the color of the samples was constant. Figure 3b,c are TEM images of samples using pure NMP as solvent. This behavior may be attributed to the different formation mechanisms of nanostructures in water/NMP and pure NMP solvents. In the water/NMP mixed solvent, water may have combined with holes generated through light excitation to produce H^+, which is related to the H^+ intercalation mechanism. As there was no H^+ production in the NMP solvent, the photo-corrosion mechanism explains the behavior of this system more accurately. The oxidation strength of NMP is limited, and the photo-generated holes cannot be consumed quickly. Owing to their strong oxidizing ability, vacancy accumulation after continuous light excitation may corrode the lattice structure of MoO_3. This is known as photo-corrosion and most often occurs in semiconductor materials used in light-related applications [20,27].

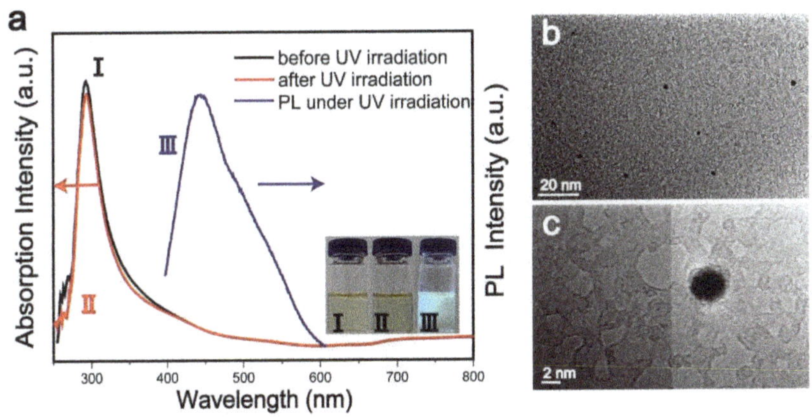

Figure 3. (**a**) UV-visible absorption spectra, photoluminescent spectra, and photographs before (I), during (III), and after UV irradiation (II). (**b**,**c**) TEM images of MoO_3 QDs with pure NMP as solvent.

In this study, the number of oxygen vacancies present in the MoO_{3-x} nanostructures was influenced by the length of the reaction time. Herein, we prepared a series of MoO_{3-x} nanostructures exhibiting varying photochromic behaviors to analyze and compare their UV-visible absorption spectra.

Figure 4a–f are the UV-visible absorption spectra generated by the MoO_{3-x} nanostructure samples, which underwent varying reaction times of 1, 3, 6, 12, 24, and 48 h. Each sample underwent UV irradiation for 5 min, which facilitated an accurate analysis of the absorption effects demonstrated in the sample solely as a function of the reaction time. As the reaction time increased, a gradual reduction in the peak intensity was confirmed; that is, the sensitivity to UV light decreased. A photograph of the color gradation exhibited by the MoO_{3-x} nanostructures was performed in Figure 4g, samples underwent the reaction for 1, 12, and 48 h before UV irradiation for 300 s.

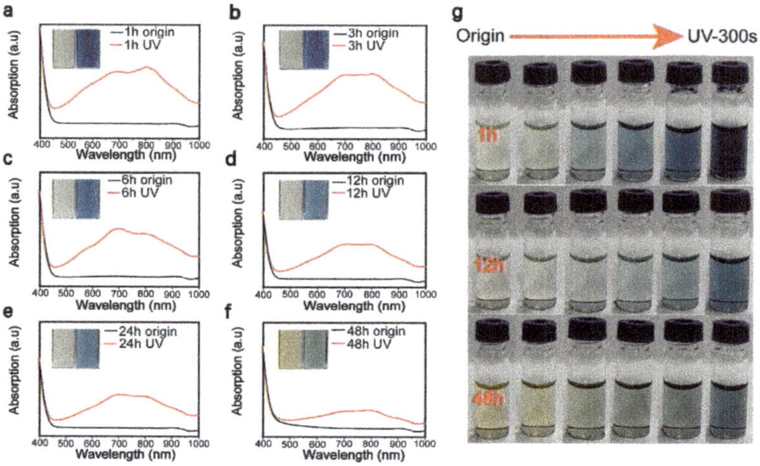

Figure 4. (**a–f**) The absorption spectrum for the MoO_{3-x} nanostructure underwent varying reaction times of 1, 3, 6, 12, 24, and 48 h, and origin samples were irradiated for 5 min by Xenon lamp to create UV samples. (**g**) Photographs of MoO_{3-x} nanostructures irradiated by Xenon lamp for 300 s.

We speculated that the reaction proceeded rapidly under hydrothermal conditions. Upon completion of the intercalation reaction, and as the reaction time increased, NMP played a significant role in controlling the size and morphology of the nanostructures. In this case, the absorption energy was greater than the forbidden band gap. Since the generated electron–hole pairs could not be completely consumed, the highly oxidizing holes oxidized the crystal of MoO_3, thus disrupting the lattice structure.

As oxygen exhibits a significant larger electronegativity than hydrogen, the majority of binary metal oxides show higher stability than their corresponding hydrides [30,35]. The hydrogenation process in this article was used to facilitate molybdenum oxide to form unstable H_xMoO_3 by ultraviolet light irradiation. This switching of the crystal structure induced a change in the optical and electrical properties of the material. The original α-MoO_3 was a semiconductor with a band gap of 3.2 eV. The charge induced upon irradiation was transferred from the newly formed Mo^{5+} state to the adjacent Mo^{6+} state, and H_XMoO_3 appeared blue. H_XMoO_3 adopts a metastable state and will spontaneously be oxidized to the corresponding oxide [36]. Under high temperatures, it can accelerate the spontaneous reduction process [37].

The MoO_{3-x} nanostructure samples prepared in this study were light yellow in color prior to UV light irradiation. Upon irradiation by sunlight or UV light, the structure of the MoO_{3-x} nanostructures changed in accordance with the element valence state, and a corresponding color change at the surface was observed. From the initial light yellow color, it turned light blue, before changing to dark blue through to black, where LSPR absorption peaks appeared in the visible light wavelength range of the absorption spectrum. This was a reversible process. When an irradiated sample was placed in an environment without UV light, its color returned to its initial light yellow slowly. Temperature imparts a considerable effect on the color change recovery process. When the dark sample was heated to 80 °C, the recovery rate increased, and the total recovery time was significantly reduced to the order of minutes.

To explore the effect of synthesis time on the photochromic behavior of MoO_{3-x} nanostructures, the reaction temperature was set at 120 °C and MoO_{3-x} nanostructures subjected to various reaction times were obtained. Their photochromic properties were analyzed by UV-visible absorption spectroscopy. The various samples were subjected to the same experimental conditions and the discoloration–recovery line chart is illustrated in Figure 5a,b. It could be concluded from Figure 5a that, as the reaction time increased, the UV-visible absorption peak intensity reduced and the sample became less sensitive to UV light. As previously discussed, this may be attributable to the different formation mechanisms exhibited by nanostructures formed in mixed water/NMP and single NMP solvent systems [20,27,38]. The trend was completely reversed when we investigated the color restoration of MoO_{3-x} nanostructure samples (Figure 5b). Samples prepared under a long reaction time formed a H_xMoO_3 metastable structure and brought about an increased rate of dehydration and reduction. This is attributed to the smaller ion size and greater number of surface sites displayed by nanostructures obtained under a longer reaction time, which are beneficial for embedding, separating hydrogen, and decomposing metastable H_xMoO_3 [36,37]. Some factors that impacted the recovery process, such as temperature and oxygen, are shown in Figure 5c,d. Briefly, high temperature and oxygen can accelerate the recovery process. During that process, the H_xMoO_3 with oxygen vacancy is oxidized and hydrogen atoms are taken away by surrounding oxygen, accompanied by a reduction in the oxygen vacancy [34,39].

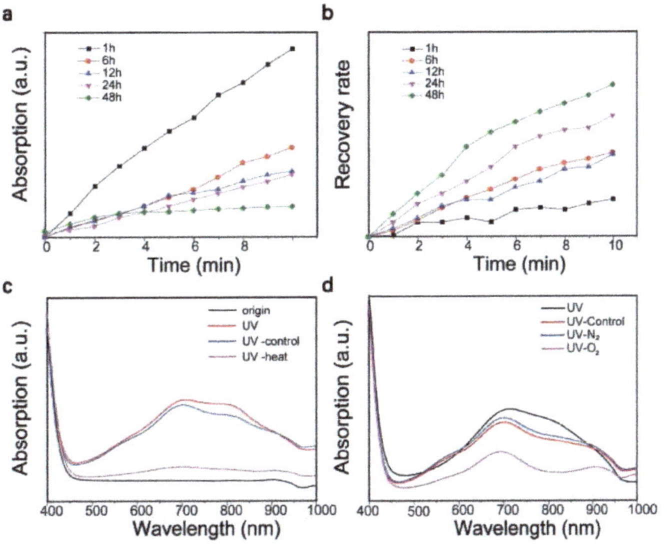

Figure 5. (**a**) The absorption peaks of MoO_{3-x} nanostructures under UV 10 min of UV irradiation using a Xenon lamp, samples were obtained at different reaction times. (**b**) The self-recovery rate of MoO_{3-x} nanostructures obtained at different reaction times after 10 min of UV irradiation using a Xenon lamp. The recovery rate is defined as the ratio of the amount of absorption intensity recovered to the absorption intensity before recovery. (**c**) The effect of heating on photochromism of MoO_{3-x} nanostructures. Origin sample was irradiated for 10 min to create a UV sample, then the UV sample was divided to two parts, treated with and without 20 min heating, marked as UV-heating and UV-control. (**d**) The effect of oxygen on photochromism of MoO_{3-x} nanostructures—the UV sample was obtained via irradiation for 10 min and samples treated with O_2, N_2, and air were marked as UV-O_2, UV-N_2, and UV-control, respectively.

In Figure 6a, a fiber optic spectrometer was used to evaluate and characterize the light transmittance of the nanostructures. The variation in the light transmittance was recorded in 100 ms intervals throughout the test. It is clear that the light intensity decreased gradually as the UV illumination time increased from 0 to 60 s in the 700–800 nm wavelength range. This suggests that the light absorption capacity of the sample increased in this wavelength band, which corresponded to an increase in the peak value recorded in UV-visible absorption spectrum. The cycle performance test was used to characterize the photochromic stability of MoO_{3-x} nanostructures. The obtained MoO_{3-x} nanostructure solution was irradiated under a solar simulator for 5 min. The UV-visible absorption spectrum was recorded at this point, and the sample was then stored in an environment without UV light. Upon completion of the recovery period, the UV-visible absorption spectrum was recorded again. The test was repeated for 20 times, as shown in Figure 6b. Photographs of samples after exposure to bright sunlight, as well as also cloudy conditions, for 0, 5, and 15 min were used as controls (Figure 6c). It was observed that the MoO_{3-x} nanostructure samples exhibited a strong discoloration effect, even when placed in an environment of low UV irradiation conditions, demonstrating that this material provides considerable potential for practical applications.

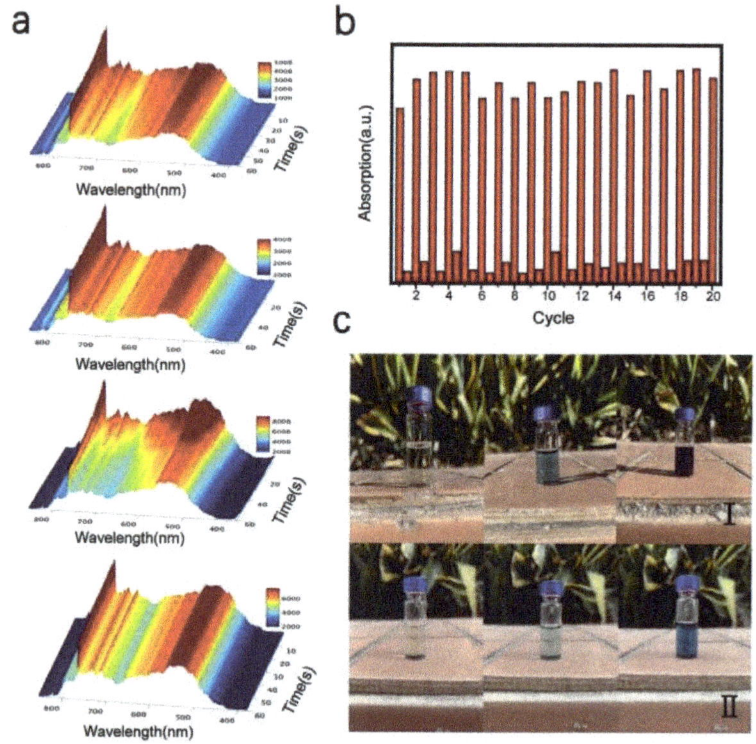

Figure 6. (**a**) Spectra of the Fiber Optic spectrometer. (**b**) Cycle stability test. (**c**) Discoloration performance of MoO_{3-x} nanostructures in sunny weather (I) and cloudy weather (II).

4. Conclusions

In summary, oxygen-vacancy-engineered MoO_x nanostructures were fabricated by a modified top-down method combining intercalation and the solvent–thermal method. In the case of the mixed solvent system of water/NMP, water was introduced to produce H^+ through photo-oxidation to facilitate intercalation into MoO_x exhibiting a layered structure. An increase in proton intercalation and oxygen vacancy in molybdenum oxide nanostructures resulted in a significant increase in their free electron concentrations and changes in their substoichiometry displayed intense LSPR in the visible light region. In the presence of oxygen, the hydrogen atoms were taken away combined with oxygen, and metastable H_XMoO_3 were oxidized, accompanied by a reduction of LSPR and color restoration. A series of MoO_x nanostructures exhibiting consistent photochromic behavior were prepared through control of the reaction time and temperature. The synthesized samples exhibited excellent photochromic properties and cycle stability. In this context, our oxygen-vacancy-engineered MoO_x nanostructures provide an important stimulus to the development of practical photochromic materials

5. Patents

There is a Chinese patent 202110009536.X resulting from the work reported in this manuscript.

Author Contributions: H.X., L.Z., and A.W.; investigation, H.X., L.Z., and A.W.; writing—original draft preparation, A.W., J.H., and X.G.; writing—review and editing, A.W., J.H., and X.G.; supervision. All authors have read and agreed to the published version of the manuscript.

Funding: This research was funded by Guangdong Provincial Department of Education Young Innovative Talents Project (2019KQNCX179) and Key Areas Scientific and Technological Research Projects in Xinjiang Production and Construction Corps (No. 2018AB004).

Institutional Review Board Statement: Not applicable.

Informed Consent Statement: Not applicable.

Data Availability Statement: Data is available on the request from the corresponding author.

Acknowledgments: We would like to extend sincere gratitude to Xue J for his instructive advice to the work.

Conflicts of Interest: The authors declare no conflict of interest.

References

1. Granqvist, C.G. *Handbook of Inorganic Electrochromic Materials*; Elsevier: Amsterdam, The Netherlands, 1995.
2. He, T.; Yao, J.N. Photochromism in composite and hybrid materials based on transition-metal oxides and polyoxometalates. *Prog. Mater. Sci.* **2006**, *51*, 810–879. [CrossRef]
3. He, T.; Yao, J.N. Photochromism in transition-metal oxides. *Res. Chem. Intermed.* **2004**, *30*, 459–488. [CrossRef]
4. He, T.; Yao, J.N. Photochromism of molybdenum oxide. *J. Photochem. Photobiol. C Photochem. Rev.* **2003**, *4*, 125–143. [CrossRef]
5. Granqvist, C.-G. Out of a niche. *Nat. Mater.* **2006**, *5*, 89–90. [CrossRef]
6. Chen, H.-G.; Wang, H.-S.; Lin, X.-C. Lateral epitaxial overgrowth of ZnO layers on hexagonally patterned buffer layers in low-temperature aqueous solution. *J. Alloys Compd.* **2017**, *707*, 341–346. [CrossRef]
7. Lampert, C.M. Chromogenic Smart Materials. *Mater. Today* **2004**, *7*, 28–35. [CrossRef]
8. Shen, Y.; Yang, Y.; Hu, F.; Xiao, Y.; Yan, P.; Li, Z. Novel coral-like hexagonal MoO_3 thin films: Synthesis and photochromic properties. *Mater. Sci. Semicond. Process.* **2015**, *29*, 250–255. [CrossRef]
9. Song, J.; Li, Y.; Zhu, X.; Zhao, S.; Hu, Y.; Hu, G. Preparation and optical properties of hexagonal and orthorhombic molybdenum trioxide thin films. *Mater. Lett.* **2013**, *95*, 190–192. [CrossRef]
10. Chithambararaj, A.; Bose, A.C. Investigation on structural, thermal, optical and sensing properties of meta-stable hexagonal MoO_3 nanocrystals of one dimensional structure. *Beilstein J. Nanotechnol.* **2011**, *2*, 585–592. [CrossRef]
11. Patil, R.S.; Uplane, M.D.; Patil, P.S. Structural and optical properties of electrodeposited molybdenum oxide thin films. *Appl. Surf. Sci.* **2006**, *252*, 8050–8056. [CrossRef]
12. Yin, X.; Han, H.; Miyamoto, A. Structure and adsorption properties of MoO_3: Insights from periodic density functional calculations. *J. Mol. Model.* **2001**, *7*, 207–215. [CrossRef]
13. Sun, Z.Z.; Yang, C.H.; Liu, G.Y.; Lu, H.B.; Zhang, R.; Wang, L.J.; Wang, H. Largely enhanced electrochemical performance in MoO_{3-x} nanobelts formed by a "sauna reaction": Importance of oxygen vacancies. *Electrochim. Acta* **2017**, *239*, 16–24. [CrossRef]
14. Song, J.M.; Ni, X.M.; Gao, L.S.; Zheng, H.G. Synthesis of metastable h-MoO_3 by simple chemical precipitation. *Mater. Chem. Phys.* **2007**, *102*, 245–248. [CrossRef]
15. Wang, M.; Koski, K.J. Reversible Chemochromic MoO_3 Nanoribbons through Zerovalent Metal Intercalation. *ACS Nano* **2015**, *9*, 3226–3233. [CrossRef]
16. Song, Y.; Zhao, Y.; Huang, Z.; Zhao, J. Aqueous synthesis of molybdenum trioxide (h-MoO_3, α-MoO_3 H_2O and h-/α-MoO_3 composites) and their photochromic properties study. *J. Alloys Compd.* **2017**, *693*, 1290–1296. [CrossRef]
17. Phuruangrat, A.; Ham, D.J.; Thongtem, S.; Lee, J.S. Electrochemical hydrogen evolution over MoO3 nanowires produced by microwave-assisted hydrothermal reaction. *Electron. Commun.* **2009**, *11*, 1740–1743. [CrossRef]
18. Wen, X.; Yang, W.; Ding, Y.; Niu, S.; Wang, Z.L. Piezoresistive effect in MoO_3 nanobelts and its application in strain-enhanced oxygen sensors. *Nano Res.* **2014**, *7*, 180–189. [CrossRef]
19. Alsaif, M.; Field, M.; Daeneke, T.; Chrimes, A.; Zhang, W.; Carey, B.; Berean, K.; Walia, S.; Embden, J.; van Zhang, B.; et al. Exfoliation Solvent Dependent Plasmon Resonances in Two Dimensional Sub-Stoichiometric Molybdenum Oxide Nanoflakes. *ACS Appl. Mater. Interfaces* **2016**, *8*, 3482–3493. [CrossRef]

20. Lu, X.; Wang, R.; Yang, F.; Jiao, W.; Liu, W.; Hao, L.; He, X. Preparation of MoO_3 QDs through combining intercalation and thermal exfoliation. *J. Mater. Chem. C* **2016**, *4*, 6720–6726. [CrossRef]
21. Soultanidis, N.; Zhou, W.; Kiely, C.J.; Wong, M.S. Solvothermal Synthesis of Ultrasmall Tungsten Oxide Nanoparticles. *Langmuir* **2012**, *28*, 17771–17777. [CrossRef]
22. Xiao, S.J.; Chu, Z.J.; Zhao, X.J.; Zhang, Z.B.; Liu, Y.H. Off-on-off detection of the activity of acetylcholine esterase and its inhibitors using MoOx quantum dots as a photoluminescent probe. *Microchim. Acta* **2017**, *184*, 4853–4860. [CrossRef]
23. Zhang, X.; Lai, Z.; Liu, Z.; Tan, C.; Huang, Y.; Li, B.; Zhao, M.; Xie, L.; Huang, W.; Zhang, H. A Facile and Universal Top-Down Method for Preparation of Monodisperse Transition-Metal Dichalcogenide Nanodots. *Angew. Chem. Int. Ed.* **2015**, *54*, 5425–5428. [CrossRef]
24. Xu, S.; Li, D.; Wu, P. One-Pot, Facile, and Versatile Synthesis of Monolayer MoS_2/WS_2 Quantum Dots as Bioimaging Probes and Efficient Electrocatalysts for Hydrogen Evolution Reaction. *Adv. Funct. Mater.* **2015**, *25*, 1127–1136. [CrossRef]
25. Huang, P.-R.; He, Y.; Cao, C.; Lu, Z.-H. Impact of Lattice Distortion and Electron Doping on R-MoO_3 Electronic Structure. *Sci. Rep.* **2014**, *4*, 7131. [CrossRef]
26. Julien, C.; Pereira-Ramos, J.-P.; Momchilov, A. *New Trends in Intercalation Compounds for Energy Storage*; Springer: Berlin, Germany, 2002; Volume 61.
27. Nicolosi, V.; Chhowalla, M.; Kanatzidis, M.G.; Strano, M.S.; Coleman, J.N. Liquid Exfoliation of Layered Materials. *Science* **2013**, *340*, 1226419. [CrossRef]
28. Shen, J.; He, Y.; Wu, J.; Gao, C.; Keyshar, K.; Zhang, X.; Yang, Y.; Ye, M.; Vajtai, R.; Lou, J.; et al. Liquid Phase Exfoliation of Two-Dimensional Materials by Directly Probing and Matching Surface Tension Components. *Nano Lett.* **2015**, *15*, 5449–5454. [CrossRef] [PubMed]
29. Hernandez, Y.; Nicolosi, V.; Lotya, M.; Blighe, F.M.; Sun, Z.; De, S.; McGovern, I.T.; Holland, B.; Byrne, M.; Gun'ko, Y.K.; et al. High-Yield Production of Graphene by Liquid-Phase Exfoliation of Graphite. *Nat. Nanotechnol.* **2008**, *3*, 563–568. [PubMed]
30. Qiao, W.; Yan, S.; He, X.; Song, X.; Li, Z.; Zhang, X.; Zhong, W.; Du, Y. Effects of Ultrasonic Cavitation Intensity on the Efficient Liquid-Exfoliation of MoS_2 Nanosheets. *RSC Adv.* **2014**, *4*, 50981–50987. [CrossRef]
31. Coleman, J.N.; Lotya, M.; O'Neill, A.; Bergin, S.D.; King, P.J.; Khan, U.; Young, K.; Gauche-r, A.; De, S.; Smith, R.J.; et al. TwoDimensional Nanosheets Produced by Liquid Exfoliation of Layered Materials. *Science* **2011**, *331*, 568–571. [CrossRef] [PubMed]
32. Hanlon, D.; Backes, C.; Higgins, T.M.; Hughes, M.; O'Neill, A.; King, P.; McEvoy, N.; Duesberg, G.S.; Mendoza Sanchez, B.; Pettersson, H.; et al. Production of Molybdenum Trioxide Nanosheets by Liquid Exfoliation and Their Application in High-Performance Supercapacitors. *Chem. Mater.* **2014**, *26*, 1751–1763. [CrossRef]
33. Alsaif, M.M.Y.A.; Latham, K.; Field, M.R.; Yao, D.D.; Medehkar, N.V.; Beane, G.A.; Kaner, R.B.; Russo, S.P.; Ou, J.Z.; Kalantar-zadeh, K. Tunable Plasmon Resonances in Two-Dimensional Molybdenum Oxide Nanoflakes. *Adv. Mater.* **2014**, *26*, 3931–3937. [CrossRef] [PubMed]
34. Cheng, H.; Qian, X.; Kuwahara, Y.; Mori, K.; Yamashita, H. A Plasmonic Molybdenum Oxide Hybrid with Reversible Tunability for Visible-Light-Enhanced Catalytic Reactions. *Adv. Mater.* **2015**, *27*, 4616–4621. [CrossRef]
35. Carey, B.J.; Daeneke, T.; Nguyen, E.P.; Wang, Y.; Zhen Ou, J.; Zhuiykov, S.; Kalantar-Zadeh, K. Two Solvent Grinding Sonication Method for the Synthesis of Two-Dimensional Tungsten Disulphide Flakes. *Chem. Comm.* **2015**, *51*, 3770–3773. [CrossRef]
36. Borgschulte, A.; Sambalova, O.; Delmelle, R.; Jenatsch, S.; Hany, R.; Nüesch, F. Hydrogen reduction of molybdenum oxide at room temperature. *Sci. Rep.* **2017**, *7*, 40761. [CrossRef]
37. Zheng, L.; Xu, Y.; Jin, D.; Xie, Y. Novel Metastable Hexagonal MoO_3 Nanobelts: Synthesis, Photochromic, and Electrochromic Properties. *Chem. Mater.* **2009**, *21*, 5681–5690. [CrossRef]
38. Alsaif, M.M.Y.A.; Balendhran, S.; Field, M.R.; Latham, K.; Wlodarski, W.; Ou, J.Z.; Kalantar-zadeh, K. Two Dimensional α- MoO_3 Nanoflakes Obtained Using Solvent-Assisted Grinding and Sonication Method: Application for H_jGas Sensing. *Sens. Actuators B* **2014**, *192*, 196–204. [CrossRef]
39. Zu, H.; Guo, Y.; Yang, H.; Huang, D.; Liu, Z.; Liu, Y.; Hu, C. Rapid room-temperature preparation of MoO3−x quantum dots by ultraviolet irradiation for photothermal treatment and glucose detection. *New J. Chem.* **2018**, *42*, 18533–18540. [CrossRef]

nanomaterials

Communication

Metal–Organic-Framework-Derived Ball-Flower-like Porous Co_3O_4/Fe_2O_3 Heterostructure with Enhanced Visible-Light-Driven Photocatalytic Activity

Qi Cao [1,*,†], Qingqing Li [2,†], Zhichao Pi [3], Jing Zhang [1], Li-Wei Sun [1], Junzhou Xu [1], Yunyi Cao [4], Junye Cheng [5,*] and Ye Bian [1,*]

1 Key Laboratory of Energy Thermal Conversion and Control of Ministry of Education, School of Energy and Environment, Wuxi Engineering Research Center of Taihu Lake Water Environment, Southeast University, Nanjing 210096, China; 213151155@seu.edu.cn (J.Z.); liwei-sun@seu.edu.cn (L.-W.S.); 213193744@seu.edu.cn (J.X.)
2 Department of Chemistry, College of Sciences, Nanjing Agricultural University, Nanjing 210095, China; 2019203072@njau.edu.cn
3 State-Operated Wuhu Machinery Plant, Wuhu 241099, China; pzc0279@163.com
4 Department of Intelligent Development Platform, Laundry Appliances Business Division of Midea Group, Wuxi 214028, China; caoyy12@midea.com
5 School of Science and Engineering, The Chinese University of Hong Kong, Shenzhen 518172, China
* Correspondence: qicao@seu.edu.cn (Q.C.); junye.cheng@polyu.edu.hk (J.C.); yebian@seu.edu.cn (Y.B.)
† These authors contributed equally to this work.

Abstract: A porous ball-flower-like Co_3O_4/Fe_2O_3 heterostructural photocatalyst was synthesized via a facile metal–organic-framework-templated method, and showed an excellent degradation performance in the model molecule rhodamine B under visible light irradiation. This enhanced photocatalytic activity can be attributed to abundant photo-generated holes and hydroxyl radicals, and the combined effects involving a porous structure, strong visible-light absorption, and improved interfacial charge separation. It is notable that the ecotoxicity of the treated reaction solution was also evaluated, confirming that an as-synthesized Co_3O_4/Fe_2O_3 catalyst could afford the sunlight-driven long-term recyclable degradation of dye-contaminated wastewater into non-toxic and colorless wastewater.

Keywords: MOF derivative; cobalt oxide; iron oxide; hierarchical heterostructure; photocatalysis

Citation: Cao, Q.; Li, Q.; Pi, Z.; Zhang, J.; Sun, L.-W.; Xu, J.; Cao, Y.; Cheng, J.; Bian, Y. Metal–Organic-Framework-Derived Ball-Flower-like Porous Co_3O_4/Fe_2O_3 Heterostructure with Enhanced Visible-Light-Driven Photocatalytic Activity. *Nanomaterials* **2022**, *12*, 904. https://doi.org/10.3390/nano12060904

Academic Editor: Stefano Agnoli

Received: 13 February 2022
Accepted: 8 March 2022
Published: 9 March 2022

1. Introduction

Various pollutants in water environments can directly cause serious harm to the lives and health of human beings, animals and plants. Organic dyes, for example rhodamine B (RhB), methylene blue and methyl orange, as one of the most common industrial pollution sources at present, have attracted tremendous attention because of their geno- and ecotoxicity [1–6]. Therefore, the development of water treatment technologies regarding dye degradation has become a top priority. Among various methods, photocatalysis is recognized as one green and efficient alternative for organic pollutant degradation, where its key issue lies in the facile preparation of highly active and stable photocatalysts [7–10].

As one of the most promising multi-functional materials, metal–organic frameworks (MOFs) are often considered to be novel photocatalysts due to their abundant and editable active sites and large surface area. However, some of their defects, such as poor light absorption and metal ion leaching due to an unstable structure, may seriously limit their practical applications [11–13]. In order to solve these problems, in this study, a flower-like cobalt 2,5-thiophenedicarboxylic coordination polymer (Co-TDC) was used as a template to synthesize a novel Co_3O_4/Fe_2O_3 heterostructural photocatalyst with improved light harvesting and photocatalytic performance. The facile preparation, structural versatility, and superior dye degradation performance of this Co_3O_4/Fe_2O_3 heterostructure provides new inspirations for the development of higher-performance photocatalysts towards water environment remediation.

2. Materials and Methods

2.1. Synthesis of Ball-Flower-like Porous Co_3O_4/Fe_2O_3 Heterostructure

Briefly, 0.1 g of Co-TDC (Sinopharm Group Co., Ltd., Shanghai, China) and 0.0482 g of $FeCl_3 \cdot 6H_2O$ (Sinopharm Group Co., Ltd., Shanghai, China) were added into 5 mL of deionized water. After 30 min of ultrasonic treatment, the mixture was dried at 60 °C for 12 h. Afterwards, the obtained powder was calcined at 550 °C for 2 h. The final Co_3O_4/Fe_2O_3 product is denoted as CF in this work for convenience.

2.2. Characterization

The chemical composition and phase structure of samples were analyzed by X-ray powder diffraction (XRD, SmartLab® by Rigaku, Tokyo, Japan). The morphology was recorded using field-emission scanning electron microscopy (SEM, JSM-7800F by JEOL, Japan) and transmission electron microscopy (TEM, JEM-2100F by JEOL, Japan). X-ray photoelectron spectroscopy (XPS, EscaLab 250Xi by Thermo Fisher Scientific, Waltham, MA, USA) was performed to investigate element distribution and valence states. The magnetism and optical properties of samples were studied using vibrating sample magnetometer (VSM, LakeShore7404 by Quantum Design, San Diego, CA, USA) and diffuse-reflection spectroscopy (DRS, Cary-5000 by Agilent, Santa Clara, CA, USA).

2.3. Photocatalysis Measurements

The adsorption and photocatalysis processes of as-prepared catalysts were evaluated by the degradation of RhB in an aqueous solution under visible light irradiation at room temperature (ca. 25 °C). A 500 W xenon lamp with a cut-off filter ($\lambda > 420$ nm) was used to generate visible light. Amounts of 0.1 g of catalyst powder and 50 mL RhB aqueous solution (initial solution pH $\approx$ 4) were added to a 100 mL quartz tube and continuously stirred during the degradation experiment. Before irradiation, the reaction solution was magnetically stirred in the dark for 30 min to reach complete adsorption/desorption equilibrium. During the photocatalytic experiment, 5 mL reaction solution was extracted every 10 min, and the concentration of residual RhB was determined by measuring its absorbance at 590 nm on a UV-visible spectrometer (UV-3600i Plus by Shimadzu, Kyoto, Japan). The 5 mL solution was added back into the reaction solution after measurement.

3. Results and Discussion

The chemical composition and crystal structure of CF were analyzed by XRD. As shown in Figure 1a, the characteristic diffraction peaks located at 19.1°, 31.2°, 36.8°, 44.7°, 59.1°, and 65.1° could be attributed to Co_3O_4 (PDF#42-1467), while the other peaks at 35.6° and 62.9° could be assigned to Fe_2O_3 (PDF#39-1346), indicating that the as-prepared sample was composed of Co_3O_4 and Fe_2O_3. The chemical states of the sample surface were further analyzed by XPS. Considering the sample preparation method, only cobalt element was studied emphatically. In Co 2p spectra (Figure 1b), the asymmetric peaks at around 780.7 eV and 796.8 eV, and shake-up type satellite peaks at 785.8 eV and 802.3 eV of Co-TDC, could be well-indexed to Co^{2+}, implying that cobalt in Co-TDC was only in the form of Co (II). On the other hand, for CF, two new peaks could be identified at around 779.4 eV and 794.3 eV, which were both ascribed to Co^{3+} [14–18]. This revealed that Co^{2+} in Co-TDC was partially oxidized to Co^{3+} during calcination, and thus Co_3O_4 was obtained as a result. Meanwhile, Fe 2p spectrum of CF was also recorded, as shown in Figure S1. It was revealed that Fe^{3+} ions were still dominant, which corresponded to the Fe_2O_3 phase. However, the minor peak at around 732.2 eV suggests that a little Fe^{3+} was reduced to Fe^{2+} along with the oxidation of Co^{2+} to Co^{3+} [19–23].

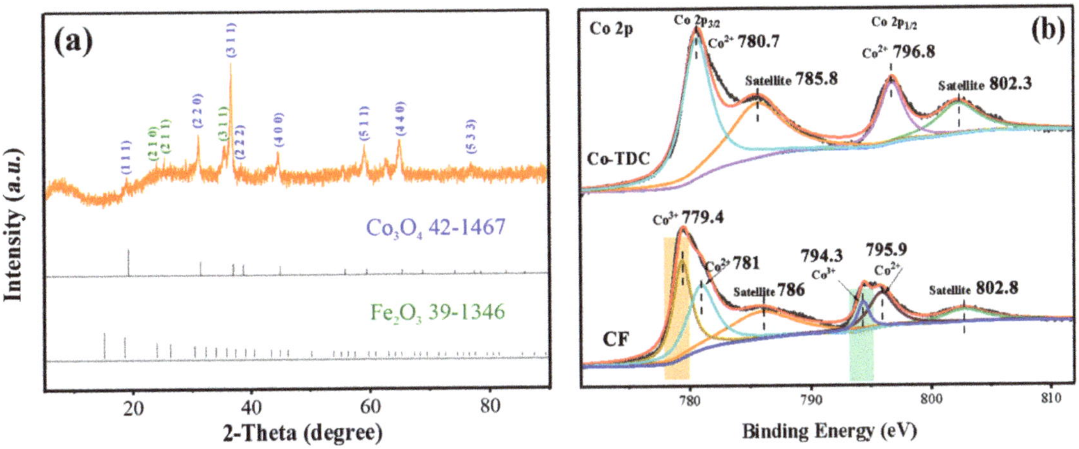

Figure 1. XRD pattern (**a**) and Co 2p XPS spectra (**b**) of as-synthesized CF catalyst.

SEM images of the CF heterostructure are shown in Figure 2a–c. It can be observed from Figure 2a,b that CF has a regular ball-flower-like morphology with a spherical size ranged at 10–20 μm, which was retained from the Co-TDC template, as shown in Figure S2 of the Supplementary Materials. It is worth noting that the sheet-like fundamental units of Co_3O_4 in CF became much more porous after calcination, with large numbers of ~200 nm Fe_2O_3 nanoparticles (Figure S3) embedded within the pores, as indicated by the yellow arrows in Figure 2c, which facilitate the adsorption and degradation of dye molecules on the surface. Moreover, the elemental mapping profiles in Figure S4 also help to verify that the distribution of Fe_2O_3 within highly porous Co_3O_4 is uniform while it is random. In order to further determine the chemical composition of the synthesized catalyst, HRTEM image was also recorded, as shown in Figure 2d. The identified two lattice fringes with an interval of 0.25 and 0.20 nm could be indexed to the (311) facet of Fe_2O_3 and (400) facet of Co_3O_4, respectively, which is in good agreement with the XRD result.

The degradation efficiencies of different samples for RhB are displayed in Figure 3a. When the catalyst was not present in the solution, RhB could hardly undergo self-degradation under visible light (i.e., black plots). The reaction solution was first stirred in the dark for 30 min for the catalyst–RhB interface to reach the adsorption/desorption equilibrium. Typically, the contribution of RhB removal by adsorption is lower than 20%, which is in proportion to the surface area of the catalyst. In photocatalysis systems, CF demonstrated a superior performance than Co-TDC and Fe_2O_3, indicating that CF possesses the highest photocatalytic activity. This could be explained by the following aspects: (i) The highly porous structure of CF provided abundant active sites, as revealed in Figure 2b,c [24–26]; (ii) The *p-n* heterojunction that formed between Fe_2O_3 and Co_3O_4 could promote the separation of photo-generated electron and hole pairs [27–30]. The promoted charge separation, and thus the inhibited charge recombination, was witnessed by the significantly decreased photoluminescence (PL) intensity of CF composites compared to pristine Fe_2O_3 particles, as displayed in Figure S5 of the Supplementary Materials [31–38]. The variation in the RhB degradation efficiency of CF in different pH conditions is presented in Figure 3b, suggesting that the catalyst could maintain a superior photocatalytic degradation activity in the pH range of 4–10, despite the fact that the degradation rate decreased to a certain extent in a strong acid environment (pH $\leq$ 2). This may be due to the dissolving of oxides by strong acid, resulting in a loss of active material in the CF catalyst for the degradation of RhB. However, considering that the actual surface water or groundwater is mostly weakly acidic or weakly alkaline, the CF catalyst is still applicable to the oxidative degradation of organic pollutants in natural water bodies.

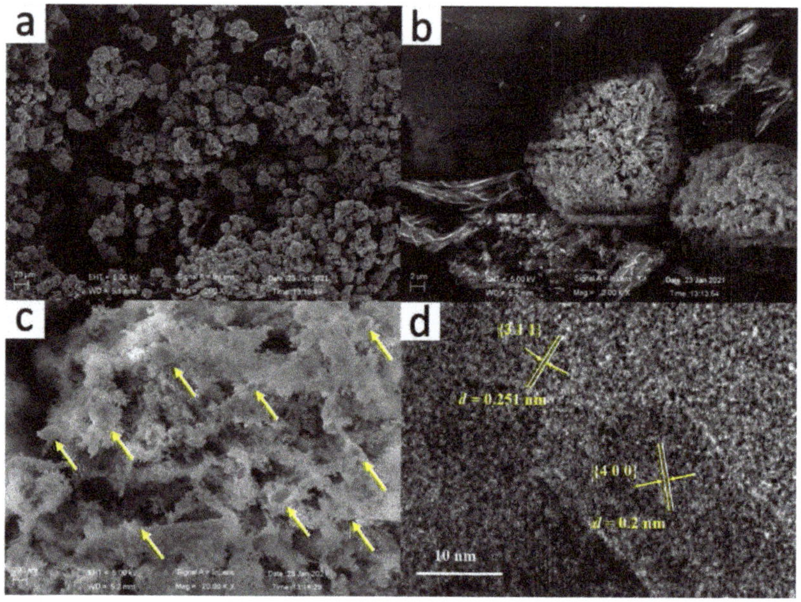

Figure 2. Representative SEM images (**a–c**), and HRTEM image (**d**) of the CF catalyst.

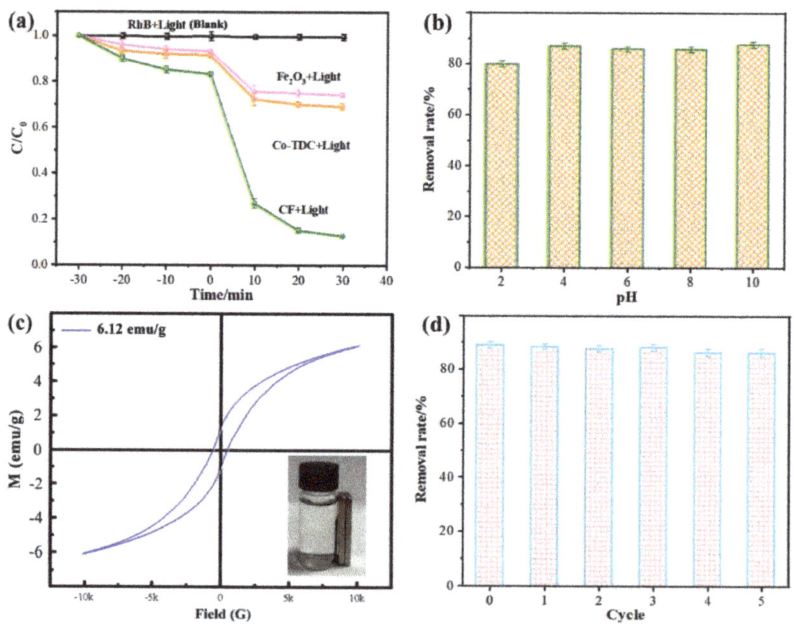

Figure 3. (**a**) RhB degradation efficiencies of different samples; (**b**) the effect of pH of reaction solution; (**c**) VSM curve of CF; (**d**) recyclability of CF photocatalyst for RhB degradation.

The service life of a catalyst is an important technical indicator for evaluating its potential for practical usage. After the reaction, the catalyst in the solution could be easily and quickly separated due to its magnetism, as revealed in Figure 3c. Then, the recycled CF catalyst was rinsed with ethanol solution to remove the residual organics on the surface.

Afterwards, it could be reused for RhB removal under the same conditions, as presented in Figure 3d. An excellent degradation efficiency of >86% was achieved after the CF catalyst was recycled and reused for five cycles, which maintained about 97% efficiency of the initial cycle (i.e., ~89.1%), confirming the recyclability of CF for long-term dye degradation in practical wastewater treatment.

Figure 4a displays the optical absorption of samples. It is observed that the absorption of Co-TDC is far lower than CF in the visible-light band. The CF catalyst maintains a superior absorption in the range of 550–750 nm, suggesting its capability for a visible-light-driven photocatalytic reaction. In addition, the threshold wavelengths of Co-TDC and CF are determined to be 619 nm and 685 nm, respectively. The corresponding bandgap and conduction band (*CB*)/valence band (*VB*) position can be calculated according to the following formulas [39–41]:

$$E_g = 1240/\lambda_g, \tag{1}$$

$$\chi(S) = \sqrt[N]{\chi_1^n \chi_2^s \cdots \chi_{n-1}^p \chi_n^q}, \tag{2}$$

$$E_{CB} = \chi(S) - E^e - \frac{1}{2}E_g, \tag{3}$$

$$E_{VB} = E_{CB} + E_g, \tag{4}$$

where E_g, λ_g, E^e, E_{CB}, and E_{VB} represent the bandgap, threshold wavelength, energy of free electrons on the hydrogen scale (~4.5 eV), and the *CB* and *VB* position, respectively. The values χ, n, and N represent the electronegativity of the constituent atom, number of species, and total number of atoms in the compound, respectively. The calculated E_g of Co-TDC and CF are 1.72 eV and 1.57 eV, indicating that the CF hybrid possesses a narrower bandgap, and thus requires less excitation energy. Thereby, Figure 4b depicts the photocatalytic mechanism of CF under visible light illumination. The photo-generated electrons in CB cannot reduce O_2 to $\cdot O_2^-$ because the E_{CB} of Co_3O_4 and Fe_2O_3 are more positive than $E(O_2/\cdot O_2^-)$ (−0.33 V vs. NHE), while the photo-generated holes are capable of oxidizing OH^- to hydroxyl radicals ($\cdot OH$) as the E_{VB} of Co_3O_4 and Fe_2O_3 are more negative than $E(\cdot OH/OH^-)$ (1.97 V vs. NHE) [42–44]. In order to further verify this perception, quenching experiments were carried out using tert-butyl alcohol (TBA), ammonium oxalate (AO) and L-ascorbic acid (L-AA) to quench the $\cdot OH$, photo-generated holes and $\cdot O_2^-$, respectively [45,46]. It can be observed from Figure S6 that the degradation efficiency of RhB clearly decreases in presence of TBA and AO. Therefore, it can be deduced that the main reactive species involved in the photocatalytic reaction are photo-generated holes and hydroxyl radicals ($\cdot OH$), which consequently degrade RhB molecules to colorless small molecules.

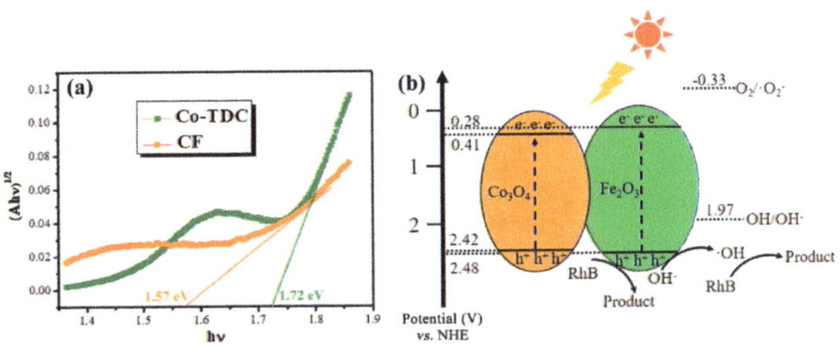

Figure 4. (**a**) Tauc plots, i.e., plots of $(\alpha h\nu)^{0.5}$ vs. photon energy ($h\nu$), derived from diffuse-reflectance spectra of the Co-TDC and CF samples; (**b**) Band alignment and photocatalytic mechanism of the CF heterojunction under visible light illumination.

In order to evaluate the ecological toxicity of the RhB solution before and after treatment, *Chlorella vulgaris* (FACHB-8) was used as the model aquatic organism being tested, and the toxicity of the residual RhB after the photocatalytic reaction was assessed according to its growth inhibition rate to *C. vulgaris*. A detailed experimental method for algae density measurement is presented in the Supplementary Materials, which could be referred to as the standard GBT 21805-2008 [47]. As exhibited in Figure 5, the growth of *C. vulgaris* was significantly suppressed in the original RhB solution, and the inhibition rate doubled as time increases. In contrast, *C. vulgaris* could grow normally in the solution after reaction, and the remaining intermediate and final products showed a neglectable influence within 24 h. Even when the incubation time was extended to 96 h, the growth inhibition rate was still about 1%, which is only 15.6% of the original RhB solution. This demonstrates that the CF catalyst can effectively degrade and mineralize RhB molecules to nearly non-toxic products.

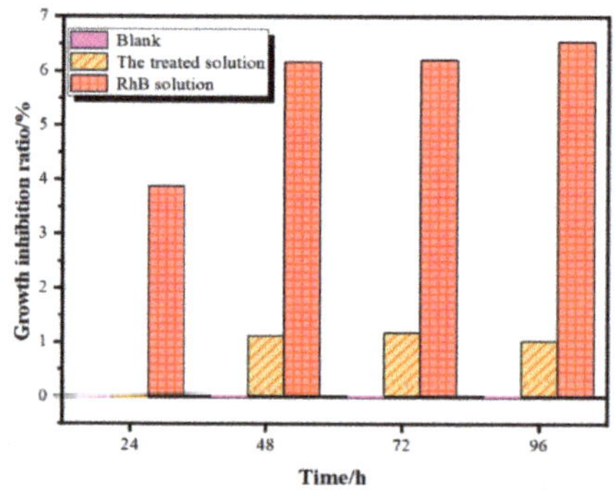

Figure 5. Time-dependent growth inhibition rates of *C. vulgaris* in different solutions.

4. Conclusions

In summary, a highly active and stable Co_3O_4/Fe_2O_3 heterostructural photocatalyst was prepared by a facile MOF-templated method, with its structure, morphology and optical properties verified by XRD, XPS, SEM and UV-visible DRS methodology. The results indicate that the CF catalyst showed a strong visible-light absorption and high photocatalytic activity towards RhB degradation. By calculating the *CB* and *VB* position, it could be inferred that hydroxyl radicals and photo-generated holes were the dominant active species in the reaction. Furthermore, the 96 h growth inhibition rate of *C. vulgaris* by the treated RhB solution was 84.4% lower than the original solution, confirming the potential of the CF photocatalyst for the sunlight-driven long-term degradation of dye molecules into non-toxic and colorless ones.

Supplementary Materials: The followings are available online at https://www.mdpi.com/article/10.3390/nano12060904/s1, Figure S1: Fe 2p XPS spectrum of as-synthesized CF catalyst, Figure S2: SEM images of pure Co-TDC at different magnifications, Figure S3: SEM images of pure Fe_2O_3 at different magnifications, Figure S4: Elemental mapping profiles of as-synthesized CF catalyst: (a) O Kα1, (b) Co Kα1, (c) Fe Kα1, Figure S5: PL spectra of CF catalyst comparing with pure Fe_2O_3 at excitation wavelength of 355 nm, Figure S6: Photocatalytic degradation efficiency of RhB with and without quenching agent, Figure S7: Photocatalytic degradation efficiency of RhB by as-synthesized CF catalyst comparing with commercial Degussa/Evonik P25-TiO_2 catalyst, Figure S8: Photocatalytic degradation efficiency of RhB by CF over a prolonged period of time; Experimental biota for toxicity test and algae density measurement.

Author Contributions: Conceptualization, Q.C. and Y.B.; Data curation, Q.L., J.Z., L.-W.S. and J.X.; Formal analysis, Q.L. and J.Z.; Funding acquisition, Q.C.; Investigation, Q.L., J.Z. and J.X.; Methodology, Q.C., Q.L., Z.P., J.Z. and L.-W.S.; Project administration, Q.C. and Y.C.; Resources, Q.C., Z.P., L.-W.S. and Y.C.; Supervision, Q.C., L.-W.S., J.C. and Y.B.; Validation, L.-W.S.; Writing—original draft, Q.L.; Writing—review and editing, Q.C., L.-W.S. and Y.B. All authors have read and agreed to the published version of the manuscript.

Funding: This research was funded by the National Natural Science Foundation of China (grant number: 52101213) and the Science and Technology Department of Jiangsu Province of China (grant number: BK20210261). The APC was funded by Southeast University of China.

Institutional Review Board Statement: Not applicable.

Informed Consent Statement: Not applicable.

Data Availability Statement: The data presented in this study are available on request from the corresponding author.

Acknowledgments: Qi Cao would like to thank the support from the "Zhi-Shan" Scholars Programme of Southeast University of China.

Conflicts of Interest: The authors declare no conflict of interest.

References

1. Han, J.; Yang, D.; Hall, D.R.; Liu, J.; Sun, J.; Gu, W.; Tang, S.; Alharbi, H.A.; Jones, P.D.; Krause, H.M.; et al. Toxicokinetics of Brominated Azo Dyes in the Early Life Stages of Zebrafish (*Danio rerio*) Is Prone to Aromatic Substituent Changes. *Environ. Sci. Technol.* **2020**, *54*, 4421–4431. [CrossRef] [PubMed]
2. Chen, M.; Jia, Y.; Li, H.; Wu, Z.; Huang, T.; Zhang, H. Enhanced pyrocatalysis of the pyroelectric $BiFeO_3/g\text{-}C_3N_4$ heterostructure for dye decomposition driven by cold-hot temperature alternation. *J. Adv. Ceram.* **2021**, *10*, 338–346. [CrossRef]
3. Cao, Q.; Liu, X.; Yuan, K.; Yu, J.; Liu, Q.; Delaunay, J.-J.; Che, R. Gold nanoparticles decorated Ag(Cl,Br) micro-necklaces for efficient and stable SERS detection and visible-light photocatalytic degradation of Sudan, I. *Appl. Catal. B Environ.* **2017**, *201*, 607–616. [CrossRef]
4. Cheng, Y.-F.; Cao, Q.; Zhang, J.; Wu, T.; Che, R. Efficient photodegradation of dye pollutants using a novel plasmonic AgCl microrods array and photo-optimized surface-enhanced Raman scattering. *Appl. Catal. B Environ.* **2017**, *217*, 37–47. [CrossRef]
5. Cao, Q.; Che, R.; Chen, N. Scalable synthesis of Cu_2S double-superlattice nanoparticle systems with enhanced UV/visible-light-driven photocatalytic activity. *Appl. Catal. B Environ.* **2015**, *162*, 187–195. [CrossRef]
6. Cao, Q.; Che, R.; Chen, N. Facile and rapid growth of Ag_2S microrod arrays as efficient substrates for both SERS detection and photocatalytic degradation of organic dyes. *Chem. Commun.* **2014**, *50*, 4931–4933. [CrossRef]
7. Lee, K.M.; Lai, C.W.; Ngai, K.S.; Juan, J.C. Recent developments of zinc oxide based photocatalyst in water treatment technology: A review. *Water Res.* **2016**, *88*, 428–448. [CrossRef]
8. Sadeghzadeh-Attar, A. Photocatalytic degradation evaluation of N–Fe codoped aligned TiO_2 nanorods based on the effect of annealing temperature. *J. Adv. Ceram.* **2020**, *9*, 107–122. [CrossRef]
9. Cao, Q.; Yuan, K.; Liu, Q.; Liang, C.; Wang, X.; Cheng, Y.-F.; Li, Q.; Wang, M.; Che, R. Porous Au–Ag Alloy Particles Inlaid AgCl Membranes As Versatile Plasmonic Catalytic Interfaces with Simultaneous, in Situ SERS Monitoring. *ACS Appl. Mater. Interfaces* **2015**, *7*, 18491–18500. [CrossRef]
10. Cao, Q.; Che, R. Tailoring Au–Ag–S composite microstructures in one-pot for both SERS detection and photocatalytic degradation of plasticizers DEHA and DEHP. *ACS Appl. Mater. Interfaces* **2014**, *6*, 7020–7027. [CrossRef]
11. Cheng, J.; Liu, K.; Li, X.; Huang, L.; Liang, J.; Zheng, G.; Shan, G. Nickel-metal-organic framework nanobelt based composite membranes for efficient Sr^{2+} removal from aqueous solution. *Environ. Sci. Ecotechnol.* **2020**, *3*, 100035. [CrossRef]
12. Cheng, J.; Liang, J.; Dong, L.; Chai, J.; Zhao, N.; Ullah, S.; Wang, H.; Zhang, D.; Imtiaz, S.; Shan, G.; et al. Self-assembly of 2D-metal–organic framework/graphene oxide membranes as highly efficient adsorbents for the removal of Cs^+ from aqueous solutions. *RSC Adv.* **2018**, *8*, 40813–40822. [CrossRef]
13. Dong, J.; Zhang, X.; Dong, X.; Ng, K.H.; Xie, Z.; Chen, I.-W.P.; Ng, Y.H.; Huang, J.; Lai, Y. Coupled porosity and heterojunction engineering: MOF-derived porous Co_3O_4 embedded on TiO_2 nanotube arrays for water remediation. *Chemosphere* **2021**, *274*, 129799. [CrossRef] [PubMed]
14. Cao, Q.; Hao, S.; Wu, Y.; Pei, K.; You, W.; Che, R. Interfacial charge redistribution in interconnected network of Ni_2P–Co_2P boosting electrocatalytic hydrogen evolution in both acidic and alkaline conditions. *Chem. Eng. J.* **2021**, *424*, 130444. [CrossRef]
15. Li, R.; Fu, Q.; Zou, X.; Zheng, Z.; Luo, W.; Yan, L. Mn–Co–Ni–O thin films prepared by sputtering with alloy target. *J. Adv. Ceram.* **2020**, *9*, 64–71. [CrossRef]
16. Zhang, L.; Liu, Y.; Tan, T.T.; Liu, Y.; Zheng, J.; Yang, Y.; Hou, X.; Feng, L.; Suo, G.; Ye, X.; et al. Thermoelectric performance enhancement by manipulation of Sr/Ti doping in two sublayers of $Ca_3Co_4O_9$. *J. Adv. Ceram.* **2020**, *9*, 769–781. [CrossRef]
17. Li, H.; Zhang, H.; Thayil, S.; Chang, A.; Sang, X.; Ma, X. Enhanced aging and thermal shock performance of $Mn_{1.95-x}Co_{0.21}Ni_{0.84}Sr_xO_4$ NTC ceramics. *J. Adv. Ceram.* **2021**, *10*, 258–270. [CrossRef]
18. Liu, L.; Huang, X.; Wei, Z.; Duan, X.; Zhong, B.; Xia, L.; Zhang, T.; Wang, H.; Jia, D.; Zhou, Y.; et al. Solvents adjusted pure phase $CoCO_3$ as anodes for high cycle stability. *J. Adv. Ceram.* **2021**, *10*, 509–519. [CrossRef]

19. Song, B.; Yuan, K.; Wei, Y.; Chen, D.; Meng, F.; Cao, Q.; Song, M.; Liu, H. In-furnace control of arsenic vapor emissions using Fe_2O_3 microspheres with good sintering resistance. *Environ. Sci. Technol.* **2021**, *55*, 8613–8621. [CrossRef]
20. Ye, F.; Dai, H.; Peng, K.; Li, T.; Chen, J.; Chen, Z.; Li, N. Effect of Mn doping on the microstructure and magnetic properties of $CuFeO_2$ ceramics. *J. Adv. Ceram.* **2020**, *9*, 444–453. [CrossRef]
21. Phor, L.; Chahal, S.; Kumar, V. Zn^{2+} substituted superparamagnetic $MgFe_2O_4$ spinel-ferrites: Investigations on structural and spin-interactions. *J. Adv. Ceram.* **2020**, *9*, 576–587. [CrossRef]
22. Chen, C.; Wang, Y.; Li, Z.; Liu, C.; Gong, W.; Tan, Q.; Han, B.; Yao, F.; Wang, K. Evolution of electromechanical properties in Fe-doped $(Pb,Sr)(Zr,Ti)O_3$ piezoceramics. *J. Adv. Ceram.* **2021**, *10*, 587–595. [CrossRef]
23. Li, J.; Tang, X.; Liu, Q.; Jiang, Y.; Tang, Z. Resistive switching and optical properties of strontium ferrate titanate thin film prepared *via* chemical solution deposition. *J. Adv. Ceram.* **2021**, *10*, 1001–1010. [CrossRef]
24. Hao, S.; Liu, J.; Cao, Q.; Zhao, Y.; Zhao, X.; Pei, K.; Zhang, J.; Chen, G.; Che, R. In-situ electrochemical pretreatment of hierarchical Ni_3S_2-based electrocatalyst towards promoted hydrogen evolution reaction with low overpotential. *J. Colloid Interface Sci.* **2020**, *559*, 282–290. [CrossRef] [PubMed]
25. Hao, S.; Cao, Q.; Yang, L.; Che, R. Morphology-optimized interconnected Ni_3S_2 nanosheets coupled with $Ni(OH)_2$ nanoparticles for enhanced hydrogen evolution reaction. *J. Alloys Compd.* **2020**, *827*, 154163. [CrossRef]
26. Cao, Q.; Yu, J.; Yuan, K.; Zhong, M.; Delaunay, J.-J. Facile and Large-Area Preparation of Porous Ag_3PO_4 Photoanodes for Enhanced Photoelectrochemical Water Oxidation. *ACS Appl. Mater. Interfaces* **2017**, *9*, 19507–19512. [CrossRef]
27. Yuan, K.; Wang, C.-Y.; Zhu, L.-Y.; Cao, Q.; Yang, J.-H.; Li, X.-X.; Huang, W.; Wang, Y.-Y.; Lu, H.-L.; Zhang, D.W. Fabrication of a Micro-Electromechanical System-Based Acetone Gas Sensor Using CeO_2 Nanodot-Decorated WO_3 Nanowires. *ACS Appl. Mater. Interfaces* **2020**, *12*, 14095–14104. [CrossRef]
28. Yuan, K.-P.; Zhu, L.-Y.; Cao, Q.; Ma, H.-P.; Tao, J.-J.; Huang, W.; Lu, H.-L. ALD-based hydrothermal facile synthesis of a dense $WO_3@TiO_2–Fe_2O_3$ nanodendrite array with enhanced photoelectrochemical properties. *J. Mater. Chem. C* **2020**, *8*, 6756–6762. [CrossRef]
29. Yuan, K.; Cao, Q.; Lu, H.-L.; Zhong, M.; Zheng, X.; Chen, H.-Y.; Wang, T.; Delaunay, J.-J.; Luo, W.; Zhang, L.; et al. Oxygen-deficient $WO_{3-x}@TiO_{2-x}$ core-shell nanosheets for efficient photoelectrochemical oxidation of neutral water solutions. *J. Mater. Chem. A* **2017**, *5*, 14697–14706. [CrossRef]
30. Yuan, K.; Cao, Q.; Li, X.; Chen, H.-Y.; Deng, Y.; Wang, Y.-Y.; Luo, W.; Lu, H.-L.; Zhang, D.W. Synthesis of $WO_3@ZnWO_4@ZnO–ZnO$ hierarchical nanocactus arrays for efficient photoelectrochemical water splitting. *Nano Energy* **2017**, *41*, 543–551. [CrossRef]
31. Lassoued, A.; Lassoued, M.S.; Dkhil, B.; Ammar, S.; Gadri, A. Synthesis, photoluminescence and magnetic properties of iron oxide (α-Fe_2O_3) nanoparticles through precipitation or hydrothermal methods. *Phys. E Low Dimens. Syst. Nanostructures* **2018**, *101*, 212–219. [CrossRef]
32. Zhang, D.; Liu, T.; Cheng, J.; Cao, Q.; Zheng, G.; Liang, S.; Wang, H.; Cao, M.S. Lightweight and high-performance microwave absorber based on 2D WS_2–RGO heterostructures. *Nano Micro Lett.* **2019**, *11*, 38. [CrossRef] [PubMed]
33. Cao, Q.; Yu, J.; Cao, Y.; Delaunay, J.-J.; Che, R. Unusual effects of vacuum annealing on large-area Ag_3PO_4 microcrystalline film photoanode boosting cocatalyst- and scavenger-free water splitting. *J. Mater.* **2021**, *7*, 929–939. [CrossRef]
34. Cao, Q.; Cheng, Y.-F.; Bi, H.; Zhao, X.; Yuan, K.; Liu, Q.; Li, Q.; Wang, M.; Che, R. Crystal defect-mediated band-gap engineering: A new strategy for tuning the optical properties of Ag_2Se quantum dots toward enhanced hydrogen evolution performance. *J. Mater. Chem. A* **2015**, *3*, 20051–20055. [CrossRef]
35. Lima, N.A.; Alencar, L.D.S.; Siu-Li, M.; Feitosa, C.A.C.; Mesquita, A.; M'peko, J.-C.; Bernardi, M.I.B. $NiWO_4$ powders prepared *via* polymeric precursor method for application as ceramic luminescent pigments. *J. Adv. Ceram.* **2020**, *9*, 55–63. [CrossRef]
36. Luchechko, A.; Shpotyuk, Y.; Kravets, O.; Zaremba, O.; Szmuc, K.; Cebulski, J.; Ingram, A.; Golovchak, R.; Shpotyuk, O. Microstructure and luminescent properties of Eu^{3+}-activated $MgGa_2O_4:Mn^{2+}$ ceramic phosphors. *J. Adv. Ceram.* **2020**, *9*, 432–443. [CrossRef]
37. Liu, N.; Mei, L.; Bin, J.; Zhang, Z.; Peng, Z. Effect of anionic group $[SiO_4]^{4-}/[PO_4]^{3-}$ on the luminescence properties of Dy^{3+}-doped tungstate structural compounds. *J. Adv. Ceram.* **2021**, *10*, 843–851. [CrossRef]
38. Cao, Q.; Che, R. Synthesis of near-infrared fluorescent, elongated ring-like Ag_2Se colloidal nanoassemblies. *RSC Adv.* **2014**, *4*, 16641–16646. [CrossRef]
39. Li, C.; Cao, Q.; Wang, F.; Xiao, Y.; Li, Y.; Delaunay, J.-J.; Zhu, H. Engineering graphene and TMDs based van der Waals heterostructures for photovoltaic and photoelectrochemical solar energy conversion. *Chem. Soc. Rev.* **2018**, *47*, 4981–5037. [CrossRef]
40. Shao, Y.; Feng, K.; Guo, J.; Zhang, R.; He, S.; Wei, X.; Lin, Y.; Ye, Z.; Chen, K. Electronic structure and enhanced photoelectrocatalytic performance of $Ru_xZn_{1-x}O/Ti$ electrodes. *J. Adv. Ceram.* **2021**, *10*, 1025–1041. [CrossRef]
41. Zhong, M.; Feng, Q.; Yuan, C.; Liu, X.; Zhu, B.; Meng, L.; Zhou, C.; Xu, J.; Wang, J.; Rao, G. Photocurrent density and electrical properties of $Bi_{0.5}Na_{0.5}TiO_3–BaNi_{0.5}Nb_{0.5}O_3$ ceramics. *J. Adv. Ceram.* **2021**, *10*, 1119–1128. [CrossRef]
42. Siahroudi, M.G.; Daryakenari, A.A.; Molamahaleh, Y.B.; Cao, Q.; Daryakenari, M.A.; Delaunay, J.-J.; Siavoshi, H.; Molaei, F. Ethylene glycol assisted solvo-hydrothermal synthesis of NGr–Co_3O_4 nanostructures for ethanol electrooxidation. *Int. J. Hydrogen Energy* **2020**, *45*, 30357–30366. [CrossRef]

43. Yu, J.; Wang, J.; Long, X.; Chen, L.; Cao, Q.; Wang, J.; Qiu, C.; Lim, J.; Yang, S. Formation of FeOOH Nanosheets Induces Substitutional Doping of CeO_{2-x} with High-Valence Ni for Efficient Water Oxidation. *Adv. Energy Mater.* **2021**, *11*, 2002731. [CrossRef]
44. Li, Z.; Dong, T.; Zhang, Y.; Wu, L.; Li, J.; Wang, X.; Fu, X. Studies on $In(OH)_yS_z$ solid solutions: Syntheses, characterizations, electronic structure, and visible-light-driven photocatalytic activities. *J. Phys. Chem. C* **2007**, *111*, 4727–4733. [CrossRef]
45. Zhao, Y.; Song, M.; Cao, Q.; Sun, P.; Chen, Y.; Meng, F. The superoxide radicals' production *via* persulfate activated with $CuFe_2O_4$@Biochar composites to promote the redox pairs cycling for efficient degradation of *o*-nitrochlorobenzene in soil. *J. Hazard. Mater.* **2020**, *400*, 122887. [CrossRef] [PubMed]
46. Meng, F.; Song, M.; Song, B.; Wei, Y.; Cao, Q.; Cao, Y. Enhanced degradation of Rhodamine B *via* α-Fe_2O_3 microspheres induced persulfate to generate reactive oxidizing species. *Chemosphere* **2020**, *243*, 125322. [CrossRef]
47. Pei, Z.-T.; Xu, R.-R.; Liu, H.-Y.; Wang, W.-Q.; Zhang, M.; Zhang, L.-L.; Zhang, J.; Wang, W.-Q.; Yu, R.; Sun, L.-W. Development and application of a novel whole sediment toxicity test using immobilized sediment and *Chlorella vulgaris*. *Ecotoxicol. Environ. Saf.* **2020**, *189*, 109979. [CrossRef]

nanomaterials

Communication

Natural Halloysite-Templated Synthesis of Highly Graphitic Boron-Doped Hollow Carbon Nanocapsule Webs

Feng Chen [1], Lulu Ma [1], Bing Li [1], Peiwen Jiang [1], Zhimin Song [1,2,*] and Lei Huang [3,*]

1 School of Environmental and Biological Engineering, Henan University of Engineering, Zhengzhou 451191, China; chenfeng871588@163.com (F.C.); malulu1001@163.com (L.M.); hngclb@126.com (B.L.); jiangpeiwen2016@163.com (P.J.)
2 School of Geosciences and Engineering, North China University of Water Resources and Electric Power, Zhengzhou 450046, China
3 School of Environmental Science and Engineering, Guangzhou University, Guangzhou 510006, China
* Correspondence: songzhimin1961@hotmail.com (Z.S.); huanglei@gzhu.edu.cn (L.H.); Tel.: +86-037162508218 (Z.S.); +86-073188830875 (L.H.)

Abstract: Hollow carbon nanocapsules have been attracting growing interest due to their fascinating characteristics and extensive potential applications. In this work, a novel natural halloysite-templated synthesis approach for highly graphitic boron-doped hollow carbon nanocapsule webs (B-HCNCWs) using glucose as the carbon source and boric acid as the heteroatom dopant was first reported. The formation process and physicochemical properties of B-HCNCWs were revealed by SEM, TEM, XRD, Raman, Brunauer–Emmett–Teller (BET), and XPS characterization techniques. The outcomes showed that the as-obtained B-HCNCWs with hollow nanocapsule network architecture had a specific surface area of 263 m^2 g^{-1}, a pore volume of 0.8 cm^3 g^{-1}, a high degree of graphitization (81.4%), graphite-like interplanar spacing (0.3370 nm), and B-containing functional groups (0.77 at%). The density function theory (DFT) calculation demonstrated that the adsorption energies of Li on B-HCNCWs were much higher than that of HCNCWs, which proved that B-doping in a carbon matrix could increase the lithium intercalation capacity.

Keywords: halloysite; carbon nanocapsule; hollow structure; boron doping; DFT calculation

Citation: Chen, F.; Ma, L.; Li, B.; Jiang, P.; Song, Z.; Huang, L. Natural Halloysite-Templated Synthesis of Highly Graphitic Boron-Doped Hollow Carbon Nanocapsule Webs. *Nanomaterials* **2022**, *12*, 2352. https://doi.org/10.3390/nano12142352

Academic Editor: Christian M. Julien

Received: 19 June 2022
Accepted: 7 July 2022
Published: 9 July 2022

Publisher's Note: MDPI stays neutral with regard to jurisdictional claims in published maps and institutional affiliations.

1. Introduction

In recent years, the design and synthesis of hollow carbon nanocapsules have aroused more and more interest due to their fascinating characteristics such as large surface area, abundant porosity, high encapsulation ability, low density, enhanced permeability, good conductivity, and excellent thermal and chemical stability [1,2]. Generally speaking, the template method (hard and soft template) is identified as a conventional and high-efficiency approach to fabricating these hollow carbon nanostructures with well-controlled morphology. Previous studies have involved a great variety of templates, such as the SBA-15, AAO, metal-organic frameworks (MOF), F127, and so on [3,4]. However, these artificial templates are pre-fabricated, which is power-wasting, time-consuming, and uneconomic [4]. With a well-defined hollow nanotubular structure, natural halloysite (Hal), a kind of aluminosilicate clay mineral with the empirical formula of $Al_2Si_2O_5(OH)_4{\cdot}2H_2O$, has merits including cost-effectiveness, environment-friendliness, huge availability, and possession of abundant mesopores, and it is distributed extensively in China, Australia, the United States, New Zealand, and Brazil [3]. Hence, it should be a promising alternative template for the controlled synthesis of hollow carbon nanomaterials.

The incorporation of heteroatoms (such as N, S, B, and P) into the carbon skeletons can further improve the physicochemical properties of hollow carbon nanocapsules and thus tremendously broaden their application [1,5,6]. Among these heteroatoms, B element has been explored as a very promising candidate for enhancing the properties of carbonaceous materials, such as conductibility, adsorption ability, surface reactivity, and lithium intercalation capacity [7,8]. In addition, increasing the degree of graphitization of carbonaceous material has also attracted the attention of many scholars all over the world, owing to its many potential applications in the fields of potassium ion batteries and photocatalytic water splitting [9,10]. However, to the best of the current authors' knowledge, a facile protocol to synthesize hollow carbon nanocapsules with a high degree of graphitization and B-doping using Hal as the template has not been reported in the literature.

Herein, we presented a natural Hal-templated synthesis technology for B-HCNCWs using glucose as the carbon precursor and boric acid as the doping agent via the hydrothermal reaction, followed by carbonization and graphitization treatment. The formation process and physicochemical properties of B-HCNCWs were then investigated in detail through a sequence of characterizations. With their hollow nanocapsule network architecture, high degree of graphitization, and B-containing functional groups, B-HCNCWs have enormous potential as electrode materials for energy storage, electrocatalytic applications, and conversion equipment. Moreover, the DFT calculation demonstrated the positive effect of B-doping in B-HCNCWs toward Li adsorption.

2. Materials and Methods

As displayed in Figure 1, the main synthetic process of B-HCNCWs comprised three steps: firstly, the HCNCWs-wrapped Hal (Hal@HCNCWs) composite was prepared by hydrothermal and carbonization treatment. Specifically, 0.6 g of Hal powder and 6 g of glucose were evenly dispersed into 70 mL of deionized water through ultrasonication for 1 h. The mixture was then moved into a 100 mL of PTFE-lined oxidation-resistant steel autoclave and maintained at 180 °C for 8 h. After cooling to indoor temperature, the as-prepared brown-black product was washed with deionized water and dehydrated ethanol a few times and then dried at 105 °C for 12 h in a vacuum drying oven, obtaining 2.1 g hydrothermal product. Afterward, the hydrothermal product (2.1 g) was put in a crucible and carbonized in a tubular furnace at 900 °C for 2 h under a nitrogen atmosphere with a heating speed of 5 °C min^{-1}, and Hal@HCNCWs (1.2 g) was obtained after cooling. Secondly, the obtained Hal@HCNCWs was exposed to 1 M HCl and HF mixed solution and stirred continuously for 24 h to remove the Hal template. The samples were filtered and washed with deionized water until the pH = 7, and then dried at 60 °C for 24 h in a drying oven to obtain HCNCWs (0.53 g). Finally, 2 g HCNCWs and 0.4143 g H_3BO_3 (3 wt% B in the sample) were mixed evenly in an agate mortar for 30 min and placed in a graphite crucible, and thermally treated at 2600 °C for 30 min with a heating speed of 20 °C min^{-1} using a graphite furnace in an argon atmosphere. When naturally cooled to room temperature, the final B-HCNCWs (1.57 g) were obtained. The information on materials (the specifications and sources of reagents) and materials characterization (SEM, TEM, XRD, Raman, BET, and XPS) are described in detail in the Supplementary Materials. The density function theory (DFT) calculation was carried out using Materials Studio. The model for calculation was also built and applied in it using the Castep model with the quality of fine. The function was GGA/PBE. Monolayer graphene sheets containing different functional groups were extracted as the simplified models and then used as the initial structures of pristine carbon and boron-doped carbon materials. Therefore, the adsorption energy (E_{ads}) of the most stable Li on the carbon materials was calculated by the following formula

$$E_{ads} = E(C \cdots Li) - E(C) - E(Li) \quad (1)$$

where $E(C\cdots Li)$ was taken as the total energy of Li adsorbed on the surface of carbon, and E(C) and E(Li) referred to the energies of the free carbon and Li, respectively.

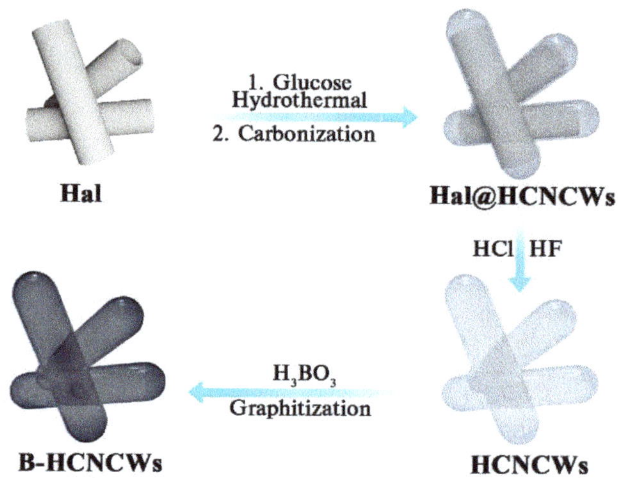

Figure 1. Schematic illustration for the synthesis of B-HCNCWs.

3. Results and Discussion

The micromorphologies and structures of natural Hal, Hal@HCNCWs, and B-HCNCWs were first characterized by scanning electron microscopy (SEM) and transmission electron microscopy (TEM) as displayed in Figures S1 and S2. From Figure S1a,b, it can be seen that natural Hal displayed the typical tubular nanostructures and the hollow lumens were readily surveyed. The length of Hal varies between a few hundred nanometers and a few micrometers, and the internal and external diameters ranged approximately from 5 to 20 and 20 to 50 nm, respectively. For Hal@HCNCWs, after the hydrothermal and high-temperature carbonization treatment, lots of carbon nanocapsules cross-linked with each other were observed as shown in Figure S1c. From Figure S1d, a thin amorphous carbon layer was found on the external surface of natural Hal, which was probably due to the adsorption of glucose molecules on the surface of Hal with the assistance of hydrogen bonding during the hydrothermal process followed by the transformation of the glucose layer into the carbon layer in the high-temperature carbonization procedure [11].

For B-HCNCWs, as displayed in Figures 2a and 2b, the tubular nanocapsule structure was surveyed and the nanocapsules interconnected with each other, forming particular porous three-dimensional webs. The length of each nanocapsule was similar to that of Hal. In addition, the TEM images of B-HCNCWs further confirmed their hollow nanocapsule structure (Figure 2c). The nanocapsules were kept very well with an outside diameter of approximately 80–120 nm after removing the Hal template. The high-resolution TEM (HRTEM) image of B-HCNCWs as shown in Figure 2d demonstrated that the carbon nanocapsules were made up of several to a dozen graphene layers, which is very similar to the microstructure of flake graphite, suggesting their high degree of graphitization [12]. The energy-dispersive spectrometer (EDS) elemental mapping of B-HCNCWs in Figure S2 verified the presence of boron (B), carbon (C), and oxygen (O) elements, which strongly proved that we had successfully synthesized boron-doped carbon materials.

Figure 3a and Figure S3 exhibited the XRD patterns of Hal, Hal@HCNCWs, HCNCWs, and B-HCNCWs. For Hal, the peaks located at 12.0°, 20.0°, 24.5°, 35.0°, 37.7°, 54.6°, and 62.4° were assigned to (001), (100), (002), (110), (003), (210), and (300) planes, respectively, which are the typical characteristic diffraction peaks of natural halloysite [13]. The diffraction peaks evident for Hal completely disappeared in the pattern of Hal@HCNCWs due to the deposition of the carbon coating layer on the surface of Hal. Two new broad peaks located in the range from 20° to 30° and from 40° to 50° were ascribed to the (002) and (100) lattice planes of graphite, respectively. The relatively broad and weak peak signals indicated the amorphous nature of the carbon coating layer, which consisted of small domains of ordered graphene sheets (Figure S3) [14]. After the removal of the Hal template, HCNCWs kept similar characteristic peaks to Hal@HCNCWs. For B-HCNCWs, four obvious and sharp characteristic peaks situated at $2\theta = 26.4°$, $42.4°$, $44.5°$, and $54.6°$ were observed, corresponding to the (002), (100), (101), and (004) planes of flake graphite [15], indicating that the amorphous carbon coating layer in HCNCWs turned into highly graphitic carbon in B-HCNCWs during the high-temperature graphitization process (2600 °C). In addition, the interplanar spacing (d_{002}) of B-HCNCWs was calculated to be 0.3370 nm according to the Bragg equation, which was very close to that of flake graphite (0.3354 nm). The graphitization degree of B-HCNCWs also reached up to 81.4% based on the calculating formula reported in the literature [16].

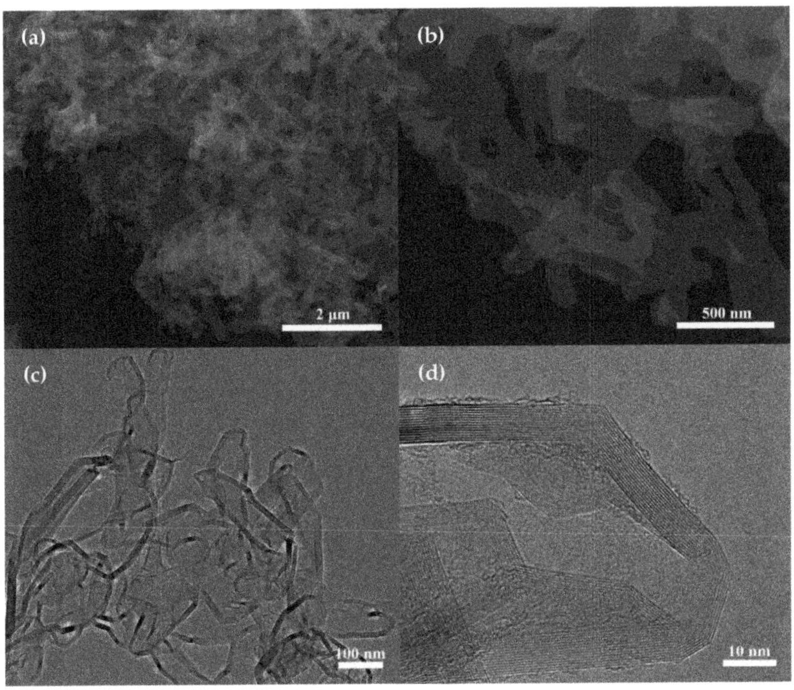

Figure 2. SEM (**a**,**b**) and TEM (**c**,**d**) images of B-HCNCWs.

The Raman spectra of Hal@HCNCWs, HCNCWs, and B-HCNCWs were shown in Figure 3b. The two distinct peaks at 1340–1350 cm^{-1} and 1565–1595 cm^{-1} in the spectra of the three samples represented the D band and G band, respectively. The D band was relevant to the defects, heteroatomic doping, and disorder induced in sp^3-bonded carbon, while the G band corresponded to the in-plane vibration of sp^2-hybridized carbon atoms of the crystalline graphite structure [3,15]. In addition, the relative strength ratio of the G band to D band ($R = I_G/I_D$) was generally used to describe the graphitization degree where a higher R value demonstrated a higher graphitic carbon content [14]. The R values of Hal@HCNCWs, HCNCWs, and B-HCNCWs were 1.06, 1.09, and 1.72, respectively. The super high R value of B-HCNCWs indicated its high graphitization degree, which was consistent with the HRTEM and XRD results.

The nitrogen adsorption–desorption isotherms of Hal, Hal@HCNCWs, HCNCWs, and B-HCNCWs were displayed in Figure 3c. The isotherms of the four specimens all belonged to the integration of type I/IV isotherms with hysteresis cycles, suggesting their microporous and mesoporous characteristics [12]. The sharp increase in N_2 adsorption at a low relative pressure of $P/P_0 < 0.1$ was normally related to the filling of micropores. The N_2 sorption volume ($P/P_0 < 0.1$) of Hal was very small, indicating that the micropores were almost negligible. Furthermore, the N_2 sorption volume ($P/P_0 < 0.1$) of Hal@HCNCWs was larger than those of HCNCWs and B-HCNCWs, suggesting the larger microporous volume of Hal@HCNCWs. The hysteresis loop at a relative pressure (P/P_0) in the scope of 0.5–1 for Hal@HCNCWs was also the most obvious, indicating its largest mesoporous volume among the four samples.

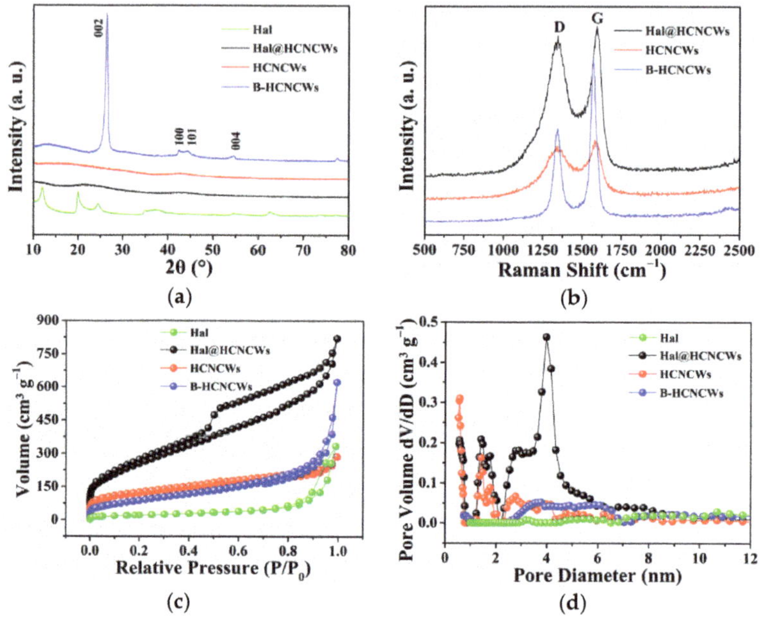

Figure 3. XRD patterns of Hal, Hal@HCNCWs, HCNCWs, and B-HCNCWs (**a**), Raman spectra (**b**) of Hal@HCNCWs, HCNCWs, and B-HCNCWs, nitrogen adsorption–desorption isotherms (**c**) and pore size distributions (**d**) of Hal, Hal@HCNCWs, HCNCWs, and B-HCNCWs.

The resulting pore size distribution patterns of Hal, Hal@HCNCWs, HCNCWs, and B-HCNCWs calculated from the N_2 adsorption data according to the DFT method were then shown in Figure 3d and Figure S4. For Hal, we can see that the pore sizes possessed the distribution between 2.5 and 25 nm and were centered at 3.2, 5.4, 8.4, 10.7, and 14.7 nm, indicating that Hal mainly contained mesoporous (Figure S4). For Hal@HCNCWs, two areas of micropores with sizes of 0.5–0.8 nm and 1.2–2.0 nm and a distinct peak centered at 4.0 nm could be surveyed, indicating that Hal@HCNCWs contained both micro- and mesopores. The HCNCWs displayed a similar area of micropores to Hal@HCNCWs except that the area of mesopores with the size of 2.0–5.6 nm disappeared, which was attributed to the etching of the Hal template. For B-HCNCWs, we can see that it contained almost negligible micropores below 1.0 nm and some mesopores between 2.4 and 7.0 nm, and few mesopores between 7.0 and 25 nm; this was because the high-temperature graphitization reduced the number of pores. Table S1 summarized the textural parameters of Hal, Hal@HCNCWs, HCNCWs, and B-HCNCWs. The corresponding specific surface areas of Hal, Hal@HCNCWs, HCNCWs, and B-HCNCWs were, respectively, 77, 920, 400, and 263 $m^2\ g^{-1}$. The pore volumes of Hal, Hal@HCNCWs, HCNCWs, and B-HCNCWs were, respectively, 0.5, 1.0, 0.3, and 0.8 $cm^3\ g^{-1}$. It is worth noting that the portion of specific surface area and porosity for B-HCNCWs was still retained even after heating at 2600 °C; such a unique textural property was conducive to the extension of its application.

Further evidence for the surface chemical compositions and states of Hal@HCNCWs, HCNCWs, and B-HCNCWs was obtained by XPS spectra as displayed in Figure 4, Figure S5 and S6. As exhibited in Figure S5a, the Hal@HCNCWs contained C, O, Si, and Al elements, and the peaks for Si and Al were not apparent due to the HCNCWs carbon coating on the outside surface of Hal. After the removal of Hal, the XPS spectrum of HCNCWs only contained C 1s and O 1s (Figure S6a). The XPS spectrum of B-HCNCWs revealed the peaks for C 1s, O 1s, and B 1s (Figure 4a), and elemental contents obtained from XPS analysis for Hal@HCNCWs, HCNCWs, and B-HCNCWs were summarized in Table S2. We can see that the contents of C, O, and B elements for B-HCNCWs were 98.58, 0.65, and 0.77 at%, respectively, indicating that B-doped carbon materials had been successfully prepared, which was in good agreement with the outcomes of EDS elemental mapping. Compared with Hal@HCNCWs and HCNCWs, the C content (98.58 at%) increased while the O content (0.65 at%) decreased for B-HCNCWs, again demonstrating its highly graphitized structure.

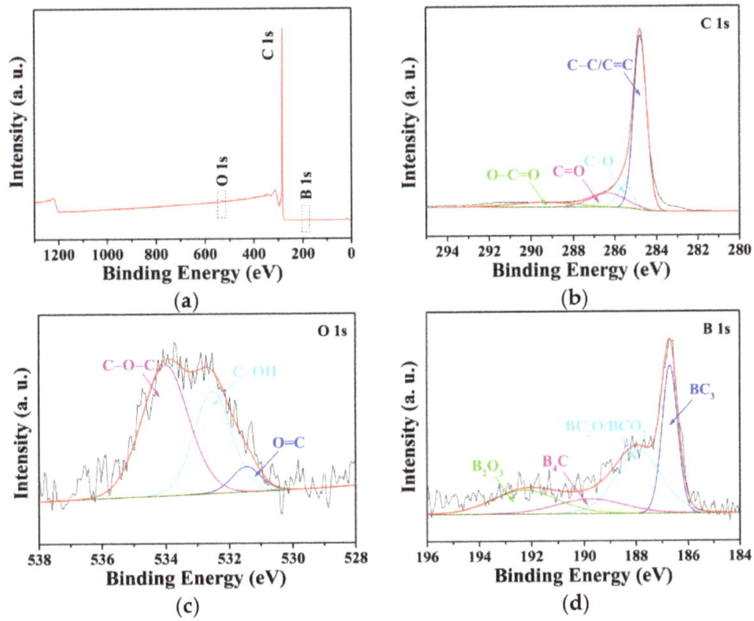

Figure 4. The XPS survey spectrum of B-HCNCWs (**a**), the high-resolution C 1s (**b**), O 1s (**c**), and B 1s (**d**) XPS spectra of B-HCNCWs.

As shown in Figure 4b, four peaks of B-HCNCWs centered at 284.8, 285.5, 286.3, and 289.4 eV were subdivided from the high-resolution C 1s spectrum, corresponding to C–C/C=C, C–O, C=O, and O–C=O, respectively [17]. The content of non-oxygenated C–C/C=C groups was as high as 66.43%, suggesting a high graphitization degree of B-HCNCWs, which was in accordance with the outcomes of the XRD and Raman analyses. The high-resolution O 1s spectrum could be fitted to three peaks located at 531.5, 532.5, and 534.0 eV, belonging to O=C, C–OH, and C–O–C, respectively (Figure 4c) [17]. The residual oxygen-containing functional groups could introduce some radicals on the surfaces of B-HCNCWs, which would greatly improve their physicochemical properties. According to the high-resolution B 1s XPS spectrum (Figure 4d), B was present in B-HCNCWs mainly as four types of B-species: BC_3 (186.7 eV), BC_2O/BCO_2 (187.9 eV), B_4C (189.8 eV), and B_2O_3 (192.3 eV) [4,18,19]. The XPS analysis suggested that B-doping has modified the surface chemistry of B-HCNCWs, which showed a good prospect in the application of the electrode materials.

In order to further verify the potential application of B-HCNCWs as electrode materials, the DFT calculation was carried out. The optimized geometries and corresponding adsorption energies of the most stable Li on HCNCWs with OH, O, BO, and B species were shown in Figure 5. We can see that the adsorption energies of Li on HCNCWs with OH, O, BO, and B species were 0.41, 1.39, 2.20, and 5.25 eV, respectively. Notably, the adsorption energies of Li on HCNCWs with BO and B species (that was B-HCNCWs) were all much higher than those of HCNCWs with OH and O species, indicating the positive effect of B-doping in the carbon matrix toward Li adsorption. Therefore, it was reasonable to conclude that B-doping in the carbon matrix could increase the lithium intercalation capacity, which also proved that B-HCNCWs have a bright prospect as the electrode materials in lithium-ion batteries. The results of the DFT calculation also verified the experimental conclusions reported in the literature [20].

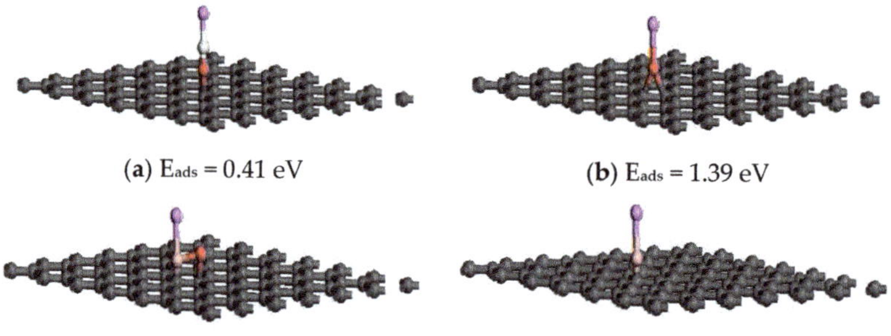

(**a**) E_{ads} = 0.41 eV (**b**) E_{ads} = 1.39 eV

(**c**) E_{ads} = 2.20 eV (**d**) E_{ads} = 5.25 eV

Figure 5. The optimized geometries and corresponding adsorption energies of the most stable Li on HCNCWs with OH (**a**), O (**b**), BO (**c**), and B (**d**) species. The gray, white, red, pink and purple balls represent C, H, O, B, and Li atoms, respectively.

4. Conclusions

In summary, B-HCNCWs were successfully synthesized via the hydrothermal reaction followed by carbonization and graphitization treatment using natural Hal as the template. The forming process and physicochemical properties of B-HCNCWs were confirmed by a sequence of characterizations. Since B-HCNCWs have hollow nanocapsule network architecture, a high degree of graphitization, graphite-like interplanar spacing, and B-containing functional groups, they could be extensively applied in the fields of electrode materials, adsorbents, catalysts, and sensors. Moreover, the DFT calculation demonstrated that B-doping in the B-HCNCWs matrix could increase the lithium intercalation capacity.

Supplementary Materials: The following are available online at https://www.mdpi.com/article/10.3390/nano12142352/s1, Figure S1: SEM (a) and TEM (b) images of natural Hal, SEM (c) and TEM (d) images of Hal@HCNCWs, Figure S2: EDS elemental maps of boron (yellow), carbon (purple), and oxygen (green) for B-HCNCWs, Figure S3: XRD patterns of Hal, Hal@HCNCWs and HCNCWs, Figure S4: pore size distributions of Hal and B-HCNCWs, Figure S5: The XPS survey spectrum of Hal@HCNCWs (a), the high-resolution C 1s (b), O 1s (c), Al 2p (d) and Si 2p (e) XPS spectra of Hal@HCNCWs, Figure S6: The XPS survey spectrum of HCNCWs (a), the high-resolution C 1s (b) and O 1s (c) XPS spectra of HCNCWs, Table S1: The textural parameters of Hal, Hal@HCNCWs, HCNCWs, and B-HCNCWs, Table S2: Elemental contents obtained from XPS analysis for Hal@HCNCWs, HCNCWs, and B-HCNCWs.

Author Contributions: Conceptualization, F.C.; software, F.C.; formal analysis, L.M., B.L. and P.J.; writing—original draft preparation, F.C.; writing—review and editing, Z.S. and L.H.; supervision, Z.S. and L.H.; funding acquisition, F.C. and Z.S. All authors have read and agreed to the published version of the manuscript.

Funding: This research was funded by NSFC-China, grant number 42172189; Natural Science Foundation of Henan province, grant number 222300420134; Key Scientific Research Projects of Henan Colleges and Universities, grant number 21A610002 and 22A170007; and Scientific Research Cultivation Fund Project of Henan University of Engineering, grant number PYXM202105.

Institutional Review Board Statement: Not applicable.

Informed Consent Statement: Not applicable.

Data Availability Statement: Not applicable.

Acknowledgments: The authors would also like to thank the workers in Shiyanjia Lab for the drafting service.

Conflicts of Interest: The authors declare no conflict of interest.

References

1. Zhu, Q.-L.; Pachfule, P.; Strubel, P.; Li, Z.; Zou, R.; Liu, Z.; Kaskel, S.; Xu, Q. Fabrication of nitrogen and sulfur co-doped hollow cellular carbon nanocapsules as efficient electrode materials for energy storage. *Energy Storage Mater.* **2018**, *13*, 72–79. [CrossRef]
2. Wang, L.; Lu, B.; Wang, S.; Cheng, W.; Zhao, Y.; Zhang, J.; Sun, X. Ultrahigh-performance of Li/Na ion batteries using N/O dual dopants porous hollow carbon nanocapsules as anode. *J. Mater. Chem. A* **2019**, *7*, 11117–11126. [CrossRef]
3. Lu, Y.; Wang, L.; Preuß, K.; Qiao, M.; Titirici, M.-M.; Varcoe, J.; Cai, Q. Halloysite-derived nitrogen doped carbon electrocatalysts for anion exchange membrane fuel cells. *J. Power Sources* **2017**, *372*, 82–90. [CrossRef]
4. Zhang, L.; Yu, Y.; Liu, B.; Liu, L.; Lv, H.; Chen, A. Synthesis of mesoporous tubular carbon using natural tubular Halloysite as template for supercapacitor. *J. Mater. Sci.: Mater. Electron.* **2018**, *29*, 12187–12194. [CrossRef]
5. Zhang, X.; Chen, Y.; Chen, M.; Yu, B.; Wang, B.; Wang, X.; Zhang, W.; Yang, D. MOF derived multi-metal oxides anchored N, P-doped carbon matrix as efficient and durable electrocatalyst for oxygen evolution reaction. *J. Colloid Interf. Sci.* **2021**, *581*, 608–618. [CrossRef] [PubMed]
6. Du, P.; Liu, L.; Dong, Y.; Li, W.; Li, J.; Liu, Z.; Wang, X. Synthesis of hierarchically porous boron-doped carbon material with enhanced surface hydrophobicity and porosity for improved supercapacitor performance. *Electrochim. Acta* **2021**, *370*, 137801. [CrossRef]
7. Chen, L.; Feng, J.; Zhou, H.; Fu, C.; Wang, G.; Yang, L.; Xu, C.; Chen, Z.; Yang, W.; Kuang, Y. Hydrothermal preparation of nitrogen, boron co-doped curved graphene nanoribbons with high dopant amounts for high-performance lithium sulfur battery cathodes. *J. Mater. Chem. A* **2017**, *5*, 7403–7415. [CrossRef]
8. Srivastava, S.; Jain, S.K.; Gupta, G.; Senguttuvan, T.D.; Gupta, B.K. Boron-doped few-layer graphene nanosheet gas sensor for enhanced ammonia sensing at room temperature. *RSC Adv.* **2020**, *10*, 1007–1014. [CrossRef] [PubMed]
9. Li, Z.; Sun, N.; Soomro, R.A.; Guan, Z.; Ma, L.; Jiang, M.; Zhu, Q.; Xu, B. Structurally engineered hollow graphitized carbon nanocages as high-performance anodes for potassium ion batteries. *ACS Nano* **2020**, *14*, 16161. [CrossRef] [PubMed]
10. Chen, S.; Zhou, X.; Liao, J.; Yang, S.; Zhou, X.; Gao, Q.; Zhang, S.; Fang, Y.; Zhong, X.; Zhang, S. FeNi intermetallic compound nanoparticles wrapped by N-doped graphitized carbon: A novel cocatalyst for boosting photocatalytic hydrogen evolution. *J. Mater. Chem. A* **2020**, *8*, 3481–3490. [CrossRef]
11. Zhang, W.; Mu, B.; Wang, A. Halloysite nanotubes template-induced fabrication of carbon/manganese dioxide hybrid nanotubes for supercapacitors. *Ionics* **2015**, *21*, 2329–2336. [CrossRef]
12. Zhou, X.; Chen, F.; Bai, T.; Long, B.; Liao, Q.; Ren, Y.; Yang, J. Interconnected highly graphitic carbon nanosheets derived from wheat stalk as high performance anode materials for lithium ion batteries. *Green Chem.* **2016**, *18*, 2078–2088. [CrossRef]
13. Dobrzański, L.A.; Tomiczek, B.; Pawlyta, M.; Nuckowski, P. TEM and XRD study of nanostructured composite materials reinforced with the halloysite particles. *Mater. Sci. Forum* **2014**, *783–786*, 1591–1596. [CrossRef]
14. Yu, K.; Wang, Y.; Wang, X.; Liu, W.; Liang, J.; Liang, C. Preparation of porous carbon anode materials for lithium-ion battery from rice husk. *Mater. Lett.* **2019**, *253*, 405–408. [CrossRef]
15. Chen, M.; Wang, Z.; Wang, A.; Li, W.; Liu, X.; Fu, L.; Huang, W. Novel self-assembled natural graphite based composite anodes with improved kinetic properties in lithium-ion batteries. *J. Mater. Chem. A* **2016**, *4*, 9865–9872. [CrossRef]
16. Greene, M.L.; Schwartz, R.W.; Treleaven, J.W. Short residence time graphitization of mesophase pitch-based carbon fibers. *Carbon* **2002**, *40*, 1217–1226. [CrossRef]
17. Ren, Y.; Chen, F.; Pan, K.; Zhao, Y.; Ma, L.; Wei, S. Studies on kinetics, isotherms, thermodynamics and adsorption mechanism of methylene blue by N and S co-doped porous carbon spheres. *Nanomaterials* **2021**, *11*, 1819. [CrossRef] [PubMed]
18. Yang, C.P.; Yin, Y.X.; Ye, H.; Jiang, K.C.; Zhang, J.; Guo, Y.G. Insight into the effect of boron doping on sulfur/carbon cathode in lithium-sulfur batteries. *ACS Appl. Mater. Interfaces* **2014**, *6*, 8789–8795. [CrossRef] [PubMed]
19. Xie, Y.; Meng, Z.; Cai, T.; Han, W.Q. Effect of boron-doping on the graphene aerogel used as cathode for the lithium-sulfur battery. *ACS Appl. Mater. Interfaces* **2015**, *7*, 25202–25210. [CrossRef] [PubMed]
20. Yin, G.; Gao, Y.; Shi, P.; Cheng, X.; Aramata, A. The effect of boron doping on lithium intercalation performance of boron-doped carbon materials. *Mater. Chem. Phys.* **2003**, *80*, 94–101. [CrossRef]

nanomaterials

Article

Photocatalytic Degradation of Organic Pollutants—Nile Blue, Methylene Blue, and Bentazon Herbicide—Using NiO-ZnO Nanocomposite

Sadaf Yasmeen [1], Luca Burratti [2,*], Leonardo Duranti [3], Emanuela Sgreccia [1] and Paolo Prosposito [1]

[1] Industrial Engineering Department, University of Rome Tor Vergata, Via del Politecnico 1, 00133 Rome, Italy; sadaf.yasmeen@students.uniroma2.eu (S.Y.); emanuela.sgreccia@uniroma2.it (E.S.); paolo.prosposito@uniroma2.it (P.P.)
[2] Department of Sciences, University of Roma Tre, Via della Vasca Navale 79, 00146 Rome, Italy
[3] Department of Chemical Science and Technologies, University of Rome Tor Vergata, Via della Ricerca Scientifica 1, 00133 Rome, Italy; leonardo.duranti@uniroma2.it
* Correspondence: luca.burratti@uniroma3.it

Abstract: Water pollution poses a significant threat to both human health and ecosystem integrity. Chemical pollutants such as dyes and pesticides affect the water quality and endanger aquatic life. Among the methods for water purification from organic pollutants, photodegradation is certainly a valid technique to decrease such contaminants. In this work, pristine NiO, ZnO, and NiO-ZnO photocatalysts were synthesized by the homogeneous co-precipitation method. X-ray diffraction confirms the formation of a photocatalyst consisting of ZnO (Hexagonal) and NiO (Cubic) structures. The crystalline size was calculated by the Scherrer formula, which is 19 nm for the NiO-ZnO photocatalyst. The band gap measurements of the prepared samples were obtained using the Tauc Plot, equation which is 2.93 eV, 3.35 eV and 2.63 eV for NiO, ZnO, and NiO-ZnO photocatalysts, respectively. The photocatalytic performance of NiO-ZnO nanocomposite was evaluated through the degradation of Methylene Blue and Nile Blue dyes under sunlight, and Bentazon herbicide under a UV light. Photocatalyst degradation efficiency was 95% and 97% for Methylene Blue and Nile Blue in 220 min under sunlight while a degradation of 70% for Bentazon after 100 min under UV light source was found.

Keywords: NiO-ZnO nanocomposite; co-precipitation method; photocatalysis; water pollutants; herbicide; dyes

Citation: Yasmeen, S.; Burratti, L.; Duranti, L.; Sgreccia, E.; Prosposito, P. Photocatalytic Degradation of Organic Pollutants—Nile Blue, Methylene Blue, and Bentazon Herbicide—Using NiO-ZnO Nanocomposite. *Nanomaterials* **2024**, *14*, 470. https://doi.org/10.3390/nano14050470

Academic Editors: Lei Huang, Junye Cheng, Hongguo Zhang and Zhenxing Wang

Received: 1 February 2024
Revised: 26 February 2024
Accepted: 1 March 2024
Published: 5 March 2024

1. Introduction

Currently, water pollution is a global issue due to its harmful effects on water species, human beings as well as animals. The release of herbicides from intensive agriculture and organic dyes from industries into freshwater reservoirs without any pretreatment has potential health effects on living beings [1,2].

Nowadays, modern agriculture uses various herbicides for the better growth of agriculture, controlling different kinds of pests and improving the food [3]. Similarly, various synthetic dyes are used in medical laboratories and industries like paint, textiles, food, and printing. The excess release of these herbicides and dyes is highly toxic for the water environment, soil fertility, aquatic creatures, and biological ecosystems. In this regard, the World Health Organization (WHO) sets the threshold levels of herbicides in drinking water at approximately 30 μg/L [4,5].

To overcome this issue, several methods have been employed to remove organic dyes and herbicides, such as coagulation, sedimentation, reverse osmosis, biological and chemical reactions, and photocatalytic activity [6]. Each method has its own advantages and limitations. In recent years, among these methods, semiconductor-mediated solar photocatalysis has been considered an efficient technique for the removal of these organic pollutants as it is an eco-friendly and sustainable approach to degrading the toxic pollutants into nontoxic molecules [7].

A literature review revealed that different semiconductor photocatalysts have been used for the degradation of dyes and herbicides. Semiconductors can be divided into two classes; n-type semiconductors, such as ZnO, CeO_2, TiO_2, SnO_2, WO_3, etc. [8–13] and p-type semiconductors that include NiO, Co_3O_4, Mn_3O_4, etc. [14–16]. Using a single metal oxide, semiconductor photocatalysts high recombination rate and poor charge carrier mobility limit the photocatalytic activity. Several techniques have been adopted, such as the mixing of two or more semiconductors [4], single doping [17], dual doping [18], and co-doping [19], to improve their charge transport properties and prevent electron-hole recombination. Metal oxides can be synthesized by different approaches, such as precipitation [20] or co-precipitation in case of two or more metal precursors [21,22], precipitation in the presence of chelating agents [23,24] or even more complex methods [25,26]. Among these approaches, the co-precipitation method ensures an easy, fast, and industrially scalable synthesis. In addition, the obtained materials have gained much attention due to their high efficiency in absorption, electron hole pair generation, and high efficiency when used for wastewater treatment and other applications [27]. In recent years, several studies have been reported on the removal of industrial dyes and herbicides using p-n heterojunction semiconductor photocatalysts such as ZnO/CdO, CuO/TiO_2, ZnO/MgO, ZnO/WO_3, NiO/ZnO [28–33] for the removal of synthetic dyes and Fe_2O_3-TiO_2 [1], ZnO/CuO [34], and Ag/TiO_2 [35] for the removal of different herbicides.

Although NiO-ZnO nanocomposite has been widely investigated for the degradation of different synthetic dyes, to the best of our knowledge, the activity of such a photocatalyst towards the degradation of Nile Blue and Bentazon has not been reported. In this work, pristine ZnO and NiO nanoparticles, along with a NiO-ZnO photocatalyst (NZP), were prepared by the co-precipitation method. The structural, morphological, photocatalytic, and optical properties were studied using X-ray diffraction (XRD), Scanning electron microscopy (SEM), Ultraviolet and visible (UV-Vis) spectroscopy, Fourier transform infrared (FTIR), and Raman spectroscopies. The degradation efficiency of the prepared photocatalyst was evaluated on Methylene Blue (MB) and Nile Blue (NB) dyes (cationic), and Bentazon (BZ) herbicides (anionic).

2. Materials and Methods

2.1. Chemicals

Nickel nitrate hexahydrate [$Ni(NO_3)_2 \cdot 6H_2O$, CAS No: 13478-00-7, purity 99%, crystals] and zinc nitrate hexahydrate [$Zn(NO_3)_2 \cdot 6H_2O$, CAS No: 10196-18-6, purity 98%, crystals] were used as precursors for the synthesis of NiO, ZnO and NZP. Sodium hydroxide (NaOH, CAS No: 1310-73-2, purity 97%, pallets) was employed as a precipitating agent. Hydrochloric acid (HCl, CAS No: 7647-01-0, concentration 37%, density 1.2 g/mL) was used for changing the pH. For the photocatalytic activity Methylene Blue ($C_{16}H_{18}ClN_3S$, CAS No: 61-73-4, dye content $\geq$ 82%, powder), Nile Blue ($C_{40}H_{40}N_6O_6S$, CAS No: 3625-57-8, dye content $\geq$ 75%, powder) and Bentazon ($C_{10}H_{12}N_2O_3S$, CAS No: 25057-89-0, purity $\geq$ 98%, powder) were used as pollutants in deionized water. All the reagents were purchased from Merck (Darmstadt, Germany) and used as received without any further refinement procedures.

2.2. Synthesis of Pristine NiO, ZnO and NiO-ZnO Photocatalyst

ZnO, NiO nanoparticles, and NZP were prepared using the homogeneous co-precipitation method. For the synthesis of NiO and ZnO nanoparticles, nickel nitrate hexahydrate (2.9079 g) and zinc nitrate hexahydrate (2.9748 g) were mixed in two separate beakers in 50 mL deionized water and magnetically stirred for 1 h. After 1 h, NaOH solution (1 M) was added dropwise to each solution until pH 9 was reached, and then the solutions were stirred for 3 h. The greenish precipitate for NiO and the white precipitate for ZnO started to form. The obtained precipitates were washed to remove impurities with distilled water and then filtered. The resulting products were dried in an oven at 60 °C for 12 h. The synthesis of NZP was based on a similar procedure, starting with a solution with a Zn:Ni mole ratio of 1:1 in 100 mL deionized water with the amount of nickel salt 2.9079 g and zinc salt 2.9748 g. Finally, dried NiO, ZnO, and NZP were grinded to obtain fine powders and then annealed at 600 °C for 2 h. The synthesis process is schematically summarized in Figure 1a.

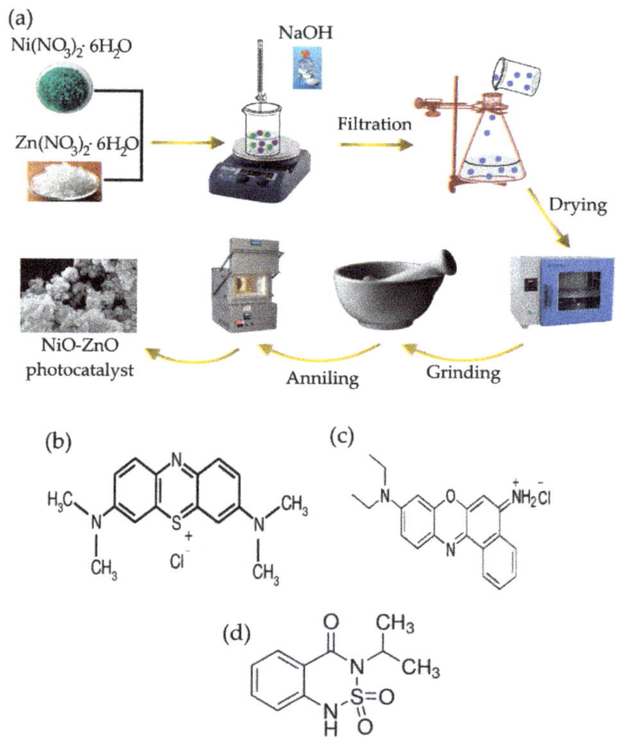

Figure 1. (**a**) Schematic representation of the synthesis process of NZP photocatalyst. Structure formula of (**b**) Methylene Blue dye, (**c**) Nile Blue dye and (**d**) Bentazon herbicide.

2.3. *Photocatalytic Test*

The photocatalytic degradation efficiency of the prepared NZP was investigated using Methylene Blue (MB) and Nile Blue (NB) as cationic dyes and Bentazon (BZ) as an herbicide pollutant. In the photocatalytic test, 30 mg of photocatalyst powder was added separately to 60 mL of a water solution of MB, NB, and BZ at a fixed concentration (5 ppm). For the adsorption–desorption equilibrium of dyes and herbicides on the surface of nanocomposite, the pollutant solutions were stirred for 1 h in the dark. The dye solutions containing NZP were exposed to solar light irradiation. In the case of BZ, the experiment was carried out under a mercury lamp (300 W) (Oriel Instruments, Newport, CA, USA), selecting the wavelength range of 220–260 with a dichroic mirror in a dark room. The pollutant solution was placed 20 cm from the lamp. The light intensity of irradiation was measured by a power meter (Thorlabs, Newton, NJ, USA, model PM100D) at wavelength 240 nm, which was about 7.8 mW during all the experiments. After regular intervals of 20 min, 2 mL of solution was taken and analyzed using a double-beam UV-Vis spectrophotometer.

The percentage degradation efficiency of the photocatalyst was calculated using the following formula [36]:

$$\% \, Degradation = (C_0 - C_t)/C_0 * 100 \tag{1}$$

with C_0 and C_t the concentration of pollutants before and after irradiation, respectively. The pH was optimized for BZ, evaluating the degradation efficiency. Three values of pH were explored: 5, 7 and 9; the values were reached using NaOH (1 M) or HCl (1 M). For the three pH values the degradation efficiency was 75%, 70% and 71% for pH 5, 7 and 9, respectively, as reported in Figure S1 of the supporting information. Due to the almost constant degradation efficiency values, for all the activities the pH was set at 7.

The self-degradation of pure organic compounds in water solution was analyzed by exposing MB and NB (concentration 5 ppm) under sunlight and BZ (5 ppm) under UV light source.

The structure of dyes and herbicide are shown in Figure 1.

2.4. *Reusability of the NZP*

The reusability of the photocatalyst was evaluated by repeating the photodegradation process for the BZ under the same reaction conditions. After completing each cycle of degradation, the photocatalyst was washed with deionized water and separated by centrifuging the samples for 10 min at 3000 rpm with a centrifuge (Thermo Fisher, Waltham, MA, USA, Megafuge 8) to collect all of the powder. After, the powder was dried for about 1 h in an oven at 70 °C and it was used again for the next cycle. Figure S5 shows the degradation efficiency of the NiO-ZnO photocatalyst for the three cycles.

2.5. *Instrumentation*

A Philips X-Pert Pro 500 (Amsterdam, The Netherlands) diffractometer X-ray diffraction (XRD) on NiO, ZnO, and NiO-ZnO powders was performed using Cu Kα radiation (λ = 1.54056 Å) in the 30–90° 2θ range, with 4 s counting time and 0.02° step size. The morphology of the samples was investigated using a Zeiss Leo SUPRA™ 35 (Oberkochen, Germany) field emission scanning electron microscope (FE-SEM). Elemental Analysis was carried out using the energy-dispersive X-ray (EDX) spectrometer. The Fourier Transform Infrared spectrophotometer (Jasco FT/IR-4X, Victoria, BC, Canada) was used to determine the functional groups. Raman data was collected using ATR8300 Raman using integral time 2000 ms and laser power 25 mW. The Optical and photocatalytic measurements were measured using a double beam UV-Vis spectrophotometer (PerkinElmer UV/VIS/NIR spectrometer Lambda 750, Shelton, CT, USA).

3. Results and Discussion

3.1. *X-ray Diffraction*

The crystal structure of the grown samples was determined using X-ray diffraction. The patterns of pristine NiO, ZnO, and NZP photocatalysts are shown in Figure 2.

Both NiO and ZnO patterns showed pure-phase samples, with peaks positions and relative intensity that closely match the reference cards of NiO rock-salt structure (JCPDS 47-1049) and ZnO hexagonal structure (JCPDS 36-1451), respectively. The diffraction peaks of both NiO and ZnO are visible in the NZP nanocomposite diffractogram; no extra peaks belonging to secondary phases are observable, indicating that NZP is only made up of NiO and ZnO. The Scherrer equation is used to determine the crystallite size "D" of the grown samples, and it can be written as [37]:

$$D = \frac{k\lambda}{\beta_{hkl} * cos\theta} \tag{2}$$

where k is a constant = 0.9, λ is the used Cu Kα radiation wavelength = 1.5406 Å, β = full width at half maximum of the peak and θ is Bragg angle [38]. The higher peaks of intensity of NZP with lower full width of half maxima shows the higher crystallinity of the photocatalyst as compared to the pristine NiO and ZnO as shown in Figure 2. The average crystallite size for NiO, ZNO, and NZP were 20 nm, 17 nm and 19 nm, respectively. The crystallite size and other XRD structural parameters such as lattice constants (a, c), unit cell volume (v), d-spacing (d), dislocation density (ρ) and strain (ε) [37,39] were calculated and listed in Table 1.

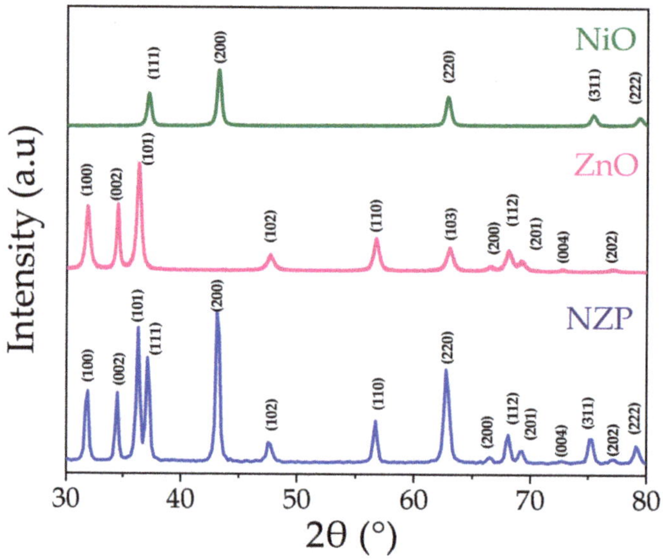

Figure 2. XRD spectra of NiO, ZnO, and NZP.

Table 1. The structural parameters of NiO, ZnO and NiO-ZnO nanocomposite.

Oxides	*a* (A)	*c* (A)	*c*/*a*	Volume (Å^3)	Micro Strain ε (×10^{-4})	d-Spacing (Å)	Dislocation Density (×10^{-3} nm^{-2})
				Individual			
NiO	4.184	-	1	72.748	2.025	1.686	3.139
ZnO	3.243	5.209	1.603	47.609	9.838	2.041	0.746
				In NZP			
NiO	4.194	-	1	73.786	12.885	1.694	1.337
ZnO	3.251	5.219	1.605	47.687	12.019	1.994	1.130

3.2. SEM Analysis and Energy Dispersive X-ray Spectroscopy

The surface morphology and chemical composition of pristine NiO, ZnO, and NZP were carried out using SEM analysis. The obtained SEM images showed that all the grown samples have nano-sized particles. Figure 3a,b reveals that ZnO nanoparticles have a rice-like structure and NiO nanoparticles have a spherical morphology with a non-homogeneous distribution. The SEM image and elemental composition of the NiO-ZnO photocatalyst are shown in Figure 3c,d. The NZP has uniform and round-shaped nanoparticles. From EDX characterization, the atomic percentage values of Ni, Zn, and O are reported in the inset of Figure 3d, indicating the presence of Nickel, Zinc, and oxygen with atomic percentages of 24%, 29%, and 46%, respectively.

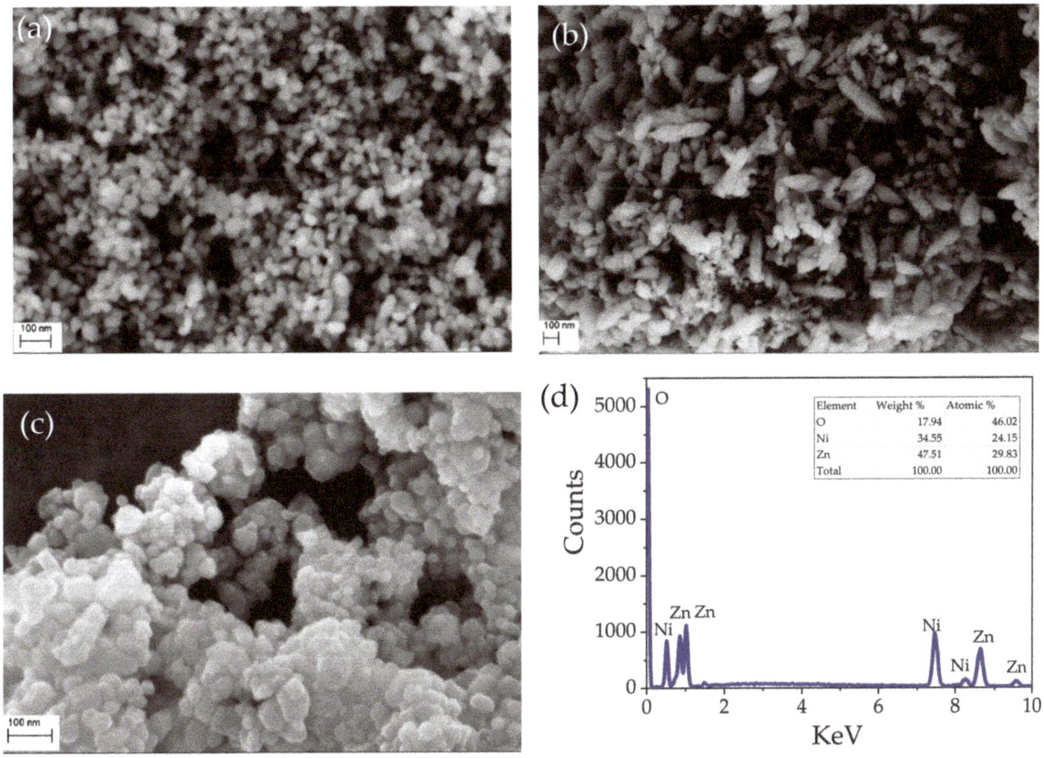

Figure 3. SEM images of: (**a**) the pristine NiO, (**b**) pristine ZnO, and (**c**) NZP; (**d**) EDS spectrum and elemental composition of NZP.

3.3. FTIR Analysis

The Fourier-transform infrared spectroscopy (FTIR) technique is used to study the major functional groups and their vibrational frequencies present in grown samples. The FTIR spectra of Pristine NiO, ZnO, and NZP are displayed in Figure 4. In the low wavenumber region (400–850 cm^{-1}), the peaks are related to metal-oxygen (M-O, M = Ni, Zn) and metal-hydroxide (M-OH) bonds [37,40]. The absorption peak at 472 cm^{-1} corresponds to the M-O vibrational mode due to the Ni-O stretching vibrations and the peaks at 447 and 503 cm^{-1} are related to ZnO stretching vibrations, while vibrations in NZP (NiO = 470 cm^{-1}, ZnO = 450 cm^{-1}) [41]. The stretching vibrations of NiO and ZnO in NZP confirm the formation of a photocatalyst [21,42]. The low-intensity peaks at 850 to 900 cm^{-1} are attributed to tetrahedral Zn^{2+} ions [43]. The peaks at 1300–1460 cm^{-1} are due to the presence of NO_3, which might not be removed well during the washing process [38].

3.4. Optical Analysis

The powder of each compound was suspended in water (concentration 5 ppm) separately, and the UV-Vis absorption spectra were recorded. Figure 5a–c shows the absorption spectra of pristine NiO, ZnO, and NZP. The absorption peaks of NiO and ZnO were observed at 279 nm and 370 nm, respectively. In the NiO-ZnO photocatalyst, the absorption peaks were centered at 320 nm and 376 nm, respectively, attributed to NiO and ZnO, which confirms the coexistence of two oxides in a single matrix. The shift in the absorption spectra of the photocatalyst might be due to the incorporation of Zn^{2+} ions into the NiO lattice. The optical energy band gaps of ZnO, NiO, and NZP were analyzed using the Tauc plot equation, which gives the correlation between the incident photon energy ($h\upsilon$) and absorption coefficient (α), as shown in Figure 5a–c (insets) [44]. The calculated energy band gaps (Eg) for NiO, ZnO, and NZP were 2.93 eV, 3.35 eV, and 2.63 eV, respectively. These measured values of band gap energy are well consistent with the literature [45–47]. The value of Eg of NZP is in the visible region, suggesting that it can enhance photocatalytic activity under sunlight.

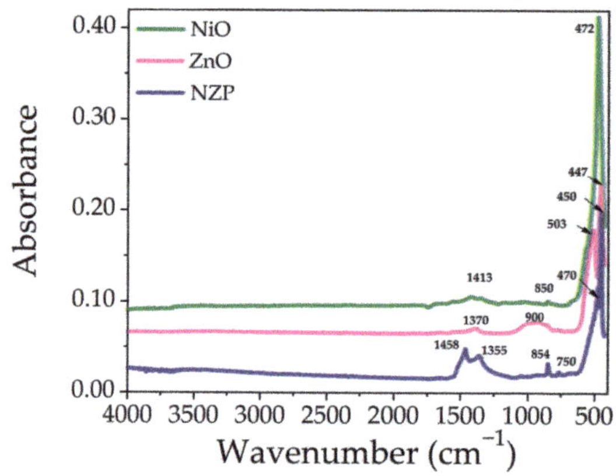

Figure 4. FTIR Spectra of pristine NiO, ZnO, and NZP in the range of 400–4000 cm^{-1}.

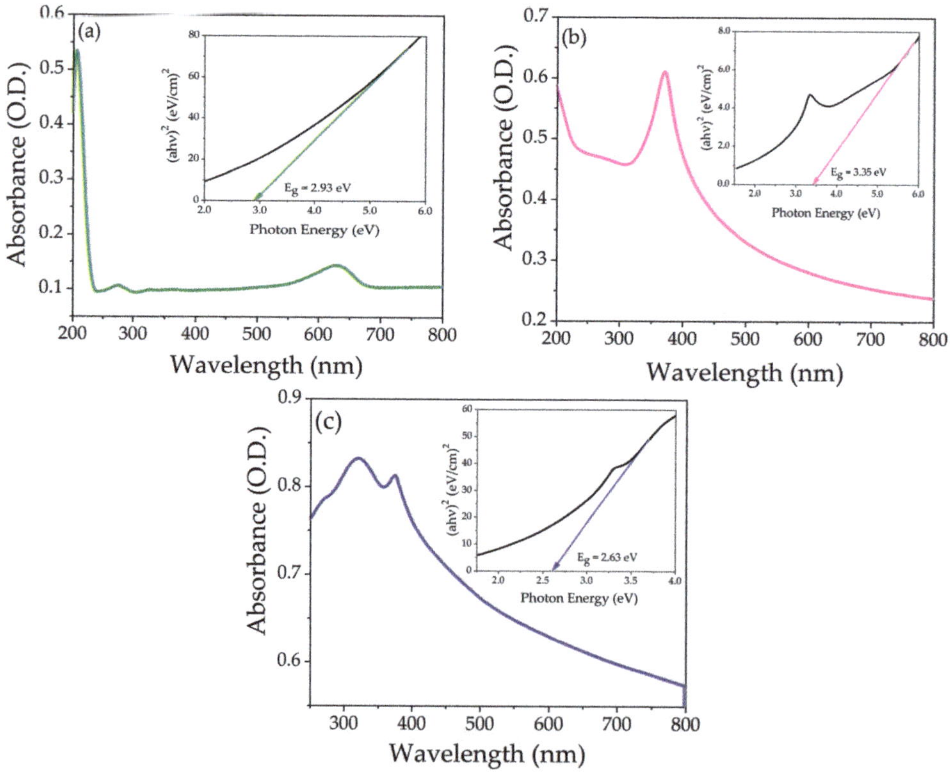

Figure 5. Absorption spectra as a function of wavelength (**a**) NiO, (**b**) ZnO, and (**c**) NZP, the insert shows Tauc's plot of energy band gap.

3.5. Raman Analysis

To study the structural and vibrational properties of prepared samples, the Raman spectroscopy technique was employed. Figure 6a–c shows the Raman spectra of ZnO, NiO nanoparticles and NZP in the spectral range of 200–700 cm^{-1}.

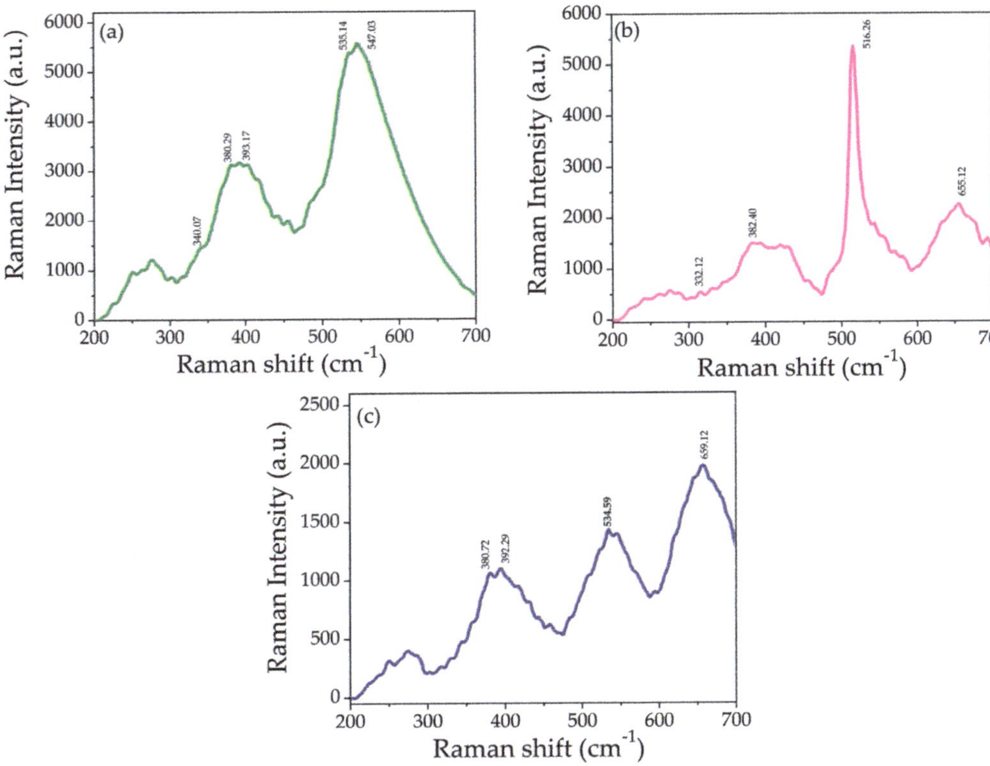

Figure 6. Raman spectra of (**a**) NiO, (**b**) ZnO, and (**c**) NZP.

The Raman spectrum of NiO showed a low-intensity band at 340.07 cm^{-1}, a medium-intensity band at 380.29 cm^{-1} and 393.17 cm^{-1}, and a high-intensity band at 535.14 cm^{-1} and 547.03 cm^{-1}, as shown in Figure 6a, and they are attributed to the active modes of cubic NiO, one-phonon (1P), one-phonon (TO), and one-phonon (LO), respectively [37,48,49].

The Raman spectrum of ZnO is shown in Figure 6b. It exhibits weak scattering peaks at 332.12 cm^{-1} and 382.20 cm^{-1} assigned to E2 (low) and E2 (high) associated with the motion of oxygen atoms in the lattice and confirmed the wurtzite crystal structure of ZnO [50]. The strong and sharp peaks at 516.26 cm^{-1} correspond to A1 (LO), and the weak peak at 655.12 cm^{-1} is an acoustic overtone with A1 symmetry, which confirmed the formation of ZnO nanoparticles [4,51]. In the Raman spectra of NZP, the NiO phase appeared at 392.29 cm^{-1} and 534.59 cm^{-1} while the ZnO phase appeared at 380.72 cm^{-1} and 659.12 cm^{-1}, which confirms the formation of the NiO-ZnO photocatalyst. The optical phonon modes of NiO and ZnO in NZP confirmed the co-existence of two phases in a single matrix. There is a slight shift in the peaks in the spectrum of NZP as shown in Figure 6c, that might be due to phonon confinement, defects (oxygen deficiency, surface impurities), and structural disorder [41,48].

3.6. Photocatalytic Activity

The photocatalytic activity of NZP was examined for MB, NB dyes under sunlight and for Bentazon herbicide under UV light at fixed concentrations of 5 ppm of contaminant at different time intervals. The absorption spectra of dyes and herbicide were measured with UV-Vis spectroscopy.

3.6.1. Degradation of Methylene Blue and Nile Blue Dyes

The photocatalytic activity of NZP was investigated using the two cationic dyes Methylene Blue and Nile Blue at fixed concentration (5 ppm) under the natural sunlight. The maximum absorption peak of MB and NB is observed at λ= 664 nm and λ= 634 nm, respectively. The photodegradation (absorbance) under direct sunlight of MB and NB dyes in the presence of NZP for various time intervals from 0 to 220 min is shown in Figure 7a,b. An evident decrease of the absorption peak as a function of time can be appreciated. In addition, after 220 min, visual degradation is reported in the inset of Figure 7a,b for MB and NB dyes before and after degradation.

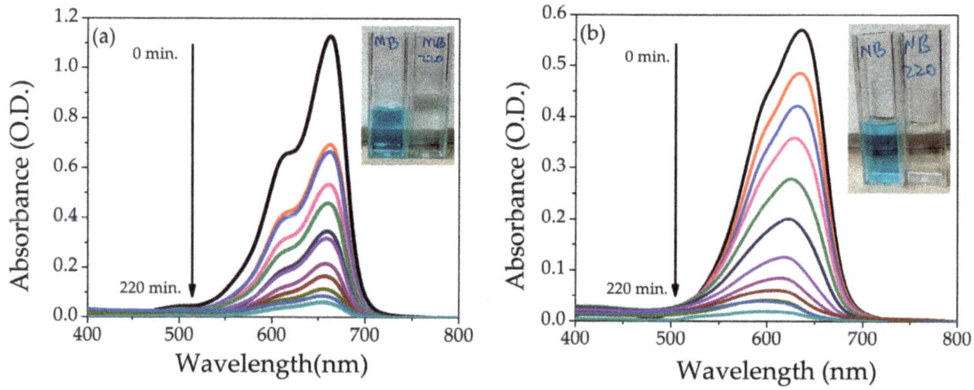

Figure 7. Absorption spectra of (**a**) MB and (**b**) NB at 5 ppm dye concentration in the presence of NZP. Insets show pictures of the cuvettes before and after sunlight irradiation of 220 min.

Figure 8a shows the percentage degradation of dyes as a function of the irradiation time. The grown photocatalyst shows higher decolorization efficiency for NB as compared to MB. The difference in degradation efficiency of both dyes may be due to the different molecular structures of MB and NB. The percentage degradation of MB and NB is 95% and 97%, respectively, in 220 min under the sunlight. The kinetic studies reveal that the photocatalytic performance of NZP can be modeled by a pseudo-first-order kinetic reaction.

$$C_t = C_0 e^{-kt} \tag{3}$$

$$ln\left(\frac{C_0}{C_t}\right) = kt \tag{4}$$

where k is the rate constant, C_0 the is initial concentration and C_t the is concentration at time t. The rate constant k is the slope of the curve obtained by plotting $ln(C_0/C_t)$ vs. irradiation time t as reported with solid line in Figure 8b. The value of k obtained for prepared NZP against MB and NB were 0.012 min^{-1} and 0.016 min^{-1}, respectively. On the other hand, the value of R^2 of the fitting were 0.970 and 0.971 for MB and NB, respectively, which also confirms the good choice of the pseudo first order reaction. The comparison of photodegradation efficiency of different metal oxide nanocomposite materials against MB and NB reported in literature is listed in Table 2. The values indicate that the photodegradation of our composite NZP against NB represents one of the best results obtained in the literature for binary composites to our knowledge.

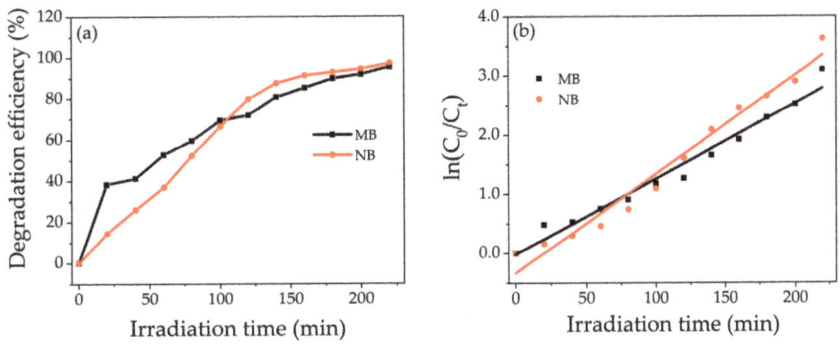

Figure 8. (**a**) Percentage degradation Efficiency of NZP for MB and NB (**b**) Degradation kinetic plot, $\ln(C_0/C_t)$ vs. irradiation time for MB (black points) and NB (red points).

Table 2. The comparison of photodegradation efficiency of different metal oxide nanocomposite materials against MB and NB.

Photocatalyst	Dyes	Source	Degradation Efficiency (%)	Ref.
ZnO-MgO	MB	Sunlight	89	[30]
ZnO-CdO	-	-	97	[28]
WO_3-ZnO	-	-	90	[52]
NiO-ZnO	-	-	95	present work
CuO-SiO_2	NB	UV-Visible	90	[53]
$FeMnO_3$	-	Sunlight	95	[54]
$CuFe_2O_4$	-	Hg lamp	93	[55]
NiO-ZnO	-	Sunlight	97	present work

For comparison, the self-degradation under sunlight of MB and NB in water solution without any catalyst was studied, and the results are shown in the supporting information. Regarding the Methylene Blue, the self-degradation was 50% in 220 min (see Figure S2), while for the NB, the self-degradation was about 26% after 220 min, as shown in Figure S3.

3.6.2. Degradation of Bentazon Herbicide

The degradation efficiency of NZP was also investigated for BZ herbicides under UV light. The photodegradation of BZ is shown in Figure 9a. The maximum degradation was obtained after 100 min under UV light. The prepared photocatalyst shows 70% degradation of BZ after 100 min under UV light, as reported in Figure 9b. As in the case of dyes previously described the value of k of the curve slope for NZP for Bentazon was obtained by plotting $\ln(C_0/C_t)$ vs. irradiation time t Figure 9c. The calculated value of k obtained by the pseudo-first-order kinetic reaction is 0.011 min^{-1}, and the value of R^2 is 0.841. The comparison study of the photodegradation efficiency of different metal oxide nanocomposite materials reported in the literature for Bentazon herbicide is listed in Table 3.

Table 3. The comparison study of photodegradation efficiency of different metal oxide nanocomposite materials against Bentazon herbicide.

Photocatalyst	Source	Irradiation Time (min)	Degradation Efficiency (%)	Ref.
Fe_2O_3-TiO_2	UV-Visible lamp	120	51	[1]
N–TiO_2–PMAA-g-PVDF/PAN	UV light	180	90	[6]
NiO-ZnO	UV light	100	70	Present work

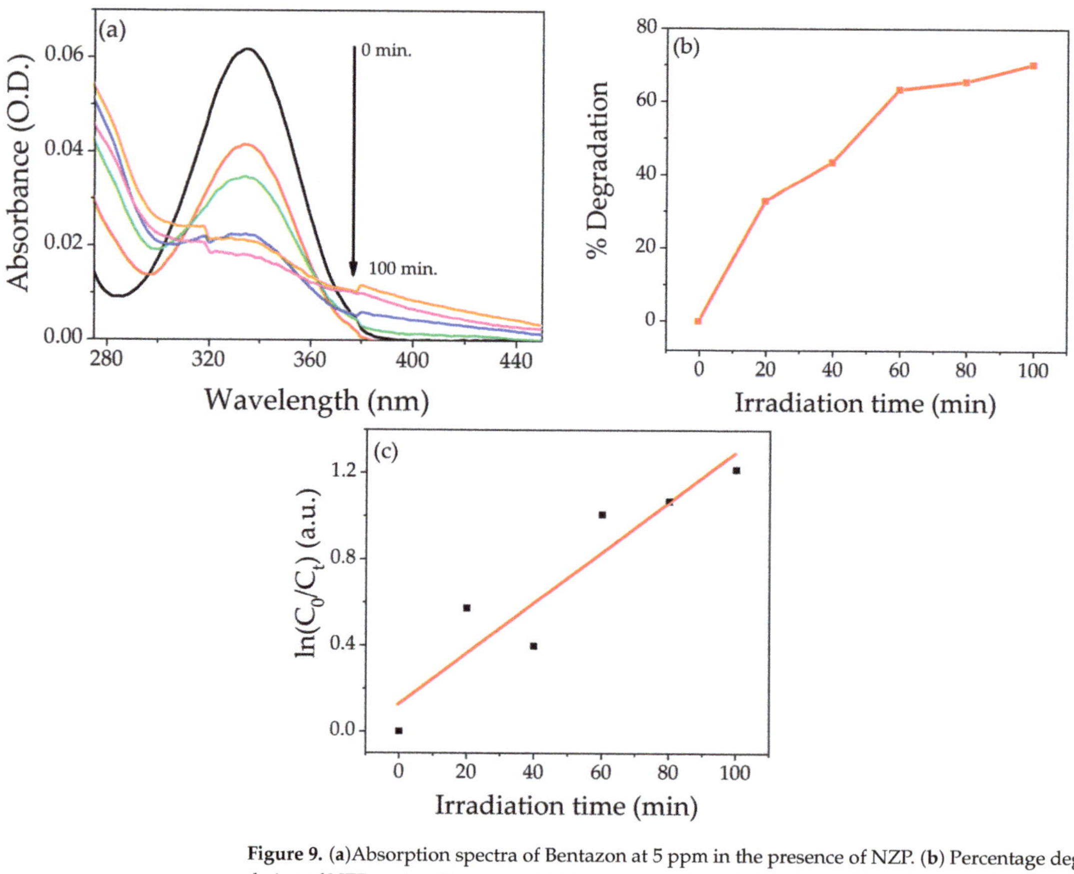

Figure 9. (**a**)Absorption spectra of Bentazon at 5 ppm in the presence of NZP. (**b**) Percentage degradation of NZP against Bentazon. (**c**) Photodegradation kinetic Plot, ln (C_0/C_t) vs. irradiation time.

Also in this case, the self-degradation of BZ was studied, and the results are shown in the supporting information in Figure S4. In the same time frame, the self-degradation reached about 38%. The presence of the catalyst is essential to boosting the degradation efficiency.

Concerning the reusability of the NZP catalyst, the efficiency in the three cycles is almost the same (~70%), as shown in Figure S5, underscoring that the composite can be reused several times.

3.7. Photodegradation Mechanism

When light strikes the NZP composite, electrons in the conduction band and holes in the valence band ($e^-{}_{CB}$ + $h^+{}_{VB}$) are generated. The oxidation and reduction processes take place at the surface of semiconductor photocatalysts. The expected photodegradation mechanism of the NiO-ZnO Photocatalyst can be summarized in four main steps:

Photo excitation:

$$\mathrm{NiO-ZnO+h\upsilon\ \rightarrow\ NiO-ZnO}\ e^-_{CB} + h^+_{VB}$$

Oxygen ion absorption:

$$\mathrm{NiO-ZnO}\ (e^-_{CB}) + \mathrm{O_2} \rightarrow \mathrm{NiO-ZnO} + {}^*\mathrm{O_2^-}$$

Ionization of water:

$$\mathrm{H_2O \rightarrow H^+ + {}^*OH}$$

$$\mathrm{{}^*O_2 + H^+ \rightarrow {}^*HO_2}$$

Protonation of superoxide:

$$\mathrm{NiO-ZnO}\ (e^-_{CB}) + *\mathrm{HO_2} + \mathrm{H^+} \rightarrow \mathrm{H_2O_2}$$

$$\mathrm{NiO-ZnO}\ \ (h^+_{VB}) + \text{dye or herbicide} \rightarrow \text{degradation products}$$

Briefly, when light is irradiated on the NiO-ZnO photocatalyst, electron-hole pairs $(e^-_{CB} + h^+_{VB})$ are generated. These photogenerated electrons react with oxygen molecules to form superoxide anion ($^*O_2{}^-$) radicals that are less toxic, while the hole reacts with hydroxyl ions to form reactive hydroxyl (*OH) radicals. These excited radicals reduce the dye and herbicide molecules, while holes oxidize the pollutants directly and cause degradation. The combination of NiO-ZnO is able to create more dynamic catalytic centers, which assist in photodegradation [56]. Figure 10 represents the schematic diagram of the action of the NiO-ZnO photocatalyst.

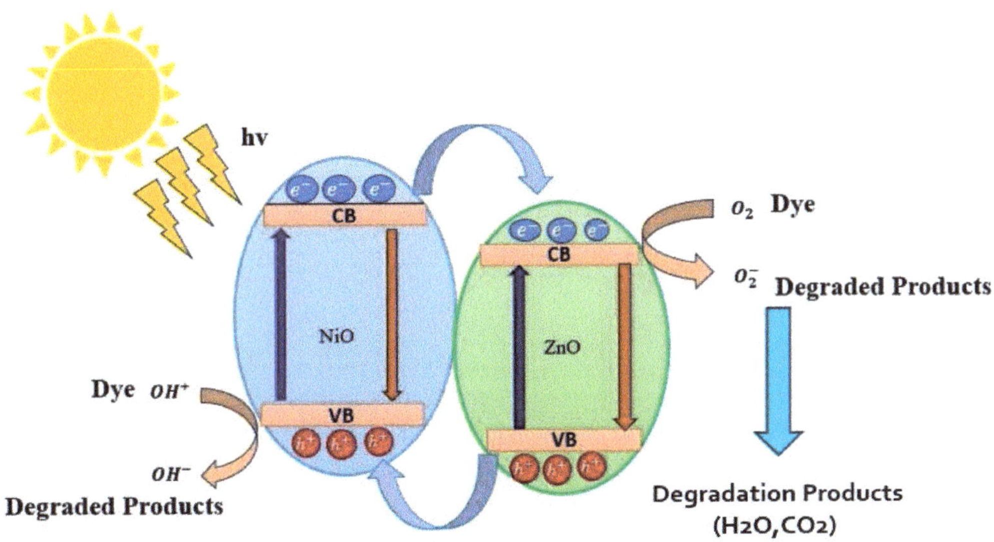

Figure 10. The schematic representation of photocatalytic mechanism for dyes in the presence of NZP.

4. Conclusions

In this study, pristine NiO, ZnO nanoparticles, and NiO-ZnO photocatalyst were synthesized and characterized for the degradation of organic pollutants. The NZP showed a smaller band gap energy (2.6 eV) compared with the pure NiO and ZiO components; consequently, this composite has a light absorption range from UV to natural light. The photocatalytic activity was investigated against MB, NB, and BZ; the degradation efficiency for dyes was 95%, 97% under sunlight, and 70% for herbicides under UV light, respectively. The photocatalyst has a recyclability of up to three cycles towards BZ without losing efficiency. Hence, this photocatalyst has great potential applications for wastewater treatment, the improvement of water quality discharge from textiles or other industries, and safeguarding the health of the ecological environment.

Supplementary Materials: The following supporting information can be downloaded at: https://www.mdpi.com/article/10.3390/nano14050470/s1, Figure S1. Degradation efficiency of BZ as a function of pH value; Figure S2. Absorption spectra of MB at 5 ppm in the absence of catalyst under sunlight irradiation; Figure S3. Absorption spectra of NB at 5 ppm in the absence of catalyst under sunlight irradiation; Figure S4. Absorption spectra of BZ at 5 ppm in the absence of catalyst under UV lamp; Figure S5: Percentage degradation of BZ under three different cycles.

Author Contributions: Conceptualization, S.Y. and L.B.; methodology, S.Y., L.B. and P.P.; validation, P.P., E.S. and L.D.; formal analysis, S.Y., E.S. and L.D.; investigation, S.Y., L.B., E.S. and L.D.; data curation, S.Y. and L.B.; writing—original draft preparation, S.Y.; writing—review and editing, L.B., P.P., E.S. and L.D.; supervision, P.P. All authors have read and agreed to the published version of the manuscript.

Funding: This research received no external funding.

Data Availability Statement: Data are contained within the article and supplementary materials.

Conflicts of Interest: The authors declare no conflicts of interests.

References

1. Braga, C.S.; Bessegato, G.G.; Maestre, K.; Espinoza-Quiñones, F.R.; Alves, H.J.; da Silva, L.C.; Eising, R.; Bariccatti, R.A. Photocatalytic Degradation of Bentazon Pesticide by a Fe_2O_3-TiO_2 Composite Catalyst Irradiated by UVA, UVB, and Natural Light. *J. Braz. Chem. Soc.* **2023**, *34*, 1284–1292. [CrossRef]
2. Jamjoum, H.A.A.; Umar, K.; Adnan, R.; Razali, M.R.; Mohamad Ibrahim, M.N. Synthesis, Characterization, and Photocatalytic Activities of Graphene Oxide/metal Oxides Nanocomposites: A Review. *Front. Chem.* **2021**, *9*, 752276. [CrossRef]
3. Shirzad-Siboni, M.; Khataee, A.; Hassani, A.; Karaca, S. Preparation, characterization and application of a CTAB-modified nanoclay for the adsorption of an herbicide from aqueous solutions: Kinetic and equilibrium studies. *Comptes Rendus Chim.* **2015**, *18*, 204–214. [CrossRef]
4. Munawar, T.; Mukhtar, F.; Yasmeen, S.; Naveed-ur-Rehman, M.; Nadeem, M.S.; Riaz, M.; Mansoor, M.; Iqbal, F. Sunlight-induced photocatalytic degradation of various dyes and bacterial inactivation using CuO–MgO–ZnO nanocomposite. *Environ. Sci. Pollut. Res.* **2021**, *28*, 42243–42260. [CrossRef]
5. Gholami, M.; Jonidi-Jafari, A.; Farzadkia, M.; Esrafili, A.; Godini, K.; Shirzad-Siboni, M. Photocatalytic removal of bentazon by copper doped zinc oxide nanorods: Reaction pathways and toxicity studies. *J. Environ. Manag.* **2021**, *294*, 112962. [CrossRef]
6. Mungondori, H.H.; Tichagwa, L.; Katwire, D.M.; Aoyi, O. Preparation of photo-catalytic copolymer grafted asymmetric membranes (N-TiO_2-PMAA-g-PVDF/PAN) and their application on the degradation of bentazon in water. *Iran. Polym. J. (Engl. Ed.)* **2016**, *25*, 135–144. [CrossRef]
7. Muhambihai, P.; Rama, V.; Subramaniam, P. Photocatalytic degradation of aniline blue, brilliant green and direct red 80 using NiO/CuO, CuO/ZnO and ZnO/NiO nanocomposites. *Environ. Nanotechnol. Monit. Manag.* **2020**, *14*, 100360. [CrossRef]
8. Uribe-López, M.C.; Hidalgo-López, M.C.; López-González, R.; Frías-Márquez, D.M.; Núñez-Nogueira, G.; Hernández-Castillo, D.; Alvarez-Lemus, M.A. Photocatalytic activity of ZnO nanoparticles and the role of the synthesis method on their physical and chemical properties. *J. Photochem. Photobiol. A Chem.* **2021**, *404*, 112866. [CrossRef]
9. Safat, S.; Buazar, F.; Albukhaty, S.; Matroodi, S. Enhanced sunlight photocatalytic activity and biosafety of marine-driven synthesized cerium oxide nanoparticles. *Sci. Rep.* **2021**, *11*, 14734. [CrossRef] [PubMed]
10. Nasikhudin; Diantoro, M.; Kusumaatmaja, A.; Triyana, K. Study on Photocatalytic Properties of TiO_2 Nanoparticle in various pH condition. *J. Phys. Conf. Ser.* **2018**, *1011*, 012069. [CrossRef]
11. Kim, S.P.; Choi, M.Y.; Choi, H.C. Photocatalytic activity of SnO_2 nanoparticles in methylene blue degradation. *Mater. Res. Bull.* **2016**, *74*, 85–89. [CrossRef]
12. Azmat, S.; Jan, T.; Ilyas, S.Z.; Hassan, A.; Habib, I.; Mahmood, Q.; Mahmood, A. Solar light triggered photocatalytic performance of WO_3 nanostructures; Waste water treatment. *Mater. Res. Express* **2018**, *5*, 115025. [CrossRef]

13. Motelica, L.; Oprea, O.-C.; Vasile, B.-S.; Ficai, A.; Ficai, D.; Andronescu, E.; Holban, A.M. Antibacterial Activity of Solvothermal Obtained ZnO Nanoparticles with Different Morphology and Photocatalytic Activity against a Dye Mixture: Methylene Blue, Rhodamine B and Methyl Orange. *Int. J. Mol. Sci.* **2023**, *24*, 5677. [CrossRef]
14. Haider, A.J.; Al-Anbari, R.; Sami, H.M.; Haider, M.J. Photocatalytic Activity of Nickel Oxide. *J. Mater. Res. Technol.* **2019**, *8*, 2802–2808. [CrossRef]
15. Althamthami, M.; Guettaf Temam, E.; Ben Temam, H.; Hasan, G.G.; Malfi, N. Influence of hole-scavenger and different withdrawn speeds on photocatalytic activity of Co3O4 thin films under sunlight irradiation. *Ceram. Int.* **2022**, *48*, 31570–31578. [CrossRef]
16. Osgouei, M.S.; Khatamian, M.; Kakili, H. Improved visible-light photocatalytic activity of Mn_3O_4-based nanocomposites in removal of methyl orange. *Mater. Chem. Phys.* **2020**, *239*, 122108. [CrossRef]
17. Jan, T.; Azmat, S.; Rahman, A.U.; Ilyas, S.Z.; Mehmood, A. Experimental and DFT study of Al doped ZnO nanoparticles with enhanced antibacterial activity. *Ceram. Int.* **2022**, *48*, 20838–20847. [CrossRef]
18. Wang, S.; Bai, L.; Ao, X. Preparation and photocatalytic application of a S, Nd double doped nano-TiO_2 photocatalyst. *RSC Adv.* **2018**, *8*, 36745–36753. [CrossRef] [PubMed]
19. Yasmeen, S.; Munawar, T.; Asghar, M.; Khan, M.A.; Hussain, A.; Iqbal, F. Synthesis and photocatalytic study of $Zn_{0.90}Co_{0.10}O$ and $Zn_{0.90}Co_{0.05}M_{0.05}O$ (M = Ca, Ba, Cr, Pb) nanocrystals: Structural, optical and electrical investigations. *J. Mater. Res. Technol.* **2020**, *9*, 4076–4096. [CrossRef]
20. Sun, Q.; Bao, S. Effects of Reaction Temperature on Microstructure and Advanced Pseudocapacitor Properties of NiO Prepared via Simple Precipitation Method. *Nano-Micro Lett.* **2013**, *5*, 289–295. [CrossRef]
21. Sharma, R.K.; Kumar, D.; Ghose, R. Synthesis of nanocrystalline ZnO-NiO mixed metal oxide powder by homogeneous precipitation method. *Ceram. Int.* **2016**, *42*, 4090–4098. [CrossRef]
22. Hameed, A.; Montini, T.; Gombac, V.; Fornasiero, P. Photocatalytic decolourization of dyes on NiO–ZnO nano-composites. *Photochem. Photobiol. Sci.* **2009**, *8*, 677–682. [CrossRef] [PubMed]
23. Liu, P.; Ng, V.M.H.; Yao, Z.; Zhou, J.; Lei, Y.; Yang, Z.; Lv, H.; Kong, L.B. Facile Synthesis and Hierarchical Assembly of Flowerlike NiO Structures with Enhanced Dielectric and Microwave Absorption Properties. *ACS Appl. Mater. Interfaces* **2017**, *9*, 16404–16416. [CrossRef]
24. Yew, Y.P.; Shameli, K.; Miyake, M.; Ahmad Khairudin, N.B.B.; Mohamad, S.E.B.; Naiki, T.; Lee, K.X. Green biosynthesis of superparamagnetic magnetite Fe_3O_4 nanoparticles and biomedical applications in targeted anticancer drug delivery system: A review. *Arab. J. Chem.* **2020**, *13*, 2287–2308. [CrossRef]
25. Al-Yunus, A.; Al-Arjan, W.; Traboulsi, H.; Schuarca, R.; Chando, P.; Hosein, I.D.; Hessien, M. Effect of Synthesis Conditions on CuO-NiO Nanocomposites Synthesized via Saponin-Green/Microwave Assisted-Hydrothermal Method. *Nanomaterials* **2024**, *14*, 308. [CrossRef]
26. Dorneanu, P.P.; Airinei, A.; Olaru, N.; Homocianu, M.; Nica, V.; Doroftei, F. Preparation and characterization of NiO, ZnO and NiO–ZnO composite nanofibers by electrospinning method. *Mater. Chem. Phys.* **2014**, *148*, 1029–1035. [CrossRef]
27. Ishaque, M.Z.; Zaman, Y.; Arif, A.; Siddique, A.B.; Shahzad, M.; Ali, D.; Aslam, M.; Zaman, H.; Faizan, M. Fabrication of ternary metal oxide (ZnO:NiO:CuO) nanocomposite heterojunctions for enhanced photocatalytic and antibacterial applications. *RSC Adv.* **2023**, *13*, 30838–30854. [CrossRef] [PubMed]
28. Warshagha, M.Z.A.; Muneer, M. Facile synthesis of CdO-ZnO heterojunction photocatalyst for rapid removal of organic contaminants from water using visible light. *Environ. Nanotechnol. Monit. Manag.* **2022**, *18*, 100728. [CrossRef]
29. Lu, D.; Zelekew, O.A.; Abay, A.K.; Huang, Q.; Chen, X.; Zheng, Y. Synthesis and photocatalytic activities of a CuO/TiO_2 composite catalyst using aquatic plants with accumulated copper as a template. *RSC Adv.* **2019**, *9*, 2018–2025. [CrossRef] [PubMed]
30. Panchal, P.; Paul, D.R.; Sharma, A.; Hooda, D.; Yadav, R.; Meena, P.; Nehra, S.P. Phytoextract mediated ZnO/MgO nanocomposites for photocatalytic and antibacterial activities. *J. Photochem. Photobiol. A Chem.* **2019**, *385*, 112049. [CrossRef]
31. Mugunthan, E.; Saidutta, M.B.; Jagadeeshbabu, P.E. Photocatalytic activity of ZnO-WO_3 for diclofenac degradation under visible light irradiation. *J. Photochem. Photobiol. A Chem.* **2019**, *383*, 111993. [CrossRef]
32. Aziz, F.; Abo-Dief, H.M.; Warsi, A.Z.; Warsi, M.F.; Shahid, M.; Ahmad, T.; Mersal, G.A.M.; Ibrahim, M.M. Facile synthesis of NiO/ZnO nano-composite by Co-precipitation, characterization and photocatalytic study of colored and colorless organic pollutants by solar irradiation. *Phys. B Condens. Matter* **2022**, *640*, 413858. [CrossRef]
33. Rogozea, E.A.; Petcu, A.R.; Olteanu, N.L.; Lazar, C.A.; Cadar, D.; Mihaly, M. Tandem adsorption-photodegradation activity induced by light on NiO-ZnO p–n couple modified silica nanomaterials. *Mater. Sci. Semicond. Process.* **2017**, *57*, 1–11. [CrossRef]
34. Aghaei, M.; Sajjadi, S.; Keihan, A.H. Sono-coprecipitation synthesis of ZnO/CuO nanophotocatalyst for removal of parathion from wastewater. *Environ. Sci. Pollut. Res.* **2020**, *27*, 11541–11553. [CrossRef] [PubMed]
35. Ma, J.; Yang, M.; Sun, Y.; Li, C.; Li, Q.; Gao, F.; Yu, F.; Chen, J. Fabrication of Ag/TiO_2 nanotube array with enhanced photo-catalytic degradation of aqueous organic pollutant. *Phys. E Low-Dimens. Syst. Nanostructures* **2014**, *58*, 24–29. [CrossRef]
36. Munawar, T.; Yasmeen, S.; Mukhtar, F.; Nadeem, M.S.; Mahmood, K.; Saqib Saif, M.; Hasan, M.; Ali, A.; Hussain, F.; Iqbal, F. $Zn_{0.9}Ce_{0.05}M_{0.05}O$ (M = Er, Y, V) nanocrystals: Structural and energy bandgap engineering of ZnO for enhancing photocatalytic and antibacterial activity. *Ceram. Int.* **2020**, *46*, 14369–14383. [CrossRef]
37. Munawar, T.; Iqbal, F.; Yasmeen, S.; Mahmood, K.; Hussain, A. Multi metal oxide NiO-CdO-ZnO nanocomposite–synthesis, structural, optical, electrical properties and enhanced sunlight driven photocatalytic activity. *Ceram. Int.* **2020**, *46*, 2421–2437. [CrossRef]

38. Yasmeen, S.; Iqbal, F.; Munawar, T.; Nawaz, M.A.; Asghar, M.; Hussain, A. Synthesis, structural and optical analysis of surfactant assisted ZnO–NiO nanocomposites prepared by homogeneous precipitation method. *Ceram. Int.* **2019**, *45*, 17859–17873. [CrossRef]
39. Munawar, T.; Yasmeen, S.; Hasan, M.; Mahmood, K.; Hussain, A.; Ali, A.; Arshad, M.I.; Iqbal, F. Novel tri-phase heterostructured ZnO–Yb_2O_3–Pr_2O_3 nanocomposite; structural, optical, photocatalytic and antibacterial studies. *Ceram. Int.* **2020**, *46*, 11101–11114. [CrossRef]
40. Subhan, M.A.; Fahim, A.M.M.; Saha, P.C.; Rahman, M.M.; Begum, K.; Azad, A.K. Structural study, photoluminescence and photocatalytic properties of La_2O_3·Fe_3O_4·ZnO, AgO·NiO·ZnO and La_2O_3·AgO·ZnO nanocomposites. *Nano-Struct. Nano-Objects* **2017**, *10*, 30–41. [CrossRef]
41. Satvekar, B.S.; Anekar, S. Synthesis of NiO-ZnO nanocomposite by sol-gel method and their characterization. *High Technol. Lett.* **2023**, *29*, 284–290.
42. Rezaei, M.; Nezamzadeh-Ejhieha, A. The ZnO-NiO nano-composite: A brief characterization, kinetic and thermodynamic study and study the Arrhenius model on the sulfasalazine photodegradation. *Int. J. Hydrogen Energy* **2020**, *45*, 24749–24764. [CrossRef]
43. Motelica, L.; Vasile, B.-S.; Ficai, A.; Surdu, A.-V.; Ficai, D.; Oprea, O.-C.; Andronescu, E.; Jinga, D.C.; Holban, A.M. Influence of the Alcohols on the ZnO Synthesis and Its Properties: The Photocatalytic and Antimicrobial Activities. *Pharmaceutics* **2022**, *14*, 2842. [CrossRef]
44. Munawar, T.; Yasmeen, S.; Hussain, A.; Akram, M.; Iqbal, F. Novel direct dual-Z-scheme ZnO-Er_2O_3-Yb_2O_3 heterostructured nanocomposite with superior photocatalytic and antibacterial activity. *Mater. Lett.* **2020**, *264*, 127357. [CrossRef]
45. Yousaf, S.; Zulfiqar, S.; Din, M.I.; Agboola, P.O.; Aly Aboud, M.F.; Warsi, M.F.; Shakir, I. Solar light irradiated photocatalytic activity of ZnO-NiO/rGO nanocatalyst. *J. Mater. Res. Technol.* **2021**, *12*, 999–1009. [CrossRef]
46. Hosny, N.M. Synthesis, characterization and optical band gap of NiO nanoparticles derived from anthranilic acid precursors via a thermal decomposition route. *Polyhedron* **2011**, *30*, 470–476. [CrossRef]
47. Weldekirstos, H.D.; Habtewold, B.; Kabtamu, D.M. Surfactant-Assisted Synthesis of NiO-ZnO and NiO-CuO Nanocomposites for Enhanced Photocatalytic Degradation of Methylene Blue Under UV Light Irradiation. *Front. Mater.* **2022**, *9*, 832439. [CrossRef]
48. Vala, M.; Pashvan, C.; Solanki, P.; Dhruv, D.; Markna, J.H.; Kataria, B. Synthesis and Characterization of ZnO: NiO Nanocomposites Prepared by Phyllanthus Emblica Fruit Extract Assisted Green Method Synthesis and Characterization of ZnO: NiO Nanocomposites Prepared by Phyllanthus Emblica Fruit Extract Assisted Green Method. *ECS J. Solid State Sci. Technol.* **2023**, *12*, 103012. [CrossRef]
49. Mironova-Ulmane, N.; Kuzmin, A.; Steins, I.; Grabis, J.; Sildos, I.; Pärs, M. Raman scattering in nanosized nickel oxide NiO. *J. Phys. Conf. Ser.* **2007**, *93*, 012039. [CrossRef]
50. Elbagermi, M.A.; Alajtal, A.I.; Edwards, H.G.M.; Sharma, A. Raman Spectroscopic Studies of Nickel-Oxide Doped ZnO Nanoparticles. *J. Mater. Sci. Appl.* **2017**, *3*, 23–27.
51. Cuscó, R.; Alarcón-Lladó, E.; Ibáñez, J.; Artús, L.; Jiménez, J.; Wang, B.; Callahan, M.J. Temperature dependence of Raman scattering in ZnO. *Phys. Rev. B* **2007**, *75*, 165202. [CrossRef]
52. Adhikari, S.; Sarkar, D.; Madras, G. Highly efficient WO_3-ZnO mixed oxides for photocatalysis. *RSC Adv.* **2015**, *5*, 11895–11904. [CrossRef]
53. Yaseen, M.; Humayun, M.; Khan, A.; Idrees, M.; Shah, N.; Bibi, S. Photo-Assisted Removal of Rhodamine B and Nile Blue Dyes from Water Using CuO–SiO_2 Composite. *Molecules* **2022**, *27*, 5343. [CrossRef]
54. Habibi, M.H.; Mosavi, V. Wet coprecipitation preparation of perovskite-type iron manganite nano powder pure phase using nitrate precursors: Structural, opto-electronic, morphological and photocatalytic activity for degradation of Nile blue dye. *J. Mater. Sci. Mater. Electron.* **2017**, *28*, 10270–10276. [CrossRef]
55. Mirzaei, M.; Habibi, M.H.; Sabzyan, H. Synthesis, characterization, and dye degradation photocatalytic activity of the nano-size copper iron binary oxide. *Environ. Sci. Pollut. Res.* **2022**, *29*, 9173–9192. [CrossRef]
56. Revathi, V.; Karthik, K. Microwave assisted CdO–ZnO–MgO nanocomposite and its photocatalytic and antibacterial studies. *J. Mater. Sci. Mater. Electron.* **2018**, *29*, 18519–18530. [CrossRef]

nanomaterials

MDPI

Article

Enhanced Adsorptivity of Hexavalent Chromium in Aqueous Solutions Using CTS@nZVI Modified Wheat Straw-Derived Porous Carbon

Tiantian Deng [1,*], Hansheng Li [1,2], Su Ding [1], Feng Chen [1], Jingbao Fu [1] and Junwei Zhao [3]

1 School of Environmental and Biological Engineering, Henan University of Engineering, Zhengzhou 451191, China; hansonlee1989@gmail.com (H.L.); sue_ding189@163.com (S.D.); chenfeng871588@163.com (F.C.); fujingbao@126.com (J.F.)
2 Faculty of Health Sciences, University of Technology MARA, Puncak Alam Campus, Puncak Alam 42300, Malaysia
3 College of Resources and Environment, Yangtze University, Wuhan 434023, China; zjw882@139.com
* Correspondence: tt18100335215@163.com or ttdeng0227@haue.edu.cn; Tel.: +86-037162508218

Abstract: Using KOH-modified wheat straw as the precursor, wheat straw biochar was produced through carbonization at 500 °C. Subsequently, a synthetic material containing nano-zero-valent iron (nZVI) was prepared via liquid phase reduction (nZVI-WSPC). To enhance its properties, chitosan (CTS) was used by crosslinking to form the new adsorbent named CTS@nZVI-WSPC. The impact of CTS on parameters such as mass ratio, initial pH value, and adsorbent dosage on the adsorption efficiency of Cr(VI) in solution was investigated through one-factor experiments. Isotherm adsorption and thermodynamic analysis demonstrated that the adsorption of Cr(VI) by CTS@nZVI-WSPC conforms to the Langmuir model, with a maximum adsorption capacity of 147.93 mg/g, and the adsorption process is endothermic. Kinetic analysis revealed that the adsorption process follows a pseudo-second-order kinetic model. The adsorption mechanism, as elucidated by SEM, FTIR, XPS, and XRD, suggests that the process may involve multiple mechanisms, including pore adsorption, electrostatic adsorption, chemical reduction, and surface chelation. The adsorption capacity of Cr(VI) by CTS@nZVI-WSPC remains high after five cycles. The adsorbent is simple to operate, economical, efficient, and reusable, making it a promising candidate for the treatment of Cr(VI) in water.

Keywords: chromium; wheat straw biochar; nZVI; chitosan; adsorption; chemical reduction

Citation: Deng, T.; Li, H.; Ding, S.; Chen, F.; Fu, J.; Zhao, J. Enhanced Adsorptivity of Hexavalent Chromium in Aqueous Solutions Using CTS@nZVI Modified Wheat Straw-Derived Porous Carbon. *Nanomaterials* **2024**, *14*, 973. https://doi.org/10.3390/nano14110973

Academic Editor: Maria Filipa Ribeiro

Received: 30 April 2024
Revised: 27 May 2024
Accepted: 31 May 2024
Published: 3 June 2024

1. Introduction

The issue of heavy metal pollution in aquatic environments has intensified, posing a significant challenge due to the indiscriminate discharge of industrial waste, mining effluents, sewage irrigation, and chemical products [1]. Among these contaminants, chromium pollution stands out as a particularly pressing concern due to its inherent toxicity and deleterious effects on aquatic ecosystems [2]. Hexavalent chromium (Cr(VI)), in particular, poses a direct threat to human health. Exposure through ingestion, inhalation, or skin contact can result in severe toxicity, potentially causing genetic defects, allergies, and long-term damage to the environment [3].

Recent investigations have revealed that biomass materials, when subjected to high-temperature pyrolysis in an oxygen-deficient environment, transform into carbon-rich materials known as biochar (BC). Biochar is characterized by a wealth of functional groups and a large specific surface area, making it highly desirable for applications in aquatic environments due to its exceptional adsorption capacity. In the context of remediating chromium-containing wastewater, various types of biochar have demonstrated superior adsorption and removal capabilities. Common agricultural waste biomasses, including oat waste [4] sawdust of beech [5], rice husk [6], bagasse [7], wheat bran [8], coconut Husk [9], bamboo [10], and rubberwood fiber [11], have all demonstrated exceptional potential for application in both chromium pollution removal and degradation.

Proper modification of biochar can significantly enhance its adsorption performance in multiple ways [12]. A range of modification techniques have been investigated to optimize the adsorption capacity of hexavalent chromium [13]. In chemical oxidation or reduction modifications, the introduction of oxygen-containing functional groups such as -COOH and -OH onto the biochar surface serves to enhance its adsorption capacity. Commonly employed agents for such modifications include HCl and HNO_3, as well as alkaline substances like KOH and NaOH [14]. Metal impregnation, characterized by the adsorption of metal ions or heteroatoms into the pores and surfaces of biochar, represents another effective modification strategy. This process increases the specific surface area of biochar, and the synergy between metal ions and adsorbents further enhances the adsorption capacity [15].

nZVI has garnered considerable attention in the scientific community due to its plethora of active sites, exceptional reactivity, and remarkable reduction efficacy in the reduction of Cr(VI) to Cr(III) [16]. However, its intrinsic magnetism, heightened reactivity, susceptibility to agglomeration, and susceptibility to oxidation pose challenges that markedly impact its performance. To surmount these challenges, there is a growing interest in investigating the amalgamation of nZVI with biochar [17,18]. The binding interactions between nanoscale zero-valent iron particles and biochar may encompass adsorption, coordination, and chelation mechanisms [19]. For instance, ferrous or iron ions can form bicrenate complexes by binding to -OH groups. Furthermore, the oxide shell encapsulating the nZVI core interacts with the surface functional groups of biochar, fostering the formation of robust and stable bonds. This interfacial interaction significantly enhances the stability and functionality of the composite material, facilitating its application in various environmental and energy-related fields [20]. These interactions diminish the strength of attraction between particles, consequently mitigating nZVI aggregation and promoting its dispersion within the biochar matrix.

In the adsorption reaction of hexavalent chromium, NH_2 has demonstrated functional reactivity [21]. Chitosan, derived from the natural polysaccharide chitin, is a high-molecular-weight product known for its non-toxic and biocompatible properties. Chitosan is rich in amino (-NH_2) and hydroxyl (-OH) functional groups, serving as robust binding sites for Cr (VI) and facilitating the adsorption of various pollutants through mechanisms such as adsorption, chelation, electrostatic attraction, and ion exchange [22,23]. When coupled with Fe-based materials, chitosan effectively addresses the inherent magnetic concerns, thereby enhancing the stability and adsorption capacity of the reaction. Notably, the synergistic combination of chitosan and biochar leads to the formation of novel carbohydrate polymers with significantly augmented adsorption capacity.

Building on the aforementioned synthesis, this study focused on the fabrication of a composite wheat straw biochar adsorbent. Chitosan was introduced for cross-linking with nanoscale zero-valent iron (nZVI) to evaluate its efficacy in adsorbing hexavalent chromium from aqueous solutions. The physical and chemical properties, along with the surface structure characteristics of the materials, were comprehensively analyzed using SEM, FTIR, XRD, BET, and XPS. Additionally, the effects of pH and adsorbent dosage on the adsorption of hexavalent chromium in water were systematically investigated to identify the critical factors influencing Cr(VI) removal. The study also delved into the dynamics of adsorption, isotherm analysis, and thermodynamic modeling to elucidate the mechanisms governing the adsorption behavior. The aim was to provide insights that can inform the development of functional materials for the treatment of chromium-containing wastewater.

2. Materials and Methods

2.1. Preparation of Materials

Wheat straw-derived porous: Wheat straws were collected from a cultivated field in Nanyang city, Henan Province, China. To eliminate impurities from the wheat straw, the materials were sieved through a 20-mesh screen and subsequently washed with distilled water before being dried in an oven at 80 °C for 24 h. An amount of 5 g of wheat straw powder was placed in 20 mL of KOH solution with a concentration of 2 mol/L and soaked for 12 h. Following the oscillation, the wheat straw powder underwent filtration and was rinsed with pure water until the pH of the filtrate reached neutrality. The resultant powder was dried at 105 °C for 3 h and stored for subsequent use. During the carbonization process, an appropriate quantity of the pre-obtained wheat straw powder was placed in a quartz boat and transferred to the quartz tube within a tube furnace. The gas tightness was ensured by nitrogen testing for 10 min, after which the heating program commenced: with a heating rate of 10 °C/min, the carbonization temperature was set to 500 °C, and the carbonization duration was 180 min. Upon completion of the carbonization process and the subsequent cooling to room temperature, the residue was retrieved and placed in a sealed beaker for preservation, denoted as the original biochar, referred to as WSPC.

nZVI-WSPC [24,25]: 2 g of wheat straw biochar (WSPC) was dispersed in 50 mL of $FeSO_4{\cdot}7H_2O$ solution with a concentration of 0.1 mol/L and stirred for 2 h. After that, 50 mL of aqueous ethanol solution with a volume ratio of 1:1 was added as a dispersant to a three-port reaction glass flask, followed by 50 mL of $NaBH_4$ solution at a concentration of 0.25 mol/L at a rate of 2 drops/s under nitrogen. Approximately 30 min later, biochar-loaded nanoscale zero-valent iron particles were successfully synthesized. The resulting carbonized products underwent a triple wash with deionized water and anhydrous ethanol, followed by vacuum drying at 60 °C for 24 h. Subsequently, the products were ground through a 0.15 mm sieve and designated as nZVI-WSPC.

CTS@nZVI-WSPC: Mix 5 g of chitosan with 100 mL of acetic acid solution with a volume fraction of 5% at 25 °C and stir vigorously until the chitosan gel dissolves. Then, nZVI-WSPC was added with a mass ratio of chitosan: nZVI-WSPC (1:1), and continuous mechanical stirring for 4 h was used to obtain a uniform CTS@nZVI-WSPC mixture. Then, the mixed gel solution is sonicated in an ultrasonic bath for 30 min to obtain a homogeneous and dispersed chitosan gel solution with nZVI-WSPC particles. The prepared solution is lyophilized and dried for 28 h to obtain complexes. The final product is crushed and passed through a 100 mesh size (150 mm) for further use.

Those three adsorbents mentioned above were obtained as shown in Figure 1.

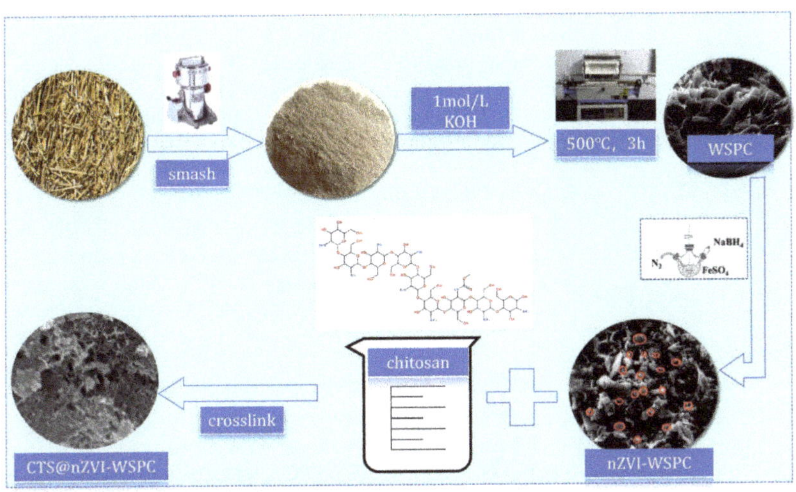

Figure 1. Schematic diagram of the preparation process of three adsorbent materials.

2.2. Adsorption Experiment

$K_2Cr_2O_7$, precisely measured at 2.835 g, was dissolved in 1000 mL of deionized water to create a stock solution with a Cr (VI) concentration of 1 g/L. This stock solution was further diluted with deionized water as necessary. The pH of the reaction solution was adjusted using varying concentrations of HCl and NaOH solutions.

2.2.1. Choice of Adsorbent

In order to find the optimal mass ratio of CTS and nZVI-WSPC, we set six different gradients, namely CTS: nZVI-WSPC = 0:1, 1:1, 1:2, 1:3, 3:1, 2:1. The experimental conditions at this time were C_0 = 50 mg/L, T = 25 °C, t = 24 h, Dose = 1 g/L, pH = 7.0. Following single-factor comparative experiments, we identified the synthesized materials resulting from a mass ratio of CTS: nZVI-WSPC = 1:3 for subsequent characterization and batch experiments.

2.2.2. Effects of Adsorbent Dose and Solution pH

Similarly, the adsorbent dose and pH value were also examined as important factors affecting the experimental results. In agreement with the experimental conditions described above, we studied the removal of Cr (VI) with different adsorbent doses and pH values. The amount of adsorbent was set to 0.2 g/L, 1 g/L, 2 g/L, 4 g/L, 8 g/L, and 10 g/L, while pH was set to 2, 3, 4, 5, 6, 7, 8, 9, 10, respectively.

2.2.3. Adsorption Isotherm Experiment

The analysis of adsorption isotherms aims to ascertain the adsorption capacity of the three adsorbents for Cr (VI). The initial concentration range of Cr (VI) was set from 5 to 200 mg/L. The mass of the adsorbent added to the 50 mL Cr (VI) solution was 0.05 g, the reaction time was 24 h, the solution maintained a pH of 2, and the temperature of the reaction system was set at 25 °C, 35 °C, and 45 °C. The data were fitted using three widely recognized isothermal adsorption models, and the experimental results of the adsorption isotherm were thoroughly analyzed.

2.2.4. Adsorption Kinetics Experiment

The adsorption kinetic experiments, along with the corresponding fitting equations, aim to elucidate the speed of the adsorption reaction and its underlying mechanism. Three distinct adsorbent materials underwent adsorption kinetics testing under identical reaction conditions. Specifically, 0.05 g of material was introduced to 50 mL of Cr (VI) solution with an initial concentration of 50 mg/L, while maintaining a reaction system pH of 2 and a temperature of 25 °C. Samples of the supernatant were extracted at specific intervals, including 5 min, 10 min, 20 min, 30 min, 50 min, 70 min, 90 min, 180 min, 240 min, 480 min, 960 min, and 1440 min. These samples were subsequently filtered to determine the concentration of Cr (VI).

2.2.5. The Reusability of CTS@nZVI-WSPC

To evaluate the reavailability of CTS@nZVI-WSPC, we carried out repeatability tests under fixed experimental conditions. The experimental conditions were set to pH = 2.0, Cr (VI) = 50.0 mg/L, and dose = 1.0 g/L. After each reaction process, we recovered the solid-phase adsorbent CTS@nZVI-WSPC composite using a magnetic material, followed by washing and drying with purified water. The experiments were repeated five times.

3. Results and Discussion

3.1. Characterization

The scanning electron microscope was employed to examine the morphological characteristics of WSPC, nZVI-WSPC, and CTS@nZVI-WSPC materials. As depicted in Figure 2a, wheat straw biochar exhibits a distinctive, regular, and porous structure, with an average pore size distribution below 5 μm. The composite biochar displays notable morphological features. On the nZVI-WSPC surface Figure 2b, a uniform distribution of heterogeneous spherical particles is evident, indicating a dispersion of nZVI nanoparticles rather than aggregation on the WSPC. This dispersion significantly enhances the adsorption sites of nZVI-WSPC [26]. Figure 2c illustrates the morphological structure of CTS@nZVI-WSPC, revealing that the introduction of CTS renders the material surface rougher. This alteration enhances the dispersion and structural stability of n-ZVI on the WSPC surface. EDS results further confirm the composite's composition, revealing the presence of C, O, and Fe. These findings affirm the successful loading of nano-zero-valent iron or iron oxide onto the biochar surface.

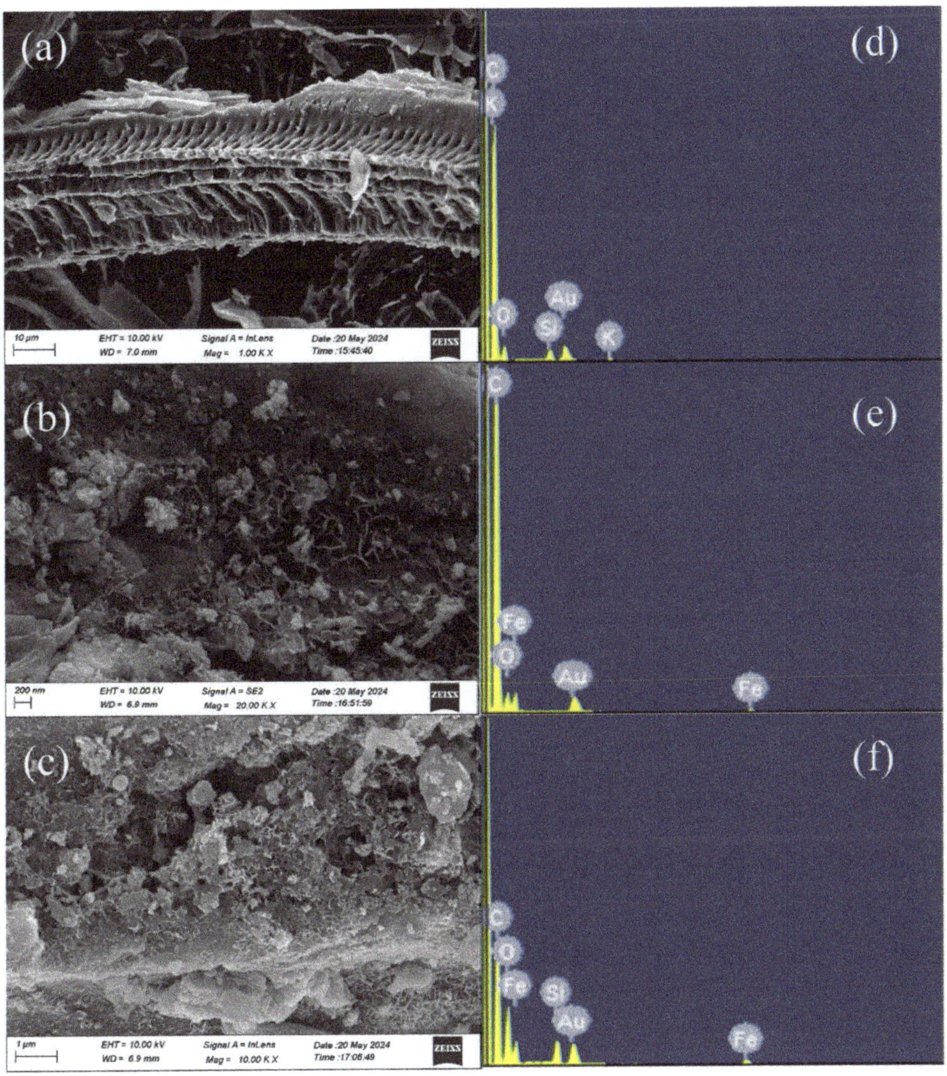

Figure 2. SEM and EDS image of three materials (**a**) SEM of WSPC; (**b**) SEM of nZVI-WSPC; (**c**) SEM of CTS@nZVI-WSPC; (**d**) EDS of WSPC; (**e**) EDS of nZVI-WSPC; (**f**) EDS of CTS@nZVI-WSPC.

Raman spectroscopy is an effective means to demonstrate the structure of carbon materials. Both the D and G peaks are recognized as Raman characteristic features of carbon atomic crystals, located at around 1350 cm^{-1} and 1580 cm^{-1}, respectively. The D peak reflects the carbon lattice defects, while the G peak signifies the material's degree of carbonization [27]. In this study, I(D)/I(G) represents the area ratio of the D peak to the G peak, with a larger ratio indicating a higher level of defects in the carbon atomic crystal.

The characteristic peaks of the three materials were fitted using the Gauss model, yielding termination coefficients (R^2) of 0.961, 0.972, and 0.963 for WSPC, nZVI-WSPC, and CTS@nZVI-WSPC, respectively. These values signify a robust degree of carbonization and prominent characteristic peaks in all three materials. As depicted in Figure 3a, the I(D)/I(G) ratios for WSPC, nZVI-WSPC, and CTS@nZVI-WSPC were 0.295, 0.415, and 0.434, respectively. This suggested an increasing degree of defects in the latter two composites, which is advantageous for providing more active sites for subsequent reactions.

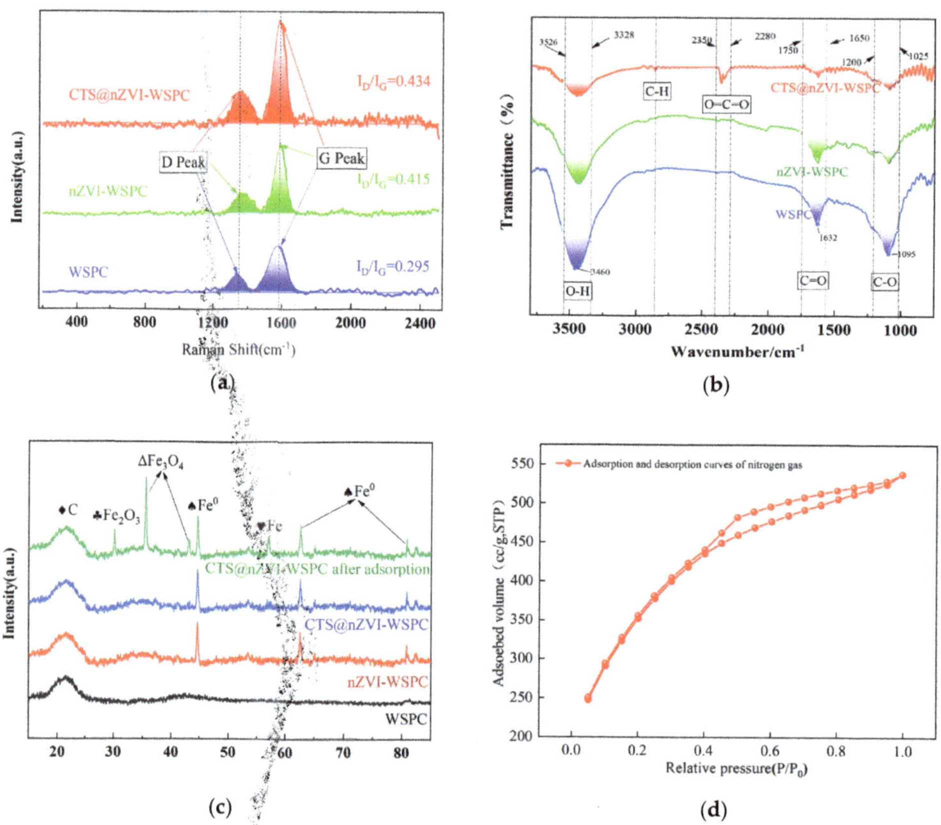

Figure 3. Characterization results of three materials: (**a**) The Raman analysis map; (**b**) FTIR curves; (**c**) XRD patterns; (**d**) N_2 adsorption-desorption isotherms of CTS@nZVI-WSPC.

The qualitative analysis of surface functional group species in three distinct materials using FTIR is presented in Figure 3b. All materials exhibit absorption peaks at 1095 cm^{-1} and 1632 cm^{-1}, which are characteristic of the functional groups inherent to biochar itself [28]. For peak around 1632 cm^{-1} is linked to the C=O stretching vibration of ester and the tensile vibration of O-H, and the wavenumber at 1095 cm^{-1} signifies the vibration absorption peak of C-O [29,30]. Nanozero-valent iron minimally influences the surface functional groups of the biochar. Upon the introduction of CTS, a contraction absorption peak emerges within the wide peak range from 3328 to 3526 cm^{-1}, indicating the presence of intramolecular and intermolecular hydrogen bonds in the CTS intramolecular hydroxyl and amino groups. The loading of CTS-nZVI preserved the structural stability of the biochar itself, allowing the C-H and C=C functional groups on the surface to directly or indirectly participate in and enhance the subsequent adsorption process.

Through the XRD characterization of the three materials, a distinctive diffraction peak appears near 20°, as illustrated in Figure 3c. Comparison with the PDF card, the characteristic peak suggested the presence of certain carbon material composition in the three materials [31]. The introduction of nZVI and CTS did not alter the structure and composition of the original biological carbon. No significant deviation was observed in the characteristic peaks of nZV-WSPC and CTS@nZVI-WSPC, indicating a minimal difference in their crystal structure. This is attributed to the fact that CTS acts as a natural polymer polysaccharide, with characteristic diffraction peaks predominantly occurring at 10–20°. The appearance of this characteristic peak is a result of intermolecular or intramolecular hydrogen bonds stemming from the presence of -OH and -NH_2 functional groups in CTS [32,33]. Many salts or oxides tend to disperse on the carrier surface to form monolayers and submonolayers. In the case where the loading amount is below a certain threshold, a monolayer dispersion state is maintained, and the active component cannot be detected by X-ray. From the XRD profile, it is evident that nZVI was successfully synthesized in both composites. The diffraction peaks are located at 30.27°, 35.68°, 44.5°, 57.4°, and 63.25°, corresponding to different forms of iron (Fe), with the peaks at 44.5° and 63.25° specifically indicating the presence of nano-zero-valent iron, as reported in various literature sources [34]. Moreover, the diffraction peaks of CTS@nZVI-WSPC at $2\theta = 29.9°$, 40.17°, and 55.08° align well with the ferric oxide standard card, which appears after the adsorption reaction [35,36]. The characteristic peak band at 35.19° represents the formation of ferric oxide, suggesting that the composite may undergo oxidation after the reduction by iron particles, which is consistent with the XPS analysis results. The comprehensive analysis showed that the introduction of nZVI and chitosan did not alter the carbon structure of the raw wheat straw. After the reaction, the material's structure remained largely unchanged, retaining a stable morphology.

As depicted in Figure 3d, the nitrogen adsorption-desorption isotherm of CTS@nZVI-WSPC exhibits a closed loop, which corresponds to I-type isotherms (IUPAC classification) and H4-type hysteresis loop [37]. According to the extended isotherm, the curve shows a steep rise followed by a plateau, indicative of micro-porous structure and monolayer reversible adsorption consistent with the results of the isothermal adsorption fitting. This adsorption generally occurs in microporous materials and mesoporous materials with pore sizes very close to micropores [18], which showed a strong agreement with the biochar pore structure observed by SEM.

3.2. Batch Experiments

3.2.1. Effect of Ratio

The standard curve for hexavalent chromium determination was shown in Figure 4a. A series of adsorbents synthesized from chitosan (CTS) and nZVI-WSPC with varying mass ratios were employed for the comparative adsorption of hexavalent chromium in aqueous solutions, and the results are presented in Figure 4b. As depicted in the figure, at a constant adsorbent mass (dose = 0.05 g), nZVI-WSPC exhibits a more pronounced impact on the adsorption of Cr (VI) in water compared to CTS alone. The composite materials demonstrated a higher adsorption capacity for Cr (VI) in water than nZVI-WSPC alone, reaching a maximum value of 39.78 mg/g at the mass ratio m(CTS):m(nZVI-WSPC) = 1:3. However, with a continuous increase in the CTS content, the adsorption capacity of the material gradually decreased. The modification of nZVI-WSPC by chitosan proved beneficial for the adsorption of Cr (VI) in water. Excessive CTS may impede the contact of nano-ZVI or biochar with hexavalent chromium, thereby reducing adsorption efficiency [38]. Based on the obtained experimental data, the optimal mass ratio of chitosan to nZVI-WSPC was determined to be 1:3, and this ratio was employed in subsequent experiments.

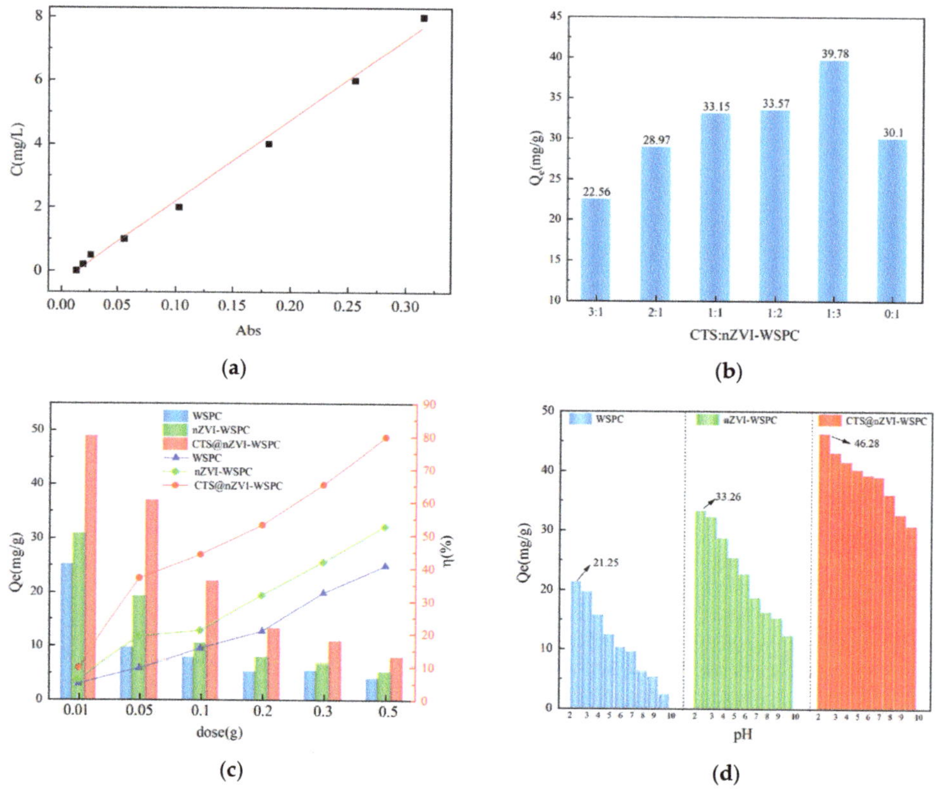

Figure 4. (**a**) Standard curve for hexavalent chromium determination; (**b**) ratio of CTS to nZVI-WSPC (C_0 = 50 mg/L, T = 25 °C, t = 24 h, dose = 0.05 g, pH = 7.0); (**c**) dose of adsorbent (C_0 = 50 mg/L, T = 25 °C, t = 24 h, pH = 7.0); (**d**) pH (C_0 = 50 mg/L, T = 25 °C, t = 24 h, dose = 0.05 g).

3.2.2. Effect of the Dose of Adsorbent

The quantity of adsorbent directly influences the adsorption effectiveness. Figure 4c illustrates the experimental results demonstrating the impact of varying material dosages on the adsorption capacity of Cr (VI) solution. Under the same reaction conditions, WSPC, nZVI-WSPC, and CTS@nZVI-WSPC all exhibited similar adsorption trends. As the dosage increased, the adsorption amount of Cr(VI) gradually decreased, while the adsorption rate steadily increased. This can be explained by the fact that when the solution concentration is constant, an increase in dosage leads to a proportional increase in the number of active adsorption sites available on the material [39]. However, as the adsorption of solutes in solution approaches saturation, an excess of active adsorption sites occurs, reducing the adsorption efficiency per unit mass of the material. Both modified materials showed improved adsorption properties to some extent. It is worth noting that the pH of the reaction system was not adjusted during these experiments, so the three adsorbents did not achieve their optimal performance.

3.2.3. Effect of Initial pH

Under the same experimental conditions, the adsorption efficiency of Cr (VI) in water was compared under different initial pH conditions. The results are presented in Figure 4d. Across solutions with pH values ranging from 2 to 10, the adsorption of Cr (VI) by all three materials exhibited a decrease with increasing pH. At pH 2, the adsorption reached its maximum, with the adsorption capacities of WSPC, nZVI-WSPC, and CTS@nZVI-WSPC reaching 21.25 mg/g, 33.26 mg/g, and 43.28 mg/g, respectively—significantly higher than under neutral and alkaline conditions. This suggests that pH is a crucial factor influencing the removal properties of the materials.

Numerous studies have shown that Cr (VI) may have different ionic forms in aqueous solutions. In the acidic environment of pH < 5.0, the primary form is hydrogen chromate ($HCrO_4^-$), and with increasing pH, Cr (VI) mainly exists as chromate ions (CrO_4^{2-}). Since the adsorption-free energy of $HCrO_4^-$ is lower than that of CrO_4^{2-}, $HCrO_4^-$ at low pH is more adsorbed to CrO_4^{2-} than at high pH [40]. In this study, WSPC underwent pretreatment modification using KOH; hence, the introduction of OH^- would lead to competitive adsorption under alkaline conditions, reducing the material's adsorption capacity [41].

Under acidic conditions, the oxide layer on the surface of nZVI was disrupted, exposing more Fe^0 sites for the reduction reaction of Cr (VI). However, under alkaline conditions, Fe tends to form a co-precipitate on the surface of nZVI-WSPC, leading to the passivation of the material surface and limiting the reaction. Furthermore, Cr (VI) has a high positive redox potential under acidic conditions (pH = 1, E0 = 1.3 V; pH = 5, E0 = 0.68 V), so the reaction between nZVI and Cr (VI) is very rapid [42]. In addition, the contact between nZVI and Cr (VI) also contributes to the iron corrosion caused by H^+ under acidic conditions. Under neutral or alkaline conditions, Cr^{3+} and Fe^{3+} will form a precipitate covering the nZVI surface to rapidly inactivate it [43].

In contrast, CTS@nZVI-WSPC was less influenced by pH than the other two materials. On the one hand, the introduction of CTS enhances the mechanical stability of the material; on the other hand, under acidic conditions, the amino group (NH_2) of chitosan is readily protonated to form (NH_3^+), acquiring a positive charge and facilitating adsorption with $HCrO_4^-$, resulting in increased anion adsorption by the adsorbent in acidic environments [44]. Under alkaline conditions, the decreased adsorption can be attributed to electrostatic repulsion on the deprotonated CTS@nZVI-WSPC surface and increased competition for more hydroxide ions. At the same time, it was obvious that CTS@nZVI-WSPC is less sensitive to pH than the other two materials, thus it can be suitable for the removal of Cr in a wide range of pH.

3.3. Adsorption Isotherm

In this study, CTS@nZVI-WSPC was employed to investigate the adsorption behavior and mechanism of Cr (VI). As depicted in Table 1 and Figure S1, the Langmuir model proved to be more effective in elucidating the adsorption behavior of Cr (VI) on the surfaces of the adsorbent, with fitting coefficients exceeding 0.99. Under the reaction conditions at pH = 2, the corresponding maximum adsorption capacities reached 111.23 mg/g, 125.00 mg/g, and 147.93 mg/g at different temperatures with 25 °C, 35 °C, and 45 °C, respectively. Langmuir model is well-suited for describing a uniform single adsorption process, where each molecule possesses constant thermal energy and an equal number of adsorption points [45]. The correlation coefficient of the Langmuir is highest at all three temperatures. This indicates that the adsorption process is mainly monolayer adsorption, and Cr (VI) is monolayer adsorption on the surface of the material. Secondly, the characteristic coefficient of the reaction adsorption degree (K_F) is positively correlated with the adsorption amount. As the temperature increases, the K_F increases, indicating that the increase in the reaction temperature is conducive to the adsorption of Cr (VI) in the solution [46].

Temperature is a critical factor influencing the adsorption capacity of the solid–liquid phase medium [47]. In this study, the impact of temperatures of 298 K, 308 K, and 318 K on Cr (VI) with CTS@nZVI-WSPC was investigated. The findings indicated that the adsorption amount rises with increasing temperature, suggesting an endothermic nature of the adsorption process. The outcomes of the thermodynamic calculations are detailed in Table S1.

Table 1. Adsorption isotherm fitting parameters of different materials.

		Langmuir			Freundlich			Temkin		
		Q_m mg/g	K_L L/mg	R^2	K_F	n	R^2	A	B	R^2
25 °C		111.23	0.0037	0.9920	12.116	2.1683	0.9769	1.3573	17.1767	0.9074
35 °C	CTS@nZVI-WSPC	125.00	0.0405	0.9956	12.965	2.2762	0.9795	1.3671	19.9112	0.9242
45 °C		147.93	0.0686	0.9821	14.532	1.8126	0.9580	1.3788	25.1972	0.9572

Based on the data, the free energy decreases as the temperature rises, indicating higher adsorption efficiency at elevated temperatures. Both the enthalpy and entropy changes are positive, signifying an endothermic adsorption process that can spontaneously occur at higher temperatures. The increased degree of disorder in the reaction enhances structural disturbance during the adsorption process, facilitating solid–liquid phase contact and thereby augmenting the adsorption capacity.

3.4. Adsorption Kinetics

The investigation into adsorption kinetics provides insights into the rate of adsorption and the primary steps involved in the confinement of the adsorbent. As depicted in Figure S2, the adsorption rate of CTS@nZVI-WSPC was notably swift in the initial stages of the reaction. The adsorption kinetics at different temperatures showed the same trend. At 90 min, the adsorption rate did not change over time. At this time, the adsorption capacity at 25 °C, 35 °C, and 45 °C reached 41.06 mg/g, 43.23 mg/g, and 48.59 mg/g, respectively. This suggests that a substantial portion of adsorption occurs in the early phase of the adsorption process. Subsequently, owing to the reduction in available adsorption sites, the adsorption capacity of the adsorbent ceases to improve, ultimately reaching saturation.

In this experiment, several kinetic models were used to analyze the adsorption behavior of Cr (VI). As can be seen, the highest R^2 was achieved by fitting data with the Pseudo-second model (Table 2), with the fitted Q_e value basically matching the actual value. Accordingly, we believed that the adsorption reaction process of Cr (VI) was mainly dominated by physical and chemical adsorption [48].

Table 2. Adsorption kinetics fitting parameters of CTS@nZVI-WSPC.

		Pseudo-First-Order			Pseudo-Second-Order			Intra-Particle	
		Q_e mg/g	K_1 L/mg	R^2	Q_e mg/g	K_2	R^2	Kp_1	Kp_2
25 °C		42.826	0.1410	0.9611	41.271	0.0047	0.9999	1.95	0.002
35 °C	CTS@nZVI-WSPC	44.511	0.1468	0.9309	44.762	0.0054	0.9999	2.23	0.015
45 °C		47.816	0.2088	0.9649	47.984	0.0089	0.9999	1.91	0.049

To elucidate the adsorption mechanism comprehensively, the intra-particulate diffusion model was employed to delineate the adsorption process in greater detail. The fitting results revealed the presence of boundary layer diffusion, intra-particle diffusion, and adsorption equilibrium throughout the adsorption of Cr (VI) by the CTS@nZVI-WSPC, signifying that internal diffusion does not singularly govern the adsorption process. A higher slope in the diffusion process within the particles indicates swifter boundary layer diffusion [49,50]. Notably, during the initial 60 min of the first stage, the slopes of the linear fitting curve were 1.95, 2.23, and 1.91, significantly larger than the subsequent two stages. This stage can be interpreted as the rapid diffusion stage, characterized by the driving force stemming from the substantial concentration difference between solid-liquid phase Cr (VI) and the porous structure of the composite material. This results in rapid adsorption under physical adsorption, driven by electrostatic action. The second stage (60–1440 min) was the slow adsorption stage, marked by a diminishing adsorption rate. Particle diffusion emerged as the predominant factor limiting the adsorption rate. As the adsorption reaction progressed, the adsorption sites approached saturation, leading to a continuous decline in the rate until equilibrium was reached [51].

3.5. Reusability of CTS@nZVI-WSPC

To investigate the reuse capacity of CTS@nZVI-WSPC, we collected the adsorbent particles using magnets at the end of each reaction, washed them with water, and dried them before subsequent use. As shown in Figure 5a, the removal rates for five consecutive runs were 95.8%, 90.36%, 85.11%, 78.37%, and 70.29%, respectively. These results demonstrate that CTS@nZVI-WSPC is effectively reusable. The removal efficiency in the first cycle is higher due to the availability of active sites. However, in subsequent runs, the number of binding sites decreases. Additionally, the removal efficiency declines in successive cycles due to the oxidation of nZVI and the formation of a passivation layer. It was noteworthy that the main form of Fe in the acidic system where the reaction proceeds was Fe (III) (Figure 5b), which may be due to the rapid oxidation of Fe (II) by reducing Cr (VI) [52,53]. Likewise, the leaching concentration of iron was relatively stable, and it may be related to the certain wrapping properties of chitosan.

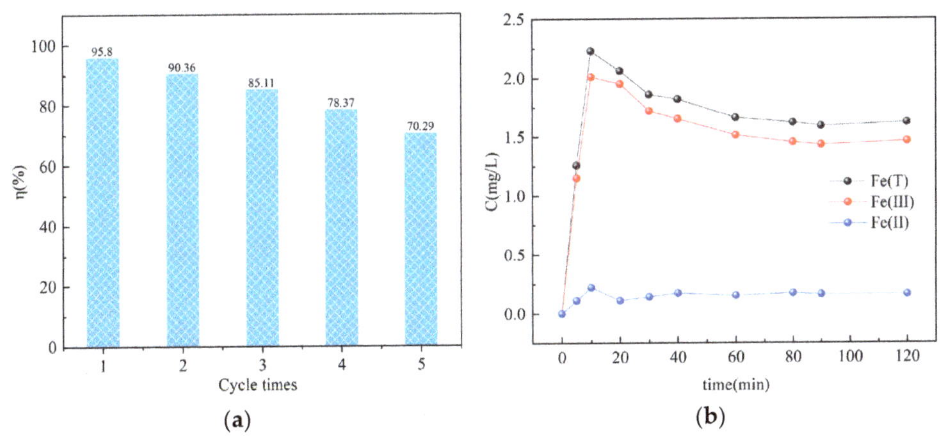

Figure 5. (**a**) Removal efficiency of CTS@nZVI-WSPC composite after five uses; (**b**) Changes in Fe concentrations during the reaction (condition: C_0 = 50 mg/L, T = 25, t = 24 h, dose = 0.05 g, pH = 2).

3.6. XPS Analysis

XPS analysis results are presented in Figure 6a–d. Comparison of the total spectra before and after the adsorption reaction reveals distinct Cr characteristic peaks, predominantly located near 577 eV and 585 eV. This observation suggests that the majority of the Cr(VI) species involved in the reaction have undergone reduction to Cr(III) [54]. Figure 6b provided an in-depth analysis and comparative study of the Fe XPS spectra pre-reaction. The electron energy spectrum exhibits original Fe characteristic peaks primarily at approximately 711 eV and 724 eV [55]. Following deconvolution, four resolved peaks emerge at 710.28 eV, 723.7 eV, 712.06 eV, and 725.51 eV, corresponding to Fe2p3/2 and Fe2p1/2 orbitals for Fe(II), as well as Fe2p3/2 and Fe2p1/2 orbitals for Fe(III). The compositional analysis based on fitted peak areas reveals Fe(II) and Fe(III) contents of 50.95% and 49.05%, respectively. After the reaction, the Fe2p3/2 orbital of Fe(II) at 710.28 eV undergoes a leftward shift to 711.06 eV, indicative of Fe(II) consumption during the reduction reaction of Cr(VI). Consequently, the Fe(II) content decreases to 36.24% post-reaction compared to its initial level, thereby corroborating the aforementioned deductions.

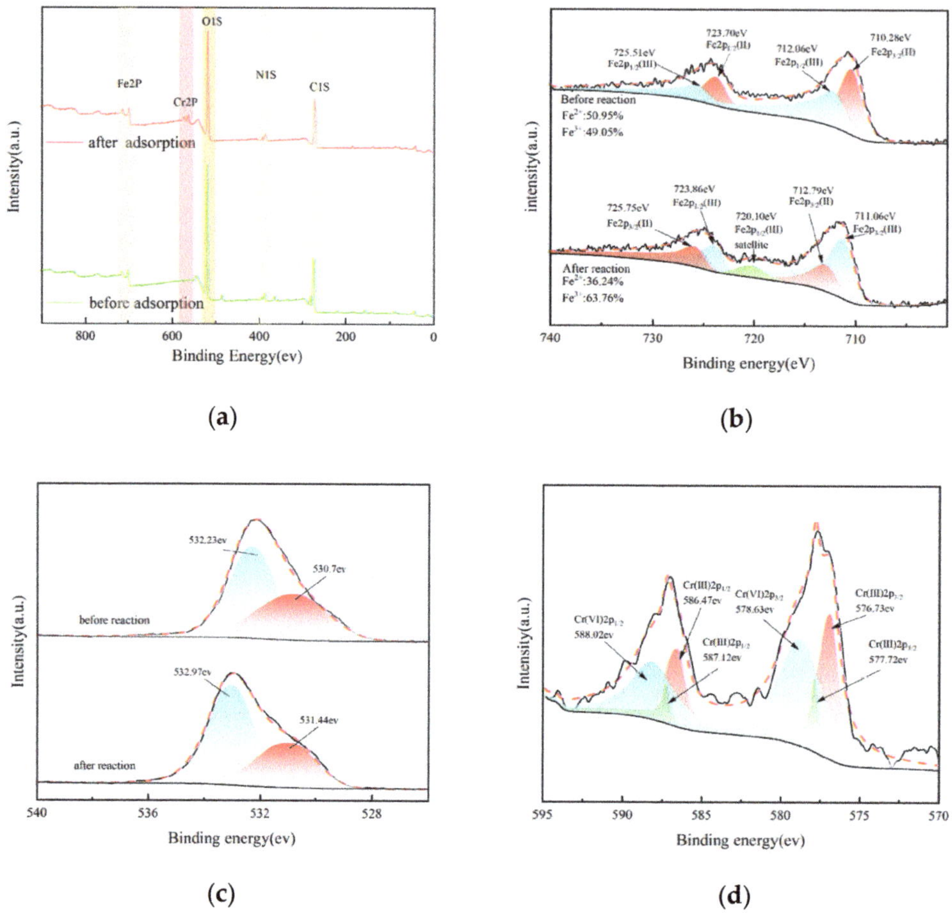

Figure 6. (**a**) Wide scan XPS analysis before and after reaction; (**b**) narrow scan of Fe before and after reaction; (**c**) narrow scan of O before and after reaction; (**d**) narrow scan of Cr after reaction.

As shown in Figure 6c, the characteristic peaks of O1s appear at the 530.7 eV and 532.2 eV binding energies, respectively, indicating that oxygen exists mainly in the form of O2-and [56]. After the adsorption reaction, the two oxygen peaks experience a shift of about 0.7 eV; this shift indicates that the oxygen-active group participates in the redox reaction between Fe and Cr. The characteristic main peaks of Cr that appeared after the reaction correspond to 577 ev and 585 ev, respectively [57]. After peak division processing, we can see that the characteristic peaks of Cr (III) 2p3/2 orbit appear near 576.73 ev and 577.72 ev, and the characteristic peaks of Cr (III) 2p1/2 orbit appear near 586.47 ev and 586.12 ev, respectively. Moreover, the peak strength signal of trivalent chromium is relatively significant, indicating that the reduction of hexavalent chromium has occurred significantly (Figure 6d).

3.7. Discussion of Adsorption Mechanism

Based on the relevant experimental results and fitting analysis, we assert that CTS@nZVI-WSPC represents a novel green biomass-modified adsorption material. The adsorption mechanism targeting Cr (VI) in water is postulated to encompass pore adsorption, chemical reduction, chelation reactions, and electrostatic interactions [58], as shown in Figure 7.

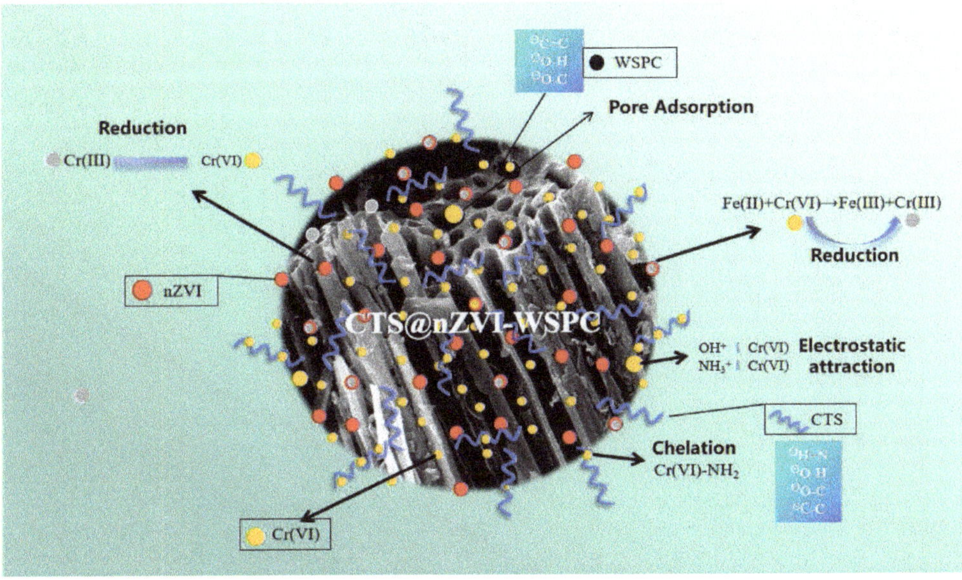

Figure 7. Diagram of the adsorption reaction mechanism.

The adsorption efficiency of wheat straw biochar (WSPC) was relatively limited, primarily attributed to the high carbon content and pronounced aromaticity developed during the high-temperature carbonization process of wheat straw. Furthermore, the wheat straw employed in this experiment underwent carbonization with the introduction of KOH as a modifier, proving advantageous not only in enhancing the biochar's pore structure but also in incorporating hydroxyl groups into the material's framework, as substantiated by Fourier infrared photoskin analysis [59]. As shown in Figure 7, CTS@nZVI-WSPC has a developed pore structure and can provide sufficient adsorption points for Cr (VI).

During the adsorption process, nanoscale zero-valent iron (nZVI) plays a crucial role in reducing the toxicity of Cr (VI) owing to its robust reduction capacity. Concurrently, Fe (II) generated in the reduction process also contributes to the reduction of Cr (VI) [60]. The newly formed Cr (III), with an atomic radius nearly identical to Fe (II), tends to precipitate onto the nanomaterial's surface in the form of complexes [61,62]. Additionally, a normalization reaction might occur between nZVI and Fe (III), generating more Fe (II), thereby enhancing the thoroughness of the reaction process and improving the material's adsorption performance to some extent. Furthermore, chitosan served as a stabilizer in the preparation of nanoscale zero-valent iron (nZVI), significantly enhancing its dispersity and preventing aggregation. Through effective modification, there was a significant increase in the adsorption capacity with the new adsorbent CTS@nZVI-WSPC, and the composite exhibited optimal adsorption performance at lower pH values. The presence of more coordination heteroatoms (-OH, -NH) provided by chitosan further improved the adsorption performance compared to the other two materials.

4. Conclusions

Utilizing wheat straw as the raw material, an amino-functionalized material was synthesized through a stepwise modification process, presenting itself as a potential alternative adsorbent for Cr (VI) in water. In comparison to a singular biochar, the modified material exhibits superior reduction and adsorption performance. Based on the above research, we can infer that the composite material of CTS@nZVI-WSPC may become a reliable adsorbent for environmental remediation due to its green, low-cost, and excellent adsorption performance. However, they still face many challenges in large-scale applications. For example, the synthesis of biochar still relies on high-temperature calcination methods, which have high energy consumption and are not conducive to large-scale use. Additionally, in the discussion of mechanism synthesis and limiting steps, more in-depth discussions are needed, which will be our focus in future efforts.

Supplementary Materials: The following supporting information can be downloaded at: https://www.mdpi.com/article/10.3390/nano14110973/s1, Equation (S1): the removal rate η for Cr (VI); Equation (S2): the equilibrium adsorption capacity Qe; Equation (S3): Langmuir isothermal adsorption equation; Equation (S4): Freudlich isothermal adsorption equation; Equation (S5): Temkin isothermal adsorption equation; Equation (S6): the pseudo-first-order mode; Equation (S7): the pseudo-second-order model; Equation (S8): intra-particle diffusion model; Equation (S9): the equation of ΔG; Equation (S10): the equation of enthalpy ΔH and ΔS; Figure S1. Adsorption isotherm models with CTS@nZVI-WSPC (a) Langmuir; (b) Freundlich; (c) Temkin; Figure S2. Effect of time and kinetic models with CTS@nZVI-WSPC (a) Effect of time on adsorptive capacity; (b) First pseudo fitting; (c) Second pseudo fitting; (d) Intra-particle fitting; Table S1. Adsorption thermodynamic parameters.

Author Contributions: Writing—original draft preparation, T.D.; writing—review and editing, H.L. and J.Z.; supervision, F.C. and H.L.; funding acquisition, T.D., J.F. and S.D. All authors have read and agreed to the published version of the manuscript.

Funding: This research was funded by NSFC-China, grant number 42377490; Natural Science Foundation of Henan, grant number 232300421343; Project for Young Key Teachers of Henan Province, grant number 2020GGJS238; Doctoral Foundation of Henan University of Engineering, grant number D2022016.

Data Availability Statement: Data are contained within the article and Supplementary Materials.

Conflicts of Interest: The authors declare no conflict of interest.

References

1. Samuel, M.S.; Selvarajan, E.; Chidambaram, R.; Patel, H.; Brindhadevi, K. Clean approach for chromium removal in aqueous environments and role of nanomaterials in bioremediation: Present research and future perspective. *Chemosphere* **2021**, *284*, 131368. [CrossRef] [PubMed]
2. Wang, H.; Cui, T.; Chen, D.; Luo, Q.; Xu, J.; Sun, R.; Zi, W.; Xu, R.; Liu, Y.; Zhang, Y. Hexavalent chromium elimination from wastewater by integrated micro-electrolysis composites synthesized from red mud and rice straw via a facile one-pot method. *Sci. Rep.* **2022**, *12*, 14242. [CrossRef] [PubMed]
3. Fan, Z.; Zhang, Q.; Gao, B.; Li, M.; Liu, C.; Qiu, Y. Removal of hexavalent chromium by biochar supported nZVI composite: Batch and fixed-bed column evaluations, mechanisms, and secondary contamination prevention. *Chemosphere* **2019**, *217*, 85–94. [CrossRef] [PubMed]
4. Gardea-Torresdey, J.; Tiemann, K.; Armendariz, V.; Bess-Oberto, L.; Chianelli, R.; Rios, J.; Parsons, J.; Gamez, G. Characterization of Cr(VI) binding and reduction to Cr(III) by the agricultural byproducts of Avena monida (Oat) biomass. *J. Hazard. Mater.* **2000**, *80*, 175–188. [CrossRef] [PubMed]
5. Witek-Krowiak, A. Application of beech sawdust for removal of heavy metals from water: Biosorption and desorption studies. *Eur. J. Wood Wood Prod.* **2013**, *71*, 227–236. [CrossRef]
6. Mehdinia, S.M. Rice Husk Silica Adsorbent for Removal of Hexavalent Chromium Pollution from Aquatic Solutions. *Iran. J. Energy Environ.* **2014**, *5*, 218–223. [CrossRef]
7. Tan, X.; Liu, Y.; Gu, Y.; Zeng, G.; Wang, X.; Hu, X.; Sun, Z.; Yang, Z. Immobilization of Cd(II) in acid soil amended with different biochars with a long term of incubation. *Environ. Sci. Pollut. Res.* **2015**, *22*, 12597–12604. [CrossRef] [PubMed]
8. Singh, K.; Hasan, S.; Talat, M.; Singh, V.; Gangwar, S. Removal of Cr (VI) from aqueous solutions using wheat bran. *Chem. Eng. J.* **2009**, *151*, 113–121. [CrossRef]
9. Osasona, I.; Onyemari, J.; Johnson, J.A.; Siyanbola, T.O. Phosphoric Acid and Cashew Bark Extract Activated Carbons derived from Coconut (*Cocos nucifera*) Shells for Cr (VI) Adsorption. *Egypt. J. Chem.* **2022**, *65*, 803–815. [CrossRef]
10. Zhang, H.; Xiao, R.; Li, R.; Ali, A.; Chen, A.; Zhang, Z. Enhanced aqueous Cr(VI) removal using chitosan-modified magnetic biochars derived from bamboo residues. *Chemosphere* **2020**, *261*, 127694. [CrossRef]
11. Mahdavi, M.; Ahmad, M.B.; Haron, J.; Rahman, M.Z.A. Adsorption of Cr(III) from aqueous solutions by polyacrylamide-grafted rubberwood fibre: Kinetics, equilibrium, and thermodynamic studies. *BioResources* **2011**, *6*, 22–33. [CrossRef]
12. Iqbal, Z.; Tanweer, M.S.; Alam, M. Recent advances in adsorptive removal of wastewater pollutants by chemically modified metal oxides: A review. *J. Water Process. Eng.* **2022**, *46*, 102641. [CrossRef]

13. Ambika, S.; Kumar, M.; Pisharody, L.; Malhotra, M.; Kumar, G.; Sreedharan, V.; Singh, L.; Nidheesh, P.; Bhatnagar, A. Modified biochar as a green adsorbent for removal of hexavalent chromium from various environmental matrices: Mechanisms, methods, and prospects. *Chem. Eng. J.* **2022**, *439*, 135716. [CrossRef]
14. Zhao, L.; Zheng, W.; Mašek, O.; Chen, X.; Gu, B.; Sharma, B.K.; Cao, X. Roles of Phosphoric Acid in Biochar Formation: Synchronously Improving Carbon Retention and Sorption Capacity. *J. Environ. Qual.* **2017**, *46*, 393–401. [CrossRef] [PubMed]
15. Chin, J.F.; Heng, Z.W.; Teoh, H.C.; Chong, W.C.; Pang, Y.L. Recent development of magnetic biochar crosslinked chitosan on heavy metal removal from wastewater—Modification, application and mechanism. *Chemosphere* **2022**, *291*, 133035. [CrossRef]
16. Sahu, U.K.; Ji, W.; Liang, Y.; Ma, H.; Pu, S. Mechanism enhanced active biochar support magnetic nano zero-valent iron for efficient removal of Cr(VI) from simulated polluted water. *J. Environ. Chem. Eng.* **2022**, *10*, 107077. [CrossRef]
17. Chen, W.-F.; Yan, C.-C.; Wang, Q.; Pan, L.; Chen, L.-F. Carbothermal synthesis of activated carbon-supported nano zero valent iron: Effects of temperature, characterization, and reactivity. *Desalination Water Treat.* **2016**, *57*, 9520–9529. [CrossRef]
18. Zhang, X.; Cao, X.-Q.; Li, G.; Yin, J.; Zhang, D.; Li, M.; Meng, N.; Dong, L.; Lyu, X.-J.; Li, L.; et al. Preparation of Novel ALRCs/nZVI Composite and Its Removal of Cr(VI) from Aqueous. *Int. J. Environ. Res.* **2020**, *14*, 123–133. [CrossRef]
19. Zhang, Y.; Jiao, X.; Liu, N.; Lv, J.; Yang, Y. Enhanced removal of aqueous Cr(VI) by a green synthesized nanoscale zero-valent iron supported on oak wood biochar. *Chemosphere* **2020**, *245*, 125542. [CrossRef]
20. Chen, H.; Gao, Y.; El-Naggar, A.; Niazi, N.K.; Sun, C.; Shaheen, S.M.; Hou, D.; Yang, X.; Tang, Z.; Liu, Z.; et al. Enhanced sorption of trivalent antimony by chitosan-loaded biochar in aqueous solutions: Characterization, performance and mechanisms. *J. Hazard. Mater.* **2022**, *425*, 127971. [CrossRef]
21. Yazdi, F.; Anbia, M.; Sepehrian, M. Recent advances in removal of inorganic anions from water by chitosan-based composites: A comprehensive review. *Carbohydr. Polym.* **2023**, *320*, 121230. [CrossRef]
22. Perera, H.M.; Rajapaksha, A.U.; Liyanage, S.; Ekanayake, A.; Selvasembian, R.; Daverey, A.; Vithanage, M. Enhanced adsorptive removal of hexavalent chromium in aqueous media using chitosan-modified biochar: Synthesis, sorption mechanism, and reusability. *Environ. Res.* **2023**, *231*, 115982. [CrossRef]
23. Huang, X.; Niu, X.; Zhang, D.; Li, X.; Li, H.; Wang, Z.; Lin, Z.; Fu, M. Fate and mechanistic insights into nanoscale zerovalent iron (nZVI) activation of sludge derived biochar reacted with Cr(VI). *J. Environ. Manag.* **2022**, *319*, 115771. [CrossRef]
24. Xu, M.; Ma, X.; Chen, Y.; Hu, L.; Wang, B.; Qiu, M. Spectroscopic investigation of Cr(VI) sorption on nZVI/biochar composites. *J. Mol. Liq.* **2022**, *366*, 120262. [CrossRef]
25. Tan, X.; Shaaban, M.; Yang, J.; Cai, Y.; Wang, B.; Peng, Q.-A. Efficient removal of hexavalent chromium from an aquatic system using nanoscale zero-valent iron supported by ramie biochar. *Nanomaterials* **2021**, *11*, 2698. [CrossRef]
26. Salama, E.; Samy, M.; Shokry, H.; El-Subruiti, G.; El-Sharkawy, A.; Hamad, H.; Elkady, M. The superior performance of silica gel supported nano zero-valent iron for simultaneous removal of Cr (VI). *Sci. Rep.* **2022**, *12*, 22443. [CrossRef]
27. Tian, H.; Huang, C.; Wang, P.; Wei, J.; Li, X.; Zhang, R.; Ling, D.; Feng, C.; Liu, H.; Wang, M.; et al. Enhanced elimination of Cr(VI) from aqueous media by polyethyleneimine modified corn straw biochar supported sulfide nanoscale zero valent iron: Performance and mechanism. *Bioresour. Technol.* **2023**, *369*, 128452. [CrossRef]
28. Chen, L.; Zhang, J.; Li, Q.; Zhang, Y.; Huangpu, J.; Shen, J.; Gu, M.; Yang, J. Removal of p-nitrophenols by BC@nZVI activated persulfate: A study of key factors and mechanisms. *J. Environ. Chem. Eng.* **2023**, *11*, 111483. [CrossRef]
29. Qian, L.; Zhang, W.; Yan, J.; Han, L.; Chen, Y.; Ouyang, D.; Chen, M. Nanoscale zero-valent iron supported by biochars produced at different temperatures: Synthesis mechanism and effect on Cr(VI) removal. *Environ. Pollut.* **2017**, *223*, 153–160. [CrossRef]
30. Xu, H.; Gao, M.; Hu, X.; Chen, Y.; Li, Y.; Xu, X.; Zhang, R.; Yang, X.; Tang, C.; Hu, X. A novel preparation of S-nZVI and its high efficient removal of Cr(VI) in aqueous solution. *J. Hazard. Mater.* **2021**, *416*, 125924. [CrossRef]
31. Luo, L.; Cheng, S.; Yue, L.; You, Z.; Cai, J. N-doped biochar from chitosan gel-like solution: Effect of hydrothermal temperature and superior aqueous Cr (VI) removal performance. *Colloids Surf. A Physicochem. Eng. Asp.* **2022**, *641*, 128426. [CrossRef]
32. Preethi, J.; Vigneshwaran, S.; Meenakshi, S. Performance of chitosan engraved iron and lanthanum mixed oxyhydroxide for the detoxification of hexavalent chromium. *Int. J. Biol. Macromol.* **2019**, *130*, 491–498. [CrossRef]
33. Wang, F.; Li, L.; Iqbal, J.; Yang, Z.; Du, Y. Preparation of magnetic chitosan corn straw biochar and its application in adsorption of amaranth dye in aqueous solution. *Int. J. Biol. Macromol.* **2022**, *199*, 234–242. [CrossRef]
34. Yi, Y.; Wang, X.; Zhang, Y.; Yang, K.; Ma, J.; Ning, P. Formation and mechanism of nanoscale zerovalent iron supported by phosphoric acid modified biochar for highly efficient removal of Cr(VI). *Adv. Powder Technol.* **2023**, *34*, 103826. [CrossRef]
35. Yin, Y.; Shen, C.; Bi, X.; Li, T. Removal of hexavalent chromium from aqueous solution by fabricating novel heteroaggregates of montmorillonite microparticles with nanoscale zero-valent iron. *Sci. Rep.* **2020**, *10*, 12137. [CrossRef]
36. Qian, L.; Long, Y.; Li, H.; Wei, Z.; Liang, C.; Liu, R.; Chen, M. Unveiling the role of biochar in simultaneous removal of hexavalent chromium and trichloroethylene by biochar supported nanoscale zero-valent iron. *Sci. Total. Environ.* **2023**, *889*, 164243. [CrossRef]
37. Sultan, M.; Miyazaki, T.; Koyama, S. Optimization of adsorption isotherm types for desiccant air-conditioning applications. *Renew. Energy* **2018**, *121*, 441–450. [CrossRef]
38. Burk, G.A.; Herath, A.; Crisler, G.B.; Bridges, D.; Patel, S.; Pittman, C.U.; Mlsna, T. Cadmium and Copper Removal From Aqueous Solutions Using Chitosan-Coated Gasifier Biochar. *Front. Environ. Sci.* **2020**, *8*, 541203. [CrossRef]
39. Roy, H.; Islam, S.; Arifin, M.T.; Firoz, S.H. Synthesis, Characterization and Sorption Properties of Biochar, Chitosan and ZnO-Based Binary Composites towards a Cationic Dye. *Sustainability* **2022**, *14*, 14571. [CrossRef]

40. Lakatos, J.; Brown, S.D.; Snape, C.E. Coals as sorbents for the removal and reduction of hexavalent chromium from aqueous waste streams. *Fuel* **2002**, *81*, 691–698. [CrossRef]
41. Zhang, X.; Lin, S.; Chen, Z.; Megharaj, M.; Naidu, R. Kaolinite-supported nanoscale zero-valent iron for removal of Pb^{2+} from aqueous solution: Reactivity, characterization and mechanism. *Water Res.* **2011**, *45*, 3481–3488. [CrossRef]
42. Chen, X.; Fan, G.; Zhu, X.; Li, H.; Li, Y.; Li, H.; Xu, X. The remediation of hexavalent chromium-contaminated soil by nanoscale zero-valent iron supported on sludge-based biochar. *J. Soils Sediments* **2023**, *23*, 1607–1616. [CrossRef]
43. Wang, X.; Zheng, M.; Qian, Y.; Chen, H.; Li, X.; Li, X.; Zhang, A.; Xue, G. Synergizing redox of zerovalent iron and singlet oxygen to remove aniline, chromium and antimony in printing and dyeing wastewater synchronously: Multifunctional effect of sludge derived biochar. *Chem. Eng. J.* **2023**, *476*, 146927. [CrossRef]
44. Chen, X.-L.; Li, F.; Xie, X.J.; Li, Z.; Chen, L. Nanoscale zero-valent iron and chitosan functionalized *Eichhornia crassipes* biochar for efficient hexavalent chromium removal. *Int. J. Environ. Res. Public Health* **2019**, *16*, 3046. [CrossRef]
45. Liu, X.; Zhang, S.; Zhang, X.; Guo, H.; Lou, Z.; Zhang, W.; Chen, Z. Cr(VI) immobilization in soil using lignin hydrogel supported nZVI: Immobilization mechanisms and long-term simulation. *Chemosphere* **2022**, *305*, 135393. [CrossRef]
46. Nong, S.; Dong, W.; Xiao, Y.; Riaz, M.S.; Dong, C.; Zhao, Y.; Liu, Z.; Wang, R.; Huang, F. Highly Hydroxylated Porous Nanozirconia for Complete Trace Cr(VI) Removal. *ACS Appl. Nano Mater.* **2020**, *3*, 3315–3322. [CrossRef]
47. Mon, P.P.; Cho, P.P.; Chanadana, L.; Kumar, K.A.; Dobhal, S.; Shashidhar, T.; Madras, G.; Subrahmanyam, C. Bio-waste assisted phase transformation of Fe_3O_4/carbon to nZVI/graphene composites and its application in reductive elimination of Cr(VI) removal from aquifer. *Sep. Purif. Technol.* **2023**, *306*, 122632. [CrossRef]
48. Vigneshwaran, S.; Sirajudheen, P.; Nikitha, M.; Ramkumar, K.; Meenakshi, S. Facile synthesis of sulfur-doped chitosan/biochar derived from tapioca peel for the removal of organic dyes: Isotherm, kinetics and mechanisms. *J. Mol. Liq.* **2021**, *326*, 115303. [CrossRef]
49. Qu, Q.; Guo, X.; Shao, Z.; Wang, X.; Zhu, M.; Qiu, L. Adsorption performance and mechanism of Fe-loaded biochar derived from waste zanthoxylum branch for removing Cr(VI) from aqueous solution. *Biomass-Convers. Biorefinery* **2022**, *14*, 10201–10215. [CrossRef]
50. Yousaf, B.; Liu, G.; Abbas, Q.; Wang, R.; Ullah, H.; Mian, M.; Amina, A.; Rashid, A. Enhanced removal of hexavalent chromium from aqueous media using a highly stable and magnetically separable rosin-biochar-coated TiO_2@C nanocomposite. *RSC Adv.* **2018**, *8*, 25983–25996. [CrossRef]
51. Yang, Y.; Zhang, Y.; Wang, G.; Yang, Z.; Xian, J.; Yang, Y.; Li, T.; Pu, Y.; Jia, Y.; Li, Y.; et al. Adsorption and reduction of Cr(VI) by a novel nanoscale FeS/chitosan/biochar composite from aqueous solution. *J. Environ. Chem. Eng.* **2021**, *9*, 105407. [CrossRef]
52. Dash, S.S.; Sahu, M.K.; Sahu, E.; Patel, R.K. Fluoride removal from aqueous solutions using cerium loaded mesoporous zirconium phosphate. *New J. Chem.* **2015**, *39*, 7300–7308. [CrossRef]
53. Kasbaji, M.; Mennani, M.; Oubenali, M.; Benhamou, A.A.; Boussetta, A.; Ablouh, E.-H.; Mbarki, M.; Grimi, N.; El Achaby, M.; Moubarik, A. Bio-based functionalized adsorptive polymers for sustainable water decontamination: A systematic review of challenges and real-world implementation. *Environ. Pollut.* **2023**, *335*, 122349. [CrossRef]
54. Bae, S.; Lee, W. Influence of Riboflavin on Nanoscale Zero-Valent Iron Reactivity during the Degradation of Carbon Tetrachloride. *Environ. Sci. Technol.* **2014**, *48*, 2368–2376. [CrossRef] [PubMed]
55. Fu, R.; Zhang, X.; Xu, Z.; Guo, X.; Bi, D.; Zhang, W. Fast and highly efficient removal of chromium (VI) using humus-supported nanoscale zero-valent iron: Influencing factors, kinetics and mechanism. *Sep. Purif. Technol.* **2017**, *174*, 362–371. [CrossRef]
56. Wang, Z.; Chen, G.; Wang, X.; Li, S.; Liu, Y.; Yang, G. Removal of hexavalent chromium by bentonite supported organosolv lignin -stabilized zero-valent iron nanoparticles from wastewater. *J. Clean. Prod.* **2020**, *267*, 122009. [CrossRef]
57. Montesinos, V.N.; Quici, N.; Halac, E.B.; Leyva, A.G.; Custo, G.; Bengio, S.; Zampieri, G.; Litter, M.I. Highly efficient removal of Cr(VI) from water with nanoparticulated zerovalent iron: Understanding the Fe(III)–Cr(III) passive outer layer structure. *Chem. Eng. J.* **2014**, *244*, 569–575. [CrossRef]
58. Liu, X.-J.; Li, M.-F.; Ma, J.-F.; Bian, J.; Peng, F. Chitosan crosslinked composite based on corncob lignin biochar to adsorb methylene blue: Kinetics, isotherm, and thermodynamics. *Colloids Surf. A Physicochem. Eng. Asp.* **2022**, *642*, 128621. [CrossRef]
59. Cao, Y.; Yang, B.; Song, Z.; Wang, H.; He, F.; Han, X. Wheat straw biochar amendments on the removal of polycyclic aromatic hydrocarbons (PAHs) in contaminated soil. *Ecotoxicol. Environ. Saf.* **2016**, *130*, 248–255. [CrossRef]
60. Cui, L.; Noerpel, M.R.; Scheckel, K.G.; Ippolito, J.A. Wheat straw biochar reduces environmental cadmium bioavailability. *Environ. Int.* **2019**, *126*, 69–75. [CrossRef]
61. Li, H.; Chen, Y.Q.; Chen, S.; Wang, X.L.; Guo, S.; Qiu, Y.F.; Di Liu, Y.; Duan, X.L.; Yu, Y.J. Wheat straw biochar-supported nanoscale zerovalent iron for removal of trichloroethylene from groundwater. *PLoS ONE* **2017**, *12*, e0172337. [CrossRef] [PubMed]
62. Kang, X.; Xiao, F.; Zhou, S.; Zhang, Q.; Qiu, L.; Wang, L. Study on the performance of sewage sludge biochar modified by nZVI to remove Cu(II) and Cr(VI) in water. *Water Sci. Technol.* **2022**, *86*, 1821–1834. [CrossRef] [PubMed]

nanomaterials

Article

Green Synthesis of Cobalt-Doped $CeFe_2O_5$ Nanocomposites Using Waste *Gossypium arboreum* L. Stalks and Their Application in the Removal of Toxic Water Pollutants

Saloni Koul [1], Mamata Singhvi [1,*] and Beom Soo Kim [2]

1 Department of Biotechnology (with Jointly Merged Institute of Bioinformatics and Biotechnology), Savitribai Phule Pune University, Pune 411007, India; n20502022@uopca.unipune.ac.in
2 Department of Chemical Engineering, Chungbuk National University, Cheongju 28644, Chungbuk, Republic of Korea; bskim@chungbuk.ac.kr
* Correspondence: mamata.singhvi@campstud.unipune.ac.in; Tel.: +91-20-25690442

Abstract: Currently, there is an increasing need to find new ways to purify water by eliminating bacterial biofilms, textile dyes, and toxic water pollutants. These contaminants pose significant risks to both human health and the environment. To address this issue, in this study, we have developed an eco-friendly approach that involves synthesizing a cobalt-doped cerium iron oxide (CCIO) nanocomposite (NC) using an aqueous extract of *Gossypium arboreum* L. stalks. The resulting nanoparticles can be used to effectively purify water and tackle the challenges associated with these harmful pollutants. Nanoparticles excel in water pollutant removal by providing a high surface area for efficient adsorption, versatile design for the simultaneous removal of multiple contaminants, catalytic properties for organic pollutant degradation, and magnetic features for easy separation, offering cost-effective and sustainable water treatment solutions. A CCIO nanocomposite was synthesized via a green co-precipitation method utilizing biomolecules and co-enzymes extracted from the aqueous solution of *Gossypium arboreum* L. stalk. This single-step synthesis process was accomplished within a 5-h reaction period. Furthermore, the synthesis of nanocomposites was confirmed by various characterization techniques such as Fourier-transform infrared (FT-IR) spectroscopy, X-ray diffraction (XRD), field emission scanning electron microscopy (FE-SEM), transmission electron microscopy (TEM), thermogravimetric analysis (TGA), dynamic light scattering (DLS), and energy dispersive X-ray (EDX) technology. CCIO NCs were discovered to have a spherical shape and an average size of 40 nm. Based on DLS zeta potential analysis, CCIO NCs were found to be anionic. CCIO NCs also showed significant antimicrobial and antioxidant activity. Overall, considering their physical and chemical properties, the application of CCIO NCs for the adsorption of various dyes (~91%) and water pollutants (chromium = ~60%) has been considered here since they exhibit great adsorption capacity owing to their microporous structure, and represent a step forward in water purification.

Keywords: green synthesis; co-precipitation; adsorption; dye removal; pollutant removal; cotton stalks; wastewater treatment; adsorption capacity

Citation: Koul, S.; Singhvi, M.; Kim, B.S. Green Synthesis of Cobalt-Doped $CeFe_2O_5$ Nanocomposites Using Waste *Gossypium arboreum* L. Stalks and Their Application in the Removal of Toxic Water Pollutants. *Nanomaterials* **2024**, *14*, 1339. https://doi.org/10.3390/nano14161339

Academic Editors: Marco Stoller, Lei Huang, Junye Cheng, Hongguo Zhang and Zhenxing Wang

Received: 29 March 2024
Revised: 15 May 2024
Accepted: 24 June 2024
Published: 12 August 2024

1. Introduction

Water pollutants pose significant environmental and public health concerns. Environmental consequences include the degradation of aquatic ecosystems, loss of biodiversity, and disruption of ecological balance. Simultaneously, the contamination of water sources increases public health risks, contributing to waterborne diseases, developmental issues, and long-term health problems for communities reliant on a source with compromised water quality. Water streams can be contaminated with a variety of toxic pollutants that include halogenated hydrocarbons, heavy metals, dyes, surfactants, organic compounds, salts, soluble bases, pesticides, and agricultural fertilizers [1]. Synthetic dyes, which are extensively employed in textiles, pharmaceuticals, and numerous other industries, exhibit remarkable color stability and versatility. Hexavalent chromium is a highly toxic heavy metal that is released into water bodies through industrial processes, including metal plating, leather tanning, and textile manufacturing [2]. The ubiquity of these synthetic colorants and metals in water bodies raises alarming concerns as once introduced into aquatic ecosystems, they resist natural degradation processes, leading to their persistence in water bodies, and they can be toxic depending on their chemical composition and concentration [2,3].

Exposure to significant levels of organic/inorganic pollutants has been associated with endocrine disruption, mutagenicity/genotoxicity, and cancer [3,4]. Consequently, nanoparticles emerge as a transformative frontier in the quest for sustainable solutions, holding promise for revolutionizing the landscape of synthetic dye and heavy metal removal [5–9]. In the case of pollutant removal techniques such as flocculation [10,11], ozonation [12], membrane filtration [11,13], activated carbon adsorption [14], electrocoagulation [15,16], sorption techniques [17,18], UV radiation [19], and biological treatment [20], their limitations in terms of efficiency, selectivity, and environmental impact underscore the need for innovative approaches.

Subsequent research focus has shifted to the expanding field of nanoparticle-mediated pollutant removal, addressing the mechanisms by which nanoparticles interact with pollutants and the factors influencing their effectiveness [21]. By reviewing the current state of knowledge, this study aims to underscore the urgency of adopting environmentally benign approaches for pollutant remediation, additionally highlighting the promising role of nanoparticles in mitigating this global environmental challenge [7,22,23]. Au NPs [24], Ag NPs [25,26], TiO_2 NPs [27], SiO_2 NPs [28,29], Fe_2O_3 NPs [30], and graphene oxide NPs (GO) [31,32] are some of the prime examples of nanomaterial-mediated pollutant remediation technologies.

Nanomaterials, with their increased surface area per unit mass, offer enhanced adsorption capacity. This mechanism involves the physical binding of pollutants to the surface of nanomaterials. The attractive forces between the nanomaterial's surface and the pollutant molecules lead to their immobilization. This process is particularly effective for the removal of organic pollutants, heavy metals, and even nanoparticles from water. Certain nanomaterials possess catalytic properties, enabling them to facilitate chemical reactions that transform pollutants into less harmful substances. This mechanism is prominent in the degradation of organic pollutants. For instance, photocatalytic nanoparticles, like titanium dioxide (TiO_2) [27] or zinc oxide (ZnO) [33], can absorb light energy and generate reactive oxygen species, initiating oxidation reactions that break down organic contaminants.

Nanomaterials can induce the precipitation of pollutants by acting as nucleation sites. This is particularly relevant for the removal of heavy metals and ions from water. Functionalized nanoparticles may promote the coagulation of suspended particles, facilitating their removal through sedimentation or filtration processes [34]. Nanomaterials can undergo ion exchange, where certain ions on the material's surface are replaced by ions from the surrounding water. This mechanism is effective for the removal of specific ions, such as heavy metal ions. Ion exchange properties are often enhanced through surface functionalization, allowing nanomaterials to selectively capture target ions [35–37].

Magnetic nanoparticles, which are endowed with magnetic properties, can be separated easily from water using external magnetic fields. This mechanism allows for the recovery and reuse of nanomaterials, contributing to their cost-effectiveness. Magnetic separation is particularly advantageous for the removal of nanoparticles and other magnetic-responsive pollutants. Nanomaterials with redox-active surfaces can participate in redox reactions, facilitating either the reduction or oxidation of pollutants. This mechanism is significant for the removal of contaminants that are susceptible to redox transformations. Redox reactions can be harnessed for the removal of various pollutants, including organic compounds and certain heavy metals [30,38].

Some nanomaterials exhibit inherent antibacterial properties, contributing to the removal of pathogenic microorganisms in water. The disruption of microbial cell membranes or interference with cellular functions by nanomaterials can effectively mitigate waterborne diseases [39,40]. In conclusion, nanomaterials offer a versatile toolkit for pollutant removal in water, employing mechanisms such as adsorption, catalysis, precipitation, ion exchange, magnetic separation, redox reactions, and antibacterial properties. The choice of nanomaterial and the optimization of its properties play a crucial role in designing effective and sustainable water treatment strategies [9]. This study deals with the green synthesis of a cobalt-doped cerium iron oxide (CCIO) nanocomposite (NC) using cotton stalks, a waste material from agriculture that serves as a highly useful, low-cost, and abundantly available material. Green chemistry is a method that aims to reduce waste products and prevent environmental pollution with the aid of sustainable and environmentally friendly materials [8]. It entails the usage of natural resources, such as vegetation, microorganisms, enzymes, and other renewable sources to synthesize various compounds and materials [40,41].

The goal of green synthesis is to develop processes that are secure and green, as well as eco-friendly [26]. In this study, we have utilized cotton stalks as they are agricultural waste; using them as a substrate for nanoparticle deposition aligns with the green chemistry principles of utilizing renewable resources and reducing waste. This approach avoids the need for synthetic or non-renewable materials as substrates. Additionally, the plant extracts are abundant, renewable, and cost-effective compared to synthetic reagents. Hence, we tested this plant waste for the synthesis of CCIO NCs, which proved its effective role in the adsorption of various dyes and toxic metals, in addition to exhibiting potent antimicrobial activity. These properties suggest promising applications for CCIO NCs in the removal of both organic and inorganic pollutants from contaminated water sources. To our knowledge, this is the first study about synthesizing CCIO NCs from *Gossypium arboreum* L. stalks, in terms of their exceptional ability to adsorb contaminants (approximately 91%) via the surface of NCs. The overall work performed in this study can be summarized as shown in Figure 1.

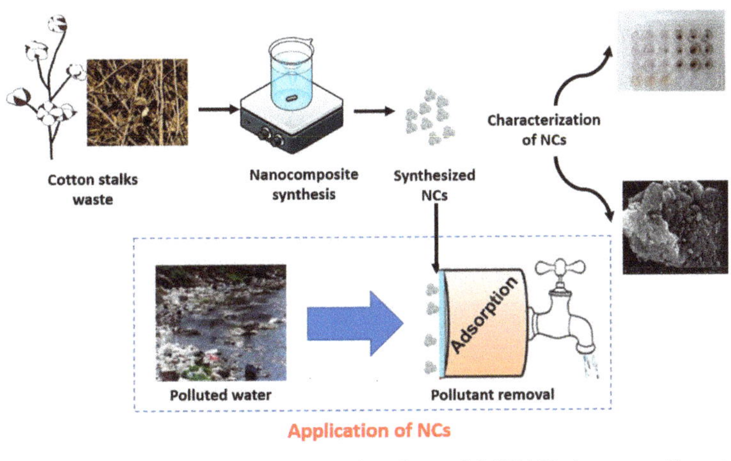

Figure 1. Illustration of the application of synthesized CCIO NCs in water pollutant removal.

2. Materials and Methods

2.1. Materials

Cerium nitrate [$Ce(NO_3)_3$], cobalt chloride [$CoCl_2$], potassium dichromate [$K_2Cr_2O_7$], 2,2-diphenyl-1-picrylhydrazyl (DPPH), malachite green (MG), and safranin (SF) were purchased from Himedia, Pune, India. Ferric chloride ($FeCl_3$) was obtained from SD Fine Chem, Mumbai, India. Methylene Blue (MB) was bought from SISCO Research Laboratories, Mumbai, India. All other reagents and chemicals were purchased locally.

2.2. Synthesis of Cobalt-Doped Cerium Iron Oxide Nanocomposites (CCIO NCs)

The stalks of *Gossypium arboreum* L. were collected from Nagpur, Maharashtra, India. After washing and heat-drying, the stalks were crushed into small pieces. First, 50 g of the crushed stalks were soaked in 500 mL of distilled water and kept in a water bath at 80 °C for 3 h [42]. All the wooden residues were removed by filtration, using a muslin cloth and centrifugation of the extract [8,26,43]. The synthesis of CCIO nanocomposites was performed by co-precipitation, a green synthesis method. In the prepared aqueous extract of *Gossypium arboreum* L. stalks (500 mL), 200 Mm each of $Ce(NO_3)_3$, $FeCl_3$, and $CoCl_2$ were added stepwise in the mentioned order and kept under stirring at 50 °C for 30 min, then overnight with stirring at 30 °C. A dark-colored solution appeared after overnight stirring and the mixture was further centrifuged at 10,000 rpm for 10 min to obtain the precipitate. The obtained precipitate was further washed in distilled water thrice, followed by ethanol washing. The obtained brown-black-colored pellet was retained for drying at 50 °C.

2.3. Characterization of Synthesized CCIO NCs

Fourier-transform infrared (FT-IR Thermo Fisher Scientific, Waltham, MA, USA) spectroscopy was conducted to observe the functional groups, stretching vibrations, and absorption peaks present on the surface of nanocomposites. Field-emission scanning electron microscopy (FE-SEM) analysis was performed to study the morphology of the CCIO NCs. Furthermore, the sizes of the CCIO NCs were estimated using transmission electron microscopy (TEM) (Carl Zeiss, Libra 120 Oberkochen, Germany) at a voltage of 120 Kv [44–46]. The elemental composition of the NCs was determined using EDX (Thermo Fisher Scientific, Waltham, MA, USA) analysis. The X-ray diffraction (XRD) [47] pattern of CCIO NCs was obtained using Cu-K_β radiation, along with a scintillation counter detector. For smoothening the data, the Savitzky–Golay (SG) digital filtering method was applied. To study the thermostability of the synthesized NCs, thermogravimetric analysis (TGA) was performed. The Malvern Panalytical zeta sizer Nano Z instrument (Malvern, UK) was used to study the zeta potential of the nanocomposites [48].

2.4. Preparation of Cationic Dyes

The catalytic properties of CCIO nanocomposites (NCs) for dye decolorization were assessed, as follows. Different concentrations (1, 2, 3, 4, 5, and 10 mg) of CCIO NCs were prepared, to which 10 mL of 100 ppm safranin, malachite green, and methylene blue dyes were added. The solution was stirred for 6 h to check the degradation rate. The dye decolorization process was analyzed by UV–vis spectrophotometer (Thermo Scientific Multiskan EX) [49]. The prepared dye solution (0.1 mg/mL) was used as the control. Eventually, the treatment solution was centrifuged at 7000 rpm for 5 min, and the absorbance (200 nm–800 nm) was measured with a Thermo Fisher Scientific microplate reader. Experiments were performed in triplicate and the mean percentage value was recorded [50,51].

2.5. Preparation of Hexavalent Chromium Solution

Analytical grade $K_2Cr_2O_7$ was dissolved in 10 mL of distilled water to prepare 0.25 mg/mL of chromium sample solution [52]. Based on the standardized concentration of CCIO NCs for dye adsorption, a 4 mg/mL concentration was selected to check its adsorption performance against hexavalent chromium ions.

2.6. Adsorption Analysis

To investigate the effect of CCIO NCs on dye as well as for heavy metal removal, the samples were checked for any residual dye/metal at a neutral pH and at room temperature using the absorbance spectra. The adsorption capacity (Q_0) and percentage of removal (RE) were calculated using the following formulas:

$$Q_0 = \frac{D_i - D_f}{W} \times V \quad (1)$$

$$RE\ (\%) = \frac{D_i - D_f}{D_i} \times 100 \quad (2)$$

where D_i and D_f are the initial and final absorbance; W is the weight of the adsorbent in g; and V is the volume of the dye solution in liters (L) [53].

2.7. Antioxidant Activity

A DPPH assay was employed to measure the scavenging activity of the synthesized NC antioxidants [26,54]. For this assay, 50 mL of 0.1 mM DPPH in methanol was prepared. Then, 0.5 mL of DPPH solution and 0.25 mL of sample solution were mixed and incubated for 30 min at 37 °C. (Additionally, the assay was performed in triplicate, after which the mean value was considered.) After 30 min, absorbance was measured at 517 nm, using the spectrophotometer. Also, the radical scavenging activity was measured using the following equation:

$$\text{Radical scavenging activity } (\%) = \left(\frac{Ac - As}{Ac}\right) \times 100 \quad (3)$$

where As = absorbance of the sample, and Ac = absorbance of the control, measured at 517 nm each.

2.8. Antimicrobial Activity

The antibacterial activity of CCIO NCs was assessed against *Staphylococcus aureus* and *Escherichia coli* bacteria, using the agar well diffusion method. CCIO NCs were sterilized under UV light for about 30 min. Then 10 mg/mL of suspension was prepared in sterile distilled water and sonicated for 5 min to prepare a homogeneous suspension. Then, 100 μL of bacterial culture was spread over nutrient agar plates. Wells were prepared on both plates using a sterile borer and 50 and 100 μL of CCIO NCs suspension was added to the wells. Then, all the plates were incubated for 24 h at 37 °C. The diameter of the zone of inhibition was then determined to calculate the rate of bacterial growth.

2.9. Recycling and Reuse of Used CCIO NCs

The NCs were recovered from the reaction mixture using sonication and Buchner filtration with Whatman filter paper and were then washed with ethanol and double-distilled water, and retained for heat drying [55]. The used NCs were washed 3 times with 70% ethanol. The samples were then dried in a hot air oven for further use. The reusability experiments were carried out using MB dye (5 mg/mL) for 4 succeeding cycles [53].

3. Results and Discussion

3.1. Synthesis and Characterization of the Synthesized CCIO NCs

During the synthesis of NCs, the plant extract (aqueous), which is added to the metal salt solution (acting as a precursor), facilitates the reduction of metal ions to their zero-valent metallic state. Furthermore, the metal ions nucleate and start to grow into nanoparticles in the presence of stabilizing agents that are present in the plant extract. To understand the structural and morphological characteristics of the synthesized CCIO NCs, SEM and TEM imaging were performed. As shown in Figure 2A, SEM analysis demonstrated that the particles were spherical in shape and were mostly in an aggregated form. The average size of the CCIO NCs was found to be 40 nm using TEM analysis, confirming that the CCIO NCs had been synthesized, as shown in Figure 2B.

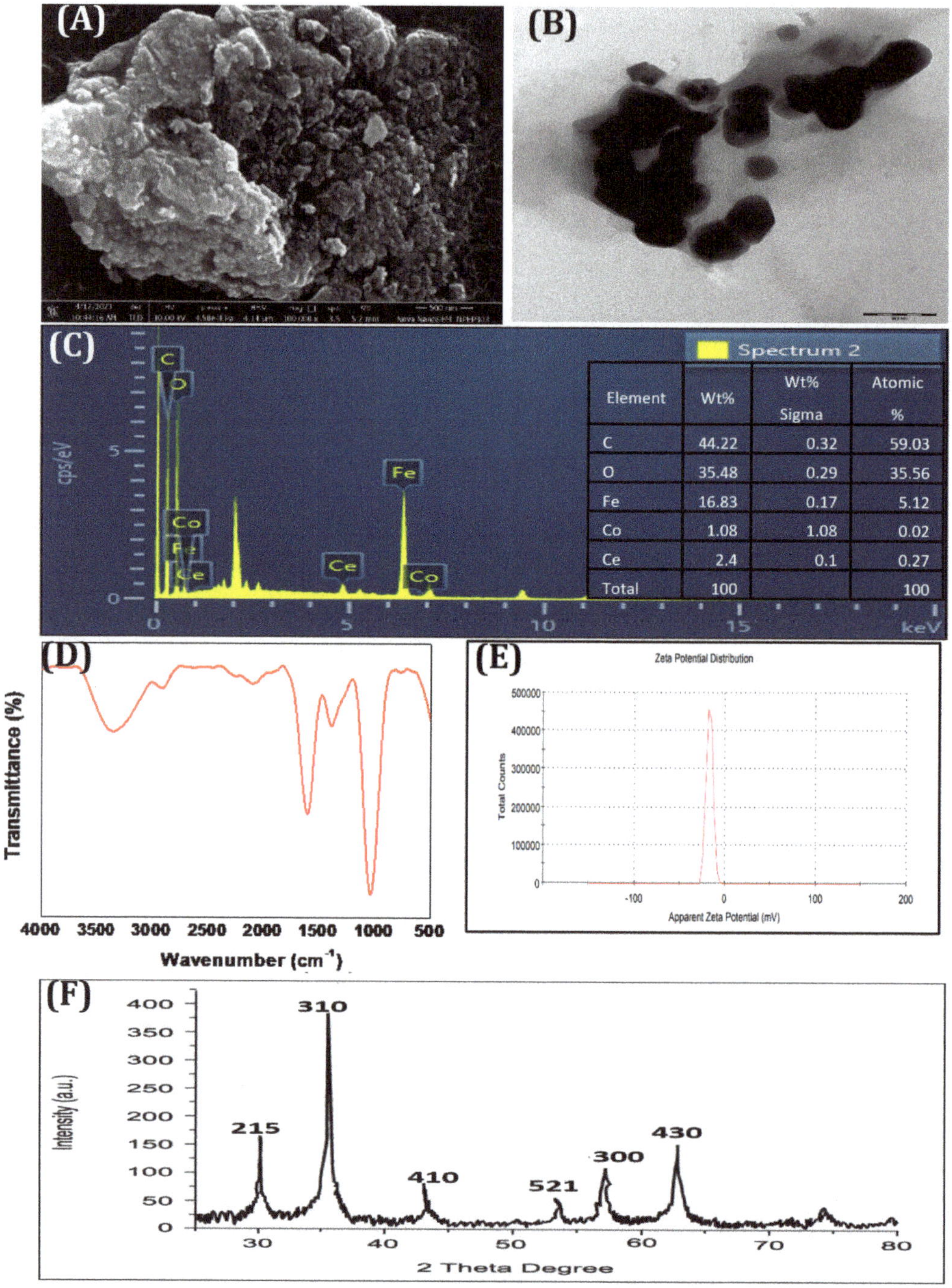

Figure 2. Characterization of CCIO NCs using (**A**) SEM, (**B**) TEM, (**C**) EDX, (**D**) FTIR, (**E**) zeta potential, and (**F**) XRD.

Then, EDX analysis was conducted to find out the percentages of elements present in the synthesized CCIO NCs. The position of the peaks from the EDX analysis confirms the elemental composition of the CCIO NCs. As depicted in Figure 2C, the peaks at 4.8 keV, 6.2 keV, and 7.1 keV confirm the presence of cerium, iron, and cobalt elements. In addition, carbon (44.22%) and oxygen (35.48%) were detected (Figure 2C). EDX analysis substantiated the finding that CCIO NCs consist of cerium, cobalt, and iron, with a higher mass percentage of iron than cerium and cobalt [56].

To validate the successful formation of CCIO NCs, FT-IR spectroscopy was administered, as seen in Figure 2D. The FT-IR spectra of CCIO showed an absorption peak at 1373 cm^{-1}, which is associated with phenol O–H stretching. The peaks rising at 1235 and 2116 cm^{-1} were attributed to the C–N and C≡C vibrations, respectively. The broad band appearing at 3384.08 cm^{-1} corresponds to the alcohol O–H stretch. The peaks at 1988.46 and 2920.32 cm^{-1} were ascribed to the presence of an aromatic compound and alkane due to the C–H stretch, whereas the peak arising at 1606 cm^{-1} was linked to an alkene (conjugated or cyclic) [43,45,57,58]. The presence of phenols, aromatic compounds, and alkanes in the NCs derived from cotton stalks is a result of the thermal decomposition (via a water bath) of cellulose, lignin, and hemicellulose into smaller molecular fragments, including phenols, aromatic compounds (such as benzene rings), and aliphatic hydrocarbons (alkanes) that are present in the cotton stalks. The zeta potential indicates the stability of the colloidal system and was observed to be -16.8 mV (Figure 2E). The synthesized NCs are anionic and stable, which proves that they have sufficient repulsive force to avoid flocculation [45].

XRD was used to analyze the degree of crystallinity of the synthesized nanocomposites. The samples were scanned in the 2θ range of 15 to 80° and the crystallite size was determined using data obtained during XRD analysis [59]. The XRD diffraction patterns of the CCIO NCs are shown in Figure 2F, and show seven peaks corresponding to the (215), (310), (410), (521), (430), and (300) planes. It is evident that all the peaks showed similarity with the standard pattern, indicating that the synthesized material was pure. The crystallite size was calculated using Scherrer's formula, which indicated a range of around 35–40 nm, indicating the high crystallinity of CCIO NCs.

TGA was performed to investigate the thermostability of the CCIO NCs. As illustrated in Figure S2B, the thermogram showed primary degradation at around 70 °C and, furthermore, degradation at around 200 °C. The weight loss percentage at the second stage of degradation was observed to be higher than the initial one.

3.2. *Adsorption Studies*

3.2.1. Adsorption Performance of CCIO NCs for Cationic Dye Removal

A green co-precipitation method was employed to synthesize CCIO NCs and assess their use for the adsorption of different cationic dyes such as safranin (SF), malachite green (MG), and methylene blue (MB) from aqueous solutions. The CCIO NCs exhibited a maximum adsorption capacity of 34.3 mg g^{-1} when using SF, followed by MB and MG, respectively, thus confirming the possibility of their application in the remediation of dye pollution [60]. The adsorbent was added to the aqueous solution of the dye and kept at room temperature with constant stirring, which allowed the cationic dye to adsorb on the anionic binding site of the CCIO NCs due to electrostatic forces, as seen in Figure 3. Different concentrations of CCIO NCs were added to the aqueous dye solution and UV-vis spectroscopy was performed at the 4th hour. Maximum dye removal was observed when a 5 mg/mL concentration of CCIO NCs was used in the case of all three dyes, as presented in Figure 3A–C. Amongst the dyes used, CCIO NCs exhibited the maximum RE (%) when using SF (92.87%) compared to MB (90.55%) and MG (89.34%), as shown in Figure 3D. Overall, chitosanbased Fe_3O_4 NCs were used as adsorbents for MB removal and showed an adsorption capacity of 0.62 to 0.95 mg/g [61]. Li et al. used silica-based chitosan NCs for studying the adsorption behavior of NCs, which demonstrated an adsorption capacity of 43.03 mg g^{-1} for MB [60]. Sadiq et al. synthesized magnetic chitosan deep eutectic solvents (MNCDES) to investigate their adsorption ability regarding MG. These studies showed that MNCDES can adsorb 92.69% of MG dye [60]. In this study, the CCIO NCs showed an almost similar adsorption performance, as reported in previous studies. However, most of the studies have been conducted using chemically synthesized nanomaterials, whereas, in this study, the CCIO NCs have been prepared using plant waste extract through a green synthesis process. Hence, green-synthesized NCs have significant application potential in the removal of dyes from wastewater sources.

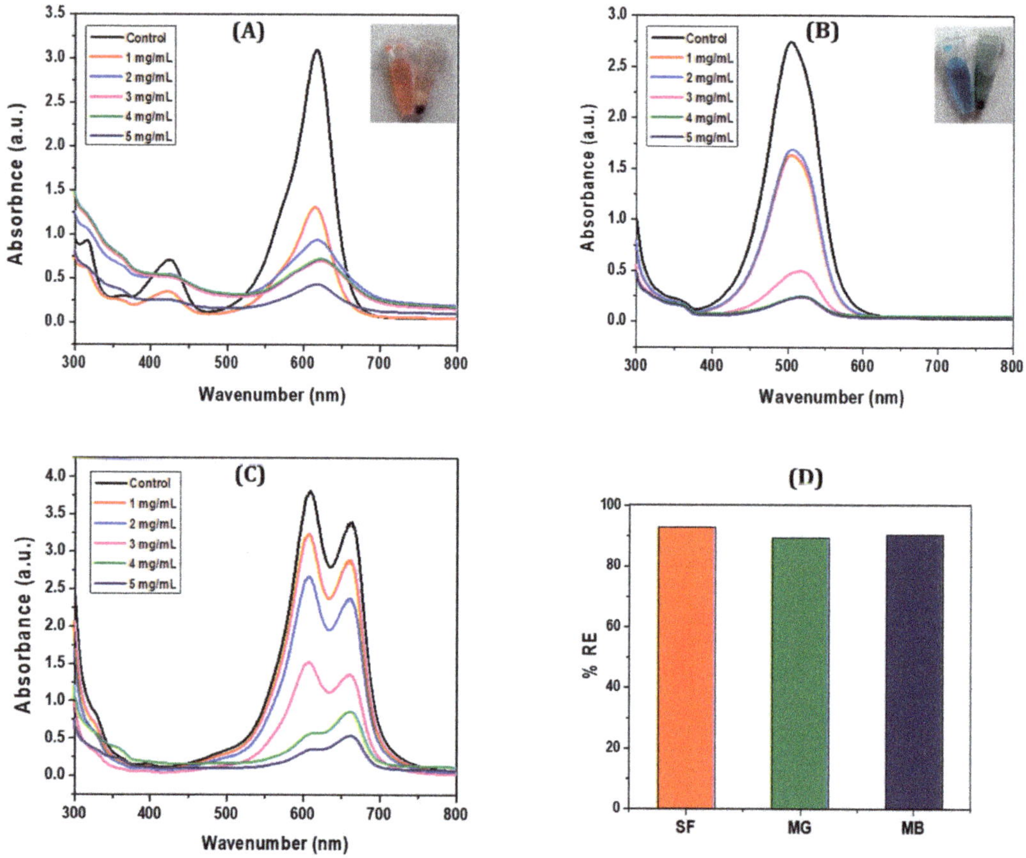

Figure 3. UV-Vis spectra of: (**A**) SF at different concentrations (1–5 mg/mL), (**B**) MG at different concentrations (1–5 mg/mL), (**C**) MB at (1–5 mg/mL), and (**D**) the regeneration efficiency (%) of CCIO NCs using SF, MG, and MB dyes.

3.2.2. Adsorption Performance of CCIO NCs for Chromium Removal

Hexavalent chromium is a tremendously toxic metal of great concern that has been found in water bodies. Several kinds of nanomaterials have been tested as adsorbents for the removal of chromium (VI) from polluted water. However, traditional adsorbents usually have a limited adsorption capacity, which limits their use in real-world applications. In this study, CCIO NCs have been synthesized and studied for chromium (VI) removal from wastewater, owing to their functionality, stability, and redox properties. As shown in Figure 4, the absorbance peak for hexavalent chromium (VI) was observed at 350 nm. The sharpness of this peak decreased significantly when the sample solution was incubated for 6 h at room temperature in the presence of CCIO NCs. CCIO NCs unveiled excellent adsorption capacity for chromium (VI) removal (~59.60%) from a solution, as displayed in Figure 4. From the FTIR analysis, it can be seen that the adsorption effect was due to electrostatic interactions between the surface of the CCIO NCs and chromium (VI). A reduction from chromium (VI) to (III) was ascribed to all the functional groups (e.g., –OH, –COOH, and –NH–) that were present on the CCIO NCs. These results corroborate the conclusion that CCIO NCs have tremendous potential as an economical and efficient adsorbent of chromium (VI) from polluted water.

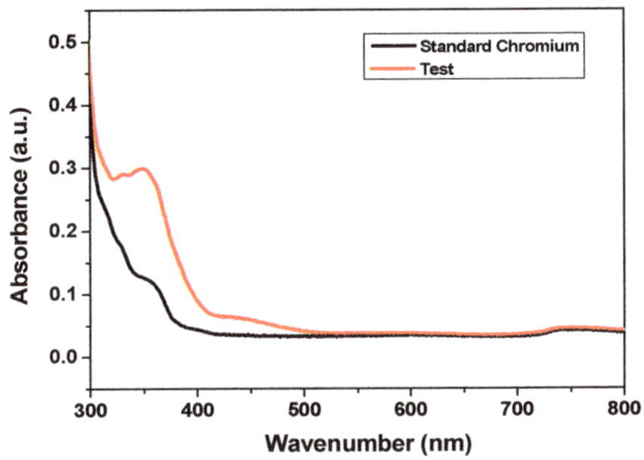

Figure 4. UV-Vis spectra for analyzing the adsorption capacity of CCIO NPs against chromium (VI).

3.2.3. Comparative Analysis

On the basis of the obtained results, maximum adsorption was obtained until 4 h into the incubation period. Comparative analysis was performed for all three dyes, i.e., SF, MB, and MG, using 10 mg/mL of CCIO concentration. The RE percentage of dye adsorption onto the adsorbent CCIO NCs in 4 h followed the order of SF > MB > MG, as seen in Figure 5. The feasibility of using CCIO NCs in wastewater treatment was investigated by preparing a reaction mixture of dye (10 mg/mL) and the adsorbent. For the adsorption capacity test, the sample was centrifuged and the supernatant was collected to determine the residual dye concentration. Taking into consideration the concentration of CCIO NCs that were utilized for the experiment, the results are highly promising when compared to other studies [52,53,62].

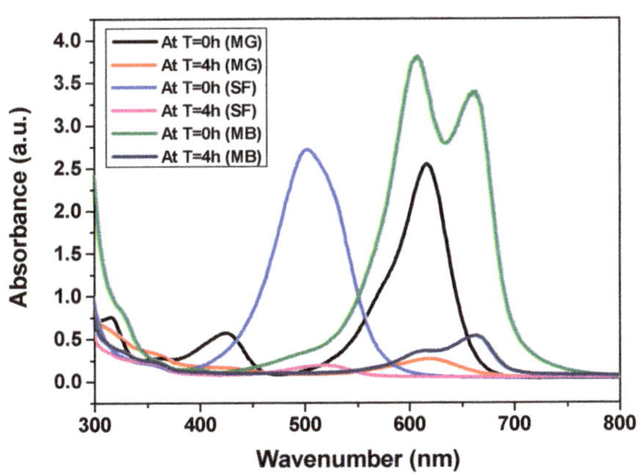

Figure 5. UV-Vis spectra of SF, MG, and MB at 10 mg/mL concentration.

3.3. Radical Scavenging Activity

The odd electron of the nitrogen atom in DPPH is reduced by the hydrogen atom that is received from the antioxidants, which reduces the color intensity (violet) (Figure 6). The degree of color reduction was measured spectrophotometrically. As shown in Figure 6, among the different concentrations of NCs, higher antioxidant activity (44%) was observed with a 1.0 mg/mL concentration of CCIO NCs.

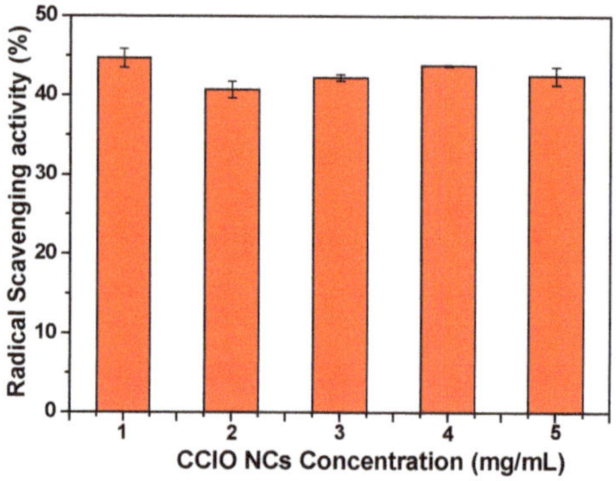

Figure 6. Antioxidant activity of CCIO NCs at different concentrations (1–5 mg/mL).

3.4. Antimicrobial Activity

As shown in Figure 7, the agar diffusion method was used to determine the antimicrobial activity of CCIO NCs. As shown in Table 1, the antimicrobial activity of CCIO NCs was measured against *E. coli* and *S. aureus*. The antimicrobial activity of CCIO NCs was found to be slightly higher in *S. aureus* (19 mm) compared to *E. coli* (16.5 mm). Once a clear zone had formed around the well containing NCs on the media plate, this exhibited proof of inhibition against a particular bacterium.

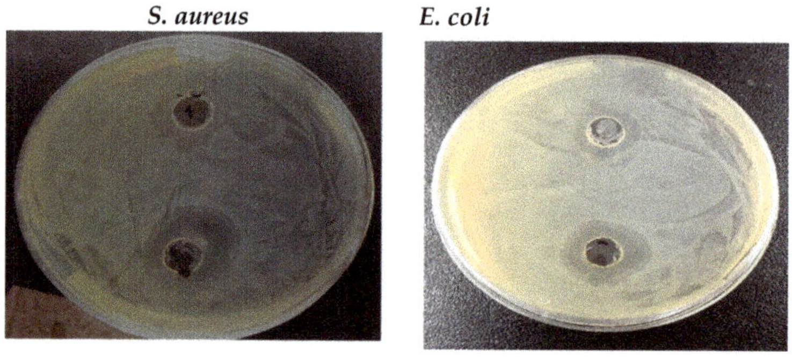

Figure 7. Antimicrobial activity of synthesized CCIO NPs against *E. coli* and *S. aureus*.

Table 1. Antimicrobial effect of CCIO NCs, using the agar well diffusion method.

Microorganisms/Sample	Antimicrobial Activity (Zone of Inhibition (dia. in mm))	
	50 µL	100 µL
E. coli	11	16.5
S. aureus	13.5	19

CCIO NCs. As per the reported studies to date, in addition to well-known antimicrobial AgNPs, other metal oxides (e.g., MgO, TiO_2, Cu_2O, ZnO, and $CoFe_2O_4$) have been found to be effective against microorganisms [63,64]. The doping of $CeFe_2O_5$ NPs with cobalt aided in improving their antimicrobial activity against both Gram-positive and Gram-negative organisms, which is evident from previous studies [65]. The excellent antimicrobial activity of CCIO NCs against *E. coli* and *S. aureus* microbes was attributed to the size, functional groups, and charge that were present on the NCs' surface [66]. The obtained results obviously validate the application of a small amount of CCIO NCs in wastewater treatment, which can reduce the numbers of various types of organisms within a short period of time.

3.5. Reusability of CCIO NCs

In addition to good adsorption performance, reusability is a crucial parameter to ensure an efficient, scalable, and economical system. To regenerate the adsorbent, ethanol and methanol were found to be good eluents. Ethanol's effectiveness in removing dyes such as safranin, methylene blue, and malachite green from nanoparticles is attributed to its ability to disrupt the electrostatic interactions between the dyes and the nanoparticle surfaces. By solvating the dye molecules, ethanol facilitates their removal, enabling the recovery and reuse of the nanoparticles. The effect of four consecutive regeneration cycles was investigated for reused CCIO NCs using SF dye, as described in Section 2.9. The reusability data are expressed with respect to regeneration efficiency (RE). As shown in Figure 8, the CCIO NCs retained almost 85–90% of their adsorption capacity, even after their fourth iteration of recycling. After each successive cycle, barely a 3.5–4% decrease in adsorption performance was observed, indicating that the obtained CCIO NCs had excellent stability, regenerability, and adsorptive properties. Therefore, CCIO can easily be recycled and can act as an excellent cost-effective material for the rapid removal of cationic dyes from an aqueous solution.

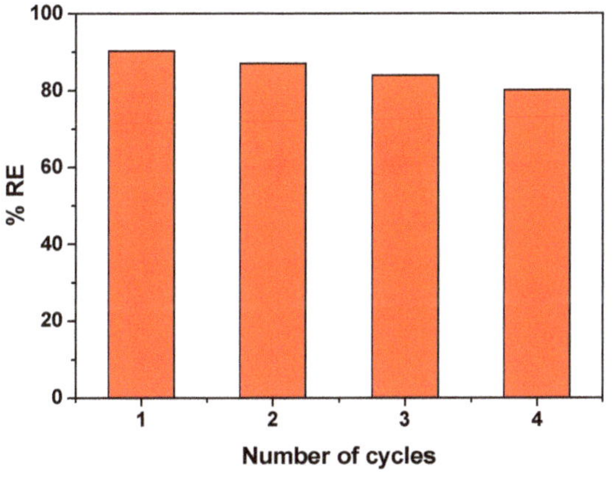

Figure 8. Regeneration efficiency of reused CCIO NCs, using SF dye.

3.6. Plausible Mechanism of Dye Removal Due to the Action of CCIO NCs

The possible action of CCIO NCs in the removal of dyes and other water pollutants is illustrated in Figure 9. The interaction between dye molecules, such as safranin, methylene blue, malachite green, and CCIO nanocomposites is primarily driven by a combination of electrostatic attraction, hydrogen bonding, and van der Waals forces. The cationic nature of these dyes, due to their dimethylamino groups, facilitates strong electrostatic interactions with negatively charged sites on the nanocomposite's surface. Additionally, hydrogen bonding plays a crucial role, particularly when the nanocomposite surface contains hydroxyl groups or other hydrogen bond donors or acceptors. The lone pairs on the nitrogen atoms in the dimethylamino groups of the dyes can form hydrogen bonds with the hydrogen atoms of hydroxyl groups on the nanocomposite. Furthermore, the high surface area and porous structure of nanocomposites provide numerous active sites for adsorption, enhancing the van der Waals interactions and allowing for the efficient capture of dye molecules. These combined interactions result in the effective adsorption and removal of dye pollutants from aqueous solutions, making such nanocomposites highly effective in environmental cleanup applications.

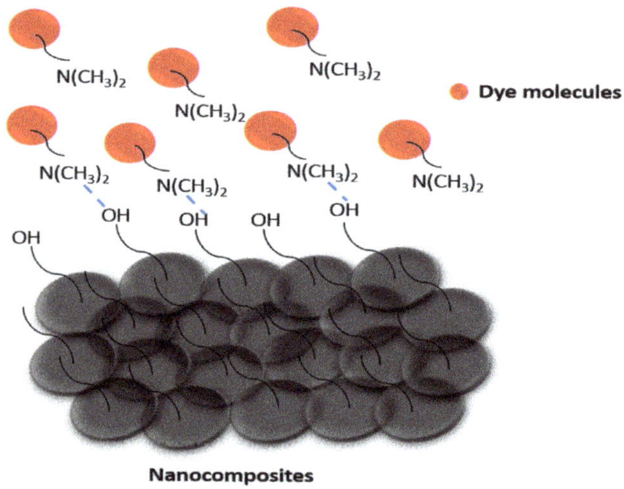

Figure 9. Plausible mechanism of action of CCIO NCs in dye adsorption.

4. Conclusions

In this study, the green synthesis of CCIO NCs was performed via the co-precipitation method after a short reaction time of 4h. As the synthesized CCIO NCs are free of secondary pollution, cost-effective, and easier to scale up, they may prove to be a better alternative to conventional wastewater treatment techniques. The characteristic features of NCs were studied using various methods, including FT-IR, XRD, TGA, FE-SEM, TEM, and DLS. The results obtained from the FT-IR spectra revealed the presence of phenolic compounds in the CCIO NCs, which established antioxidant activity. The dynamic light scattering (DLS) graph for zeta potential confirms the CCIO NCs to be anionic, which justifies their ability to adsorb cationic dyes like SF, MB, and MG. The TGA curve demonstrates that the synthesized CCIO NPs offer dry heat resistance of up to ± 200 °C. The maximum adsorption capacity of the CCIO NCs was found to be 34.3 mg/g against SF dye. Also, the reused NCs retained the maximum level of 85–90% of their adsorption capacity, even after 4 successive recycling iterations, indicating its reusability. An adsorption capability of CCIO NCs against cationic dyes such as SF, MG, and MB, along with the removal of hexavalent chromium, was determined by using UV-Vis spectroscopy. The results demonstrated a ~91% adsorption of various dyes and ~60% of water pollutants. Hence, considering the environmental factors, CCIO NCs can provide a cost-effective solution against multiple pollutants.

Supplementary Materials: The following supporting information can be downloaded at: https://www.mdpi.com/article/10.3390/nano14161339/s1, Figure S1. UV-Vis spectrometry analysis of synthesized CCIO NCs; Figure S2. TGA analysis of synthesized CCIO NCs; Table S1. Biocompositional analysis of raw cotton stalk waste powder. References [55,67] are cited in Supplementary Materials.

Author Contributions: Conceptualization, M.S.; methodology, M.S., S.K. and B.S.K.; software, M.S. and S.K.; validation, M.S. formal analysis, M.S., S.K. and B.S.K.; investigation, S.K.; resources, M.S.; data curation, M.S. and S.K.; writing—original draft preparation, S.K.; writing—review and editing, M.S.; visualization, M.S.; supervision, M.S.; project administration, M.S.; funding acquisition, M.S. All authors have read and agreed to the published version of the manuscript.

Funding: This work is funded by the Ramalingaswami Re-entry Fellowship (BT/RLF/Re-entry/27/2021) Department of Biotechnology (DBT), Government of India.

Institutional Review Board Statement: Not applicable.

Data Availability Statement: The raw data supporting the conclusions of this article will be made available by the authors upon request.

Acknowledgments: The authors thank Ayesha Shafiq (from Chungbuk National University, Republic of Korea) for assistance with the EDX and TEM analysis.

Conflicts of Interest: The authors declare no conflicts of interest.

References

1. Loux, N.T.; Su, Y.S.; Hassan, S.M. Issues in Assessing Environmental Exposures to Manufactured Nanomaterials. *Int. J. Environ. Res. Public Health* **2011**, *8*, 3562–3578. [CrossRef]
2. Balali-Mood, M.; Naseri, K.; Tahergorabi, Z.; Khazdair, M.R.; Sadeghi, M. Toxic Mechanisms of Five Heavy Metals: Mercury, Lead, Chromium, Cadmium, and Arsenic. *Front. Pharmacol.* **2021**, *12*, 643972. [CrossRef]
3. Wani, K.A.; Jangid, N.K.; Bhat, A.R. (Eds.) *Impact of Textile Dyes on Public Health and the Environment*; IGI Global: Hershey, PA, USA, 2019; ISBN 978-1-79980-313-3.
4. Mudgal, V.; Madaan, N.; Mudgal, A.; Singh, R.B.; Mishra, S. Effect of Toxic Metals on Human Health. *Open Nutraceuticals J.* **2010**, *3*, 94–99. [CrossRef]
5. Anastopoulos, I.; Hosseini-Bandegharaei, A.; Fu, J.; Mitropoulos, A.C.; Kyzas, G.Z. Use of Nanoparticles for Dye Adsorption: Review. *J. Dispers. Sci. Technol.* **2018**, *39*, 836–847. [CrossRef]
6. Baig, N.; Kammakakam, I.; Falath, W. Nanomaterials: A Review of Synthesis Methods, Properties, Recent Progress, and Challenges. *Mater. Adv.* **2021**, *2*, 1821–1871. [CrossRef]
7. Hlongwane, G.N.; Sekoai, P.T.; Meyyappan, M.; Moothi, K. Simultaneous Removal of Pollutants from Water Using Nanoparticles: A Shift from Single Pollutant Control to Multiple Pollutant Control. *Sci. Total Environ.* **2019**, *656*, 808–833. [CrossRef] [PubMed]
8. Hussain, I. Green Synthesis of Nanoparticles and Its Potential Application. *Biotechnol. Lett.* **2016**, *38*, 545–560. [CrossRef] [PubMed]
9. Savage, N.; Diallo, M.S. Nanomaterials and Water Purification: Opportunities and Challenges. *J. Nanopart Res.* **2005**, *7*, 331–342. [CrossRef]
10. Lee, C.S.; Robinson, J.; Chong, M.F. A Review on Application of Flocculants in Wastewater Treatment. *Process Saf. Environ. Prot.* **2014**, *92*, 489–508. [CrossRef]
11. Leiknes, T. The Effect of Coupling Coagulation and Flocculation with Membrane Filtration in Water Treatment: A Review. *J. Environ. Sci.* **2009**, *21*, 8–12. [CrossRef]
12. Wang, J.; Chen, H. Catalytic Ozonation for Water and Wastewater Treatment: Recent Advances and Perspective. *Sci. Total Environ.* **2020**, *704*, 135249. [CrossRef] [PubMed]
13. Hube, S.; Eskafi, M.; Hrafnkelsdóttir, K.F.; Bjarnadóttir, B.; Bjarnadóttir, M.Á.; Axelsdóttir, S.; Wu, B. Direct Membrane Filtration for Wastewater Treatment and Resource Recovery: A Review. *Sci. Total Environ.* **2020**, *710*, 136375. [CrossRef] [PubMed]
14. Perrich, J.R. *Activated Carbon Adsorption for Wastewater Treatment*; CRC Press: Boca Raton, FL, USA, 2018; ISBN 978-1-351-07791-0.
15. Shahedi, A.; Darban, A.K.; Taghipour, F.; Jamshidi-Zanjani, A. A Review on Industrial Wastewater Treatment via Electrocoagulation Processes. *Curr. Opin. Electrochem.* **2020**, *22*, 154–169. [CrossRef]
16. Butler, E.; Hung, Y.-T.; Yeh, R.Y.-L.; Suleiman Al Ahmad, M. Electrocoagulation in Wastewater Treatment. *Water* **2011**, *3*, 495–525. [CrossRef]
17. Rashid, R.; Shafiq, I.; Akhter, P.; Iqbal, M.J.; Hussain, M. A State-of-the-Art Review on Wastewater Treatment Techniques: The Effectiveness of Adsorption Method. *Environ. Sci. Pollut. Res.* **2021**, *28*, 9050–9066. [CrossRef]
18. Rajendran, S.; Priya, A.K.; Senthil Kumar, P.; Hoang, T.K.A.; Sekar, K.; Chong, K.Y.; Khoo, K.S.; Ng, H.S.; Show, P.L. A Critical and Recent Developments on Adsorption Technique for Removal of Heavy Metals from Wastewater-A Review. *Chemosphere* **2022**, *303*, 135146. [CrossRef] [PubMed]
19. Brahmi, M.; Hassen, A. Ultraviolet Radiation for Microorganism Inactivation in Wastewater. *J. Environ. Prot.* **2012**, *2012*. [CrossRef]
20. Oller, I.; Malato, S.; Sánchez-Pérez, J.A. Combination of Advanced Oxidation Processes and Biological Treatments for Wastewater Decontamination—A Review. *Sci. Total Environ.* **2011**, *409*, 4141–4166. [CrossRef] [PubMed]

21. Prasse, C.; Ternes, T. Removal of Organic and Inorganic Pollutants and Pathogens from Wastewater and Drinking Water Using Nanoparticles—A Review. In *Nanoparticles in the Water Cycle: Properties, Analysis and Environmental Relevance*; Frimmel, F.H., Niessner, R., Eds.; Springer: Berlin/Heidelberg, Germany, 2010; pp. 55–79. ISBN 978-3-642-10318-6.
22. Aragaw, T.A.; Bogale, F.M.; Aragaw, B.A. Iron-Based Nanoparticles in Wastewater Treatment: A Review on Synthesis Methods, Applications, and Removal Mechanisms. *J. Saudi Chem. Soc.* **2021**, *25*, 101280. [CrossRef]
23. Mondal, P.; Nandan, A.; Ajithkumar, S.; Siddiqui, N.A.; Raja, S.; Kola, A.K.; Balakrishnan, D. Sustainable Application of Nanoparticles in Wastewater Treatment: Fate, Current Trend & Paradigm Shift. *Environ. Res.* **2023**, *232*, 116071. [CrossRef]
24. Theerthagiri, J.; Lee, S.J.; Karuppasamy, K.; Park, J.; Yu, Y.; Kumari, M.L.A.; Chandrasekaran, S.; Kim, H.-S.; Choi, M.Y. Fabrication Strategies and Surface Tuning of Hierarchical Gold Nanostructures for Electrochemical Detection and Removal of Toxic Pollutants. *J. Hazard. Mater.* **2021**, *420*, 126648. [CrossRef] [PubMed]
25. Singh, J.; Kumar, V.; Singh Jolly, S.; Kim, K.-H.; Rawat, M.; Kukkar, D.; Tsang, Y.F. Biogenic Synthesis of Silver Nanoparticles and Its Photocatalytic Applications for Removal of Organic Pollutants in Water. *J. Ind. Eng. Chem.* **2019**, *80*, 247–257. [CrossRef]
26. Melkamu, W.W.; Bitew, L.T. Green Synthesis of Silver Nanoparticles Using Hagenia Abyssinica (Bruce) J.F. Gmel Plant Leaf Extract and Their Antibacterial and Anti-Oxidant Activities. *Heliyon* **2021**, *7*, e08459. [CrossRef] [PubMed]
27. Gopinath, K.P.; Madhav, N.V.; Krishnan, A.; Malolan, R.; Rangarajan, G. Present Applications of Titanium Dioxide for the Photocatalytic Removal of Pollutants from Water: A Review. *J. Environ. Manag.* **2020**, *270*, 110906. [CrossRef] [PubMed]
28. A Novel and Facile Green Synthesis of SiO_2 Nanoparticles for Removal of Toxic Water Pollutants | Applied Nanoscience. Available online: https://link.springer.com/article/10.1007/s13204-021-01898-1 (accessed on 29 December 2023).
29. Jadhav, S.A.; Garud, H.B.; Patil, A.H.; Patil, G.D.; Patil, C.R.; Dongale, T.D.; Patil, P.S. Recent Advancements in Silica Nanoparticles Based Technologies for Removal of Dyes from Water. *Colloid Interface Sci. Commun.* **2019**, *30*, 100181. [CrossRef]
30. Bhateria, R.; Singh, R. A Review on Nanotechnological Application of Magnetic Iron Oxides for Heavy Metal Removal. *J. Water Process Eng.* **2019**, *31*, 100845. [CrossRef]
31. Wang, J.; Zhang, J.; Han, L.; Wang, J.; Zhu, L.; Zeng, H. Graphene-Based Materials for Adsorptive Removal of Pollutants from Water and Underlying Interaction Mechanism. *Adv. Colloid Interface Sci.* **2021**, *289*, 102360. [CrossRef]
32. Yap, P.L.; Nine, M.J.; Hassan, K.; Tung, T.T.; Tran, D.N.H.; Losic, D. Graphene-Based Sorbents for Multipollutants Removal in Water: A Review of Recent Progress. *Adv. Funct. Mater.* **2021**, *31*, 2007356. [CrossRef]
33. Sanjeev, N.O.; Valsan, A.E.; Zachariah, S.; Vasu, S.T. Synthesis of Zinc Oxide Nanoparticles from *Azadirachta Indica* Extract: A Sustainable and Cost-Effective Material for Wastewater Treatment. *J. Hazard. Toxic Radioact. Waste* **2023**, *27*, 04023027. [CrossRef]
34. Shabani, N.; Javadi, A.; Jafarizadeh-Malmiri, H.; Mirzaie, H.; Sadeghi, J. Potential Application of Iron Oxide Nanoparticles Synthesized by Co-Precipitation Technology as a Coagulant for Water Treatment in Settling Tanks. *Min. Metall. Explor.* **2021**, *38*, 269–276. [CrossRef]
35. Bashir, A.; Malik, L.A.; Ahad, S.; Manzoor, T.; Bhat, M.A.; Dar, G.N.; Pandith, A.H. Removal of Heavy Metal Ions from Aqueous System by Ion-Exchange and Biosorption Methods. *Environ. Chem. Lett.* **2019**, *17*, 729–754. [CrossRef]
36. Ebrahimi, M.; Van der Bruggen, B.; Hosseini, S.M.; Askari, M.; Nemati, M. Improving Electrochemical Properties of Cation Exchange Membranes by Using Activated Carbon-Co-Chitosan Composite Nanoparticles in Water Deionization. *Ionics* **2019**, *25*, 1199–1214. [CrossRef]
37. Hosseini, S.M.; Sohrabnejad, S.; Nabiyouni, G.; Jashni, E.; Van der Bruggen, B.; Ahmadi, A. Magnetic Cation Exchange Membrane Incorporated with Cobalt Ferrite Nanoparticles for Chromium Ions Removal via Electrodialysis. *J. Membr. Sci.* **2019**, *583*, 292–300. [CrossRef]
38. Farahbakhsh, J.; Vatanpour, V.; Ganjali, M.R.; Saeb, M.R. 21-Magnetic Nanoparticles in Wastewater Treatment. In *Magnetic Nanoparticle-Based Hybrid Materials*; Ehrmann, A., Nguyen, T.A., Ahmadi, M., Farmani, A., Nguyen-Tri, P., Eds.; Woodhead Publishing Series in Electronic and Optical Materials; Woodhead Publishing: Sawston, UK, 2021; pp. 547–589, ISBN 978-0-12-823688-8.
39. Naseem, T.; Durrani, T. The Role of Some Important Metal Oxide Nanoparticles for Wastewater and Antibacterial Applications: A Review. *Environ. Chem. Ecotoxicol.* **2021**, *3*, 59–75. [CrossRef]
40. Jamzad, M. Green Synthesis of Iron Oxide Nanoparticles by the Aqueous Extract of *Laurus nobilis* L. Leaves and Evaluation of the Antimicrobial Activity | SpringerLink. Available online: https://link.springer.com/article/10.1007/s40097-020-00341-1 (accessed on 20 May 2023).
41. Ying, S.; Guan, Z.; Ofoegbu, P.C.; Clubb, P.; Rico, C.; He, F.; Hong, J. Green Synthesis of Nanoparticles: Current Developments and Limitations. *Environ. Technol. Innov.* **2022**, *26*, 102336. [CrossRef]
42. Mittal, A.K.; Chisti, Y.; Banerjee, U.C. Synthesis of Metallic Nanoparticles Using Plant Extracts. *Biotechnol. Adv.* **2013**, *31*, 346–356. [CrossRef] [PubMed]
43. Manogar, P.; Esther Morvinyabesh, J.; Ramesh, P.; Dayana Jeyaleela, G.; Amalan, V.; Ajarem, J.S.; Allam, A.A.; Seong Khim, J.; Vijayakumar, N. Biosynthesis and Antimicrobial Activity of Aluminium Oxide Nanoparticles Using Lyngbya Majuscula Extract. *Mater. Lett.* **2022**, *311*, 131569. [CrossRef]
44. Joudeh, N.; Linke, D. Nanoparticle Classification, Physicochemical Properties, Characterization, and Applications: A Comprehensive Review for Biologists. *J. Nanobiotechnol.* **2022**, *20*, 262. [CrossRef]
45. Das, P.; Das, M.K. Chapter 4-Production and Physicochemical Characterization of Nanocosmeceuticals. In *Nanocosmeceuticals*; Das, M.K., Ed.; Academic Press: Cambridge, MA, USA, 2022; pp. 95–138. ISBN 978-0-323-91077-4.

46. Petcharoen, K.; Sirivat, A. Synthesis and Characterization of Magnetite Nanoparticles via the Chemical Co-Precipitation Method. *Mater. Sci. Eng. B* **2012**, *177*, 421–427. [CrossRef]
47. Ali, M. Qualitative Analyses of Thin Film-Based Materials Validating New Structures of Atoms. *Mater. Today Commun.* **2023**, *36*, 106552. [CrossRef]
48. Kim, M.; Singhvi, M.S.; Kim, B.S. Eco-Friendly and Rapid One-Step Fermentable Sugar Production from Raw Lignocellulosic Biomass Using Enzyme Mimicking Nanomaterials: A Novel Cost-Effective Approach to Biofuel Production. *Chem. Eng. J.* **2023**, *465*, 142879. [CrossRef]
49. Ghanbary, F.; Jafarnejad, E. Removal of Malachite Green from the Aqueous Solutions Using Polyimide Nanocomposite Containing Cerium Oxide as Adsorbent. *Inorg. Nano-Met. Chem.* **2017**, *47*, 1675–1681. [CrossRef]
50. Tkaczyk, A.; Mitrowska, K.; Posyniak, A. Synthetic Organic Dyes as Contaminants of the Aquatic Environment and Their Implications for Ecosystems: A Review. *Sci. Total Environ.* **2020**, *717*, 137222. [CrossRef] [PubMed]
51. Hammad, E.N.; Salem, S.S.; Mohamed, A.A.; El-Dougdoug, W. Environmental Impacts of Ecofriendly Iron Oxide Nanoparticles on Dyes Removal and Antibacterial Activity. *Appl. Biochem. Biotechnol.* **2022**, *194*, 6053–6067. [CrossRef] [PubMed]
52. Behera, S.K.; Sahni, S.; Tiwari, G.; Rai, A.; Mahanty, B.; Vinati, A.; Rene, E.R.; Pugazhendhi, A. Removal of Chromium from Synthetic Wastewater Using Modified Maghemite Nanoparticles. *Appl. Sci.* **2020**, *10*, 3181. [CrossRef]
53. Singh, N.; Riyajuddin, S.; Ghosh, K.; Mehta, S.K.; Dan, A. Chitosan-Graphene Oxide Hydrogels with Embedded Magnetic Iron Oxide Nanoparticles for Dye Removal. *ACS Appl. Nano Mater.* **2019**, *2*, 7379–7392. [CrossRef]
54. Kedare, S.B.; Singh, R.P. Genesis and Development of DPPH Method of Antioxidant Assay. *J. Food Sci. Technol.* **2011**, *48*, 412–422. [CrossRef] [PubMed]
55. Singhvi, M.S.; Deshmukh, A.R.; Kim, B.S. Cellulase Mimicking Nanomaterial-Assisted Cellulose Hydrolysis for Enhanced Bioethanol Fermentation: An Emerging Sustainable Approach. *Green Chem.* **2021**, *23*, 5064–5081. [CrossRef]
56. Ning, J.; Shi, P.; Jiang, M.; Liu, C.; Jia, Z. Synthesis and Characterization of Cerium Oxide/Iron Oxide Nanocomposite and Its Surface Acid-Base Characteristics. *J. Environ. Chem. Eng.* **2021**, *9*, 105540. [CrossRef]
57. IR Spectrum Table. Available online: https://www.sigmaaldrich.com/IN/en/technical-documents/technical-article/analytical-chemistry/photometry-and-reflectometry/ir-spectrum-table (accessed on 3 June 2023).
58. Mourdikoudis, S.; Pallares, R.M.; Thanh, N.T. Characterization Techniques for Nanoparticles: Comparison and Complementarity upon Studying Nanoparticle Properties. *Nanoscale* **2018**, *10*, 12871–12934. [CrossRef]
59. Thorat, M.N.; Dastager, S.G. High Yield Production of Cellulose by a Komagataeibacter Rhaeticus PG2 Strain Isolated from Pomegranate as a New Host. *RSC Adv.* **2018**, *8*, 29797–29805. [CrossRef]
60. Azari, A.; Noorisepehr, M.; Dehghanifard, E.; Karimyan, K.; Hashemi, S.Y.; Kalhori, E.M.; Norouzi, R.; Agarwal, S.; Gupta, V.K. Experimental Design, Modeling and Mechanism of Cationic Dyes Biosorption on to Magnetic Chitosan-Lutaraldehyde Composite. *Int. J. Biol. Macromol.* **2019**, *131*, 633–645. [CrossRef]
61. Rahmi; Ishmaturrahmi; Mustafa, I. Methylene Blue Removal from Water Using H2SO4 Crosslinked Magnetic Chitosan Nanocomposite Beads. *Microchem. J.* **2019**, *144*, 397–402. [CrossRef]
62. Burakov, A.E.; Galunin, E.V.; Burakova, I.V.; Kucherova, A.E.; Agarwal, S.; Tkachev, A.G.; Gupta, V.K. Adsorption of Heavy Metals on Conventional and Nanostructured Materials for Wastewater Treatment Purposes: A Review. *Ecotoxicol. Environ. Saf.* **2018**, *148*, 702–712. [CrossRef]
63. Naik, C.C.; Gaonkar, S.K.; Furtado, I.; Salker, A.V. Effect of Cu^{2+} Substitution on Structural, Magnetic and Dielectric Properties of Cobalt Ferrite with Its Enhanced Antimicrobial Property. *J. Mater. Sci. Mater. Electron.* **2018**, *29*, 14746–14761. [CrossRef]
64. Abou Hammad, A.B.; Abd El-Aziz, M.E.; Hasanin, M.S.; Kamel, S. A Novel Electromagnetic Biodegradable Nanocomposite Based on Cellulose, Polyaniline, and Cobalt Ferrite Nanoparticles. *Carbohydr. Polym.* **2019**, *216*, 54–62. [CrossRef]
65. Vitta, Y.; Figueroa, M.; Calderon, M.; Ciangherotti, C. Synthesis of Iron Nanoparticles from Aqueous Extract of *Eucalyptus Robusta* Sm and Evaluation of Antioxidant and Antimicrobial Activity. *Mater. Sci. Energy Technol.* **2020**, *3*, 97–103. [CrossRef]
66. Ramanavičius, S.; Žalnėravičius, R.; Niaura, G.; Drabavičius, A.; Jagminas, A. Shell-Dependent Antimicrobial Efficiency of Cobalt Ferrite Nanoparticles. *Nano-Struct. Nano-Objects* **2018**, *15*, 40–47. [CrossRef]
67. Rajak, R.C.; Saha, P.; Singhvi, M.; Kwak, D.; Kim, D.; Lee, H.; Deshmukh, A.R.; Bu, Y.; Kim, B.S. An eco-friendly biomass pretreatment strategy utilizing reusable enzyme mimicking nanoparticles for lignin depolymerization and biofuel production. *Green Chem.* **2021**, *23*, 5584–5599. [CrossRef]

nanomaterials

Article

Comprehensive Utilization of Formation Water Scale to Prepare Controllable Size $CaCO_3$ Nanoparticles: A New Method to Improve Oil Recovery

Bo Huang [1,2], Shengzhen Hou [1,2,*], Zhao Hua [1,2], Jian Zhang [1,2], Huan Yang [3,*], Yuejun Zhu [1,2], Yumiao Tang [3] and Benru Wang [3]

1 State Key Laboratory of Offshore Oil and Gas Exploitation, Beijing 102209, China; huangbo@cnooc.com.cn (B.H.); huazhao@cnooc.com.cn (Z.H.); zhangjian@cnooc.com.cn (J.Z.); zhuyj3@cnooc.com.cn (Y.Z.)
2 CNOOC Research Institute Co., Ltd., Beijing 100028, China
3 School of Chemistry and Environmental Engineering, Yangtze University, Jingzhou 434020, China; 2022710239@yangtzeu.edu.cn (Y.T.); 2023710233@yangtzeu.edu.cn (B.W.)
* Correspondence: houshzh2@cnooc.com.cn (S.H.); yanghuan@yangtzeu.edu.cn (H.Y.)

Abstract: Formation water scale blocks pipelines and results in oil/gas production decreasing and energy consumption increasing. Many methods have been developed to inhibit scale formation. However, these previous methods are limited by their complications and low efficiency. A new method is proposed in this paper that uses the scale in formation water as a nanomaterial to improve oil recovery via controlling particle size. A series of ligands were synthesized and characterized. Micrometer-$CaCO_3$ was formed and accumulated to form scale of a large size under uncontrolled conditions. The tetradentate ligands (L^4) exhibited an excellent capturing yield of Ca^{2+} (87%). The particle size was very small, but they accumulated to form large particles (approximately 1300 nm) in the presence of Na_2CO_3. The size of the $CaCO_3$ could be further controlled by poly(aspartic acid) to form sizes of about 700 nm. The flooding test showed that this material effectively improved oil recovery from 55.2% without nano $CaCO_3$ to 61.5% with nano $CaCO_3$. This paves a new pathway for the utilization of Ca^{2+} in formation water.

Keywords: formation water; scale; enhanced oil recovery; calcium carbonate nanoparticles

Citation: Huang, B.; Hou, S.; Hua, Z.; Zhang, J.; Yang, H.; Zhu, Y.; Tang, Y.; Wang, B. Comprehensive Utilization of Formation Water Scale to Prepare Controllable Size $CaCO_3$ Nanoparticles: A New Method to Improve Oil Recovery. *Nanomaterials* **2024**, *14*, 1452. https://doi.org/10.3390/nano14171452

Academic Editor: Alexey Pestryakov

Received: 22 July 2024
Revised: 4 September 2024
Accepted: 4 September 2024
Published: 6 September 2024

1. Introduction

With the continuous development of modern society, the demand for oil and gas resources is gradually increasing [1–3]. Upon entering the middle and later stages of oilfield maturity, the composition of formation water becomes complicated. High-valent metal ions in formation water, such as Ca^{2+} and Mg^{2+}, can form scale on the surface of pipelines and facilities and lead to blockages [4–6]. Moreover, these high-valent ions can also react with pipelines, which leads to severe corrosion in water injection systems [7–9]. The formation of scale is very complex, involving thermodynamics, dynamics, fluid dynamics, etc. The scale can block the pipelines, resulting in oil and gas production decreasing, energy consumption increasing, unnormal continuous production, or even the stopping of production. In order to inhibit scale formation in oilfields, many methods have been developed, including chemical anti-scaling agents and physical anti-scaling methods [10,11]. However, owing to their complications, the effect of these methods on inhibiting scale formation is limited [12].

In addition, in different oilfields, the types and contents of ions are also different, leading to many uncertain factors and unsatisfactory results when using traditional chemical methods for anti-scaling [12]. Normally, one needs to add much more agent to achieve the anticipated aim. Physical methods, such as ultrasound method for anti-scaling, are as of now immature and easily affected by on-site conditions [13,14]. Most importantly, physical methods must be coupled with chemical methods to treat scale formation. Therefore, although chemical methods are not universally applicable, their low cost, mature technology, fast effectiveness, and good results make them the main means of removing scale [15–18]. However, chemical agents may remain in formation water after removing scale and gradually accumulate, leading to more serious side effects [19]. Therefore, new methods need to be developed to remove or treat scale.

Owing to their decreasing interfacial tension, changing wetting, enhancing the stability of emulsions, and decreasing injection pressure, many nanomaterials have been extensively utilized in oilfields to enhance oil recovery [20–22]. Polyacrylamide (PAM) microsphere and its emulsions have been demonstrated in some oilfields, such as the Changqing Oilfield in China, not only to have excellent dispersion performance, but also to be capable of increasing flow resistance and forcing flow steering to increase swept volume in deep or large channels when the nanospheres flow into the reservoir [23]. Scale formation is a slow process in formation water that involves crystal growth, including homogeneous and heterogeneous crystallization [24]. Therefore, it suggests the probability of controlling the size of scale particles during scale formation in nanometers or micrometers as an inorganic material to enhance oil recovery. This can not only solve the scale problem, but also achieve Ca^{2+} and Mg^{2+} sustainability in formation water [25,26].

Normally, $CaCO_3$ and $MgCO_3$ are precipitated to form scale in formation water owing to their relatively high content of Ca^{2+} and Mg^{2+}. To prevent the rapid precipitation of $CaCO_3$ and $MgCO_3$, the fast capture of the above ions in formation water is a critical step. For instance, chelated agents such as ethylenediamine tetra-acetic acid (EDTA) can be used to capture Ca^{2+} and Mg^{2+} ions and form a very stable complex [27,28], which would be beneficial to the formation of nano- or micrometer $CaCO_3$ and $MgCO_3$. Moreover, the stability of the chelation should not be too strong, so as to facilitate the fast release metal ions and the in situ formation of nano- or micro-nanoparticles of $CaCO_3$ and $MgCO_3$ [29–31]. Therefore, a suitable ligand is a key factor in the formation of micro-nanoparticles of $CaCO_3$ and $MgCO_3$. To this end, different kinds of compounds have been designed and synthesized (Figure 1) to capture Ca^{2+} and Mg^{2+}. The objectives of this paper include: (1) preparing compounds to capture Ca^{2+} and Mg^{2+}; (2) forming micro-nanoparticles of $CaCO_3$ and enhancing oil recovery; and (3) elucidating a possible mechanism under present conditions.

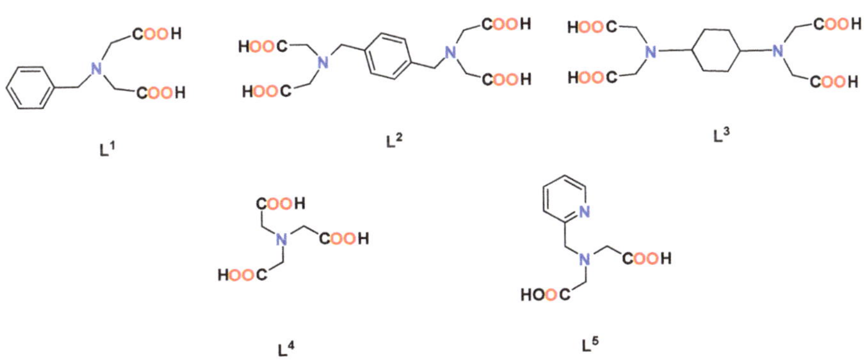

Figure 1. The structures of different ligands.

2. Experimental Section

2.1. Materials

Chloroacetic acid (moderate toxicity); sodium hydroxide (NaOH); benzylamine; hydrochloric acid; sodium carbonate; butanediamine; p-xylylenediamine; ammonium chloride (low toxicity); iminodiacetic acid (low toxicity); absolute ethanol; 2-picolyl chloride-HCl (low toxicity); methanol (trace toxicity); poly (aspartic acid) (PASP) (trace toxicity);

$CaCl_2$; and $NaHCO_3$ were purchased from Acros Organic Chemical Company (Waltham, MA, USA). They are analytical grade and without further purification.

2.2. Instruments

Fourier-transform infrared spectroscopy (FTIR): VERTEX 70 FT-IR spectrometer Thermo Fisher Technologies (Waltham, MA, USA); ^{1}H nuclear magnetic resonance spectroscopy: 400 MHz AVANCE III 600 M (Bruker, Berlin, Germany); X-ray diffraction (XRD): D8 Advance (Dutch Panaco, Almelo, The Netherlands); S3500 laser particle size analyzer(Microtrac, Montgomeryville, PA, USA); Nano ZS Malvern laser particle sizer (Masterizer, Micro, Buckinghamshire, UK); scanning electron microscopy (SEM-EDS): Merlin (Zeiss, Jena, Germany); mass spectrometer: Agilent 6520 Q-TOF LC/MS (Agilent Technologies, Santa Clara, CA, USA); atomic absorption spectrophotometer: AA-7003 (East-west Analytical Instruments Co., Ltd., Beijing, China).

2.3. Preparation of Ion Regulators

N-Benzyl iminodiacetic acid (L^1): Chloroacetic acid (9.5 g, 100 µmol) in H_2O (35 mL) was cooled to 50 °C and neutralized with NaOH solution (0.17 g/mL). The alkali was added slowly to avoid the solution temperature exceeding 20 °C. Then, benzylamine (50 µmol, 5.5 mL) was added and the mixture was heated at reflux for one hour. The NaOH solution was added to mixture dropwise, and the mixture was further heated under reflux for one hour. After stopping heating, the mixture's pH value was adjusted to neutral with hydrochloric acid. The precipitation was formed and collected with a 64% yield.

2,2′,2″,2‴-((1,4-phenylenebis(methylene))bis(azanetriyl))tetraacetic acid (L^2): 5.25 g of chloroacetic acid was weighed and dissolved in water. The mixture was stirred and neutralized with sodium carbonate until the pH was 6.8–7. According to the ratio of chloroacetic acid:butanediamine = 6.7:1 (mass ratio), the sodium chloroacetate solution was first transferred into a three-necked flask, then stirred. Next, 906 µL (0.78 g) of butanediamine and 25% NaOH mixture was slowly added to the three-necked flask using a drip funnel. After addition, the pH was adjusted to 9–9.5 using a 25% NaOH solution, and the reaction temperature was maintained at 90–95 °C until the pH value of the solution did not decrease (at the end of the reaction, the pH was maintained at around 9). The reaction solution was cooled down to 40–45 °C, then continuously stirred while 1:1 hydrochloric acid was added for acidification. When the pH of the solution reached 1, no more acid was added and the solution was stirred until precipitation formed. The yield was 64%.

Trans-2,2′,2″,2‴-(cyclohexane-1,4-diylbis(azanetriyl))tetraacetic acid (L^3): An aqueous solution of chloroacetic acid (4.75 g; 50 µmol in 17.5 mL) was cooled in an ice bath to 5 °C and neutralized with an aqueous sodium hydroxide solution (2.0 g; 50 µmol in 10 mL). The alkali was added at such a rate that the temperature of the reaction mixture never reached 20 °C. Then, 3.40 g (25 µmol) of p-xylylenediamine was added. The reaction mixture was refluxed and a solution of 2.0 g NaOH in 10 mL of water was added dropwise to it over one hour. A clear colorless solution was formed, and the refluxing was continued for one additional hour. The reaction mixture was poured into a large beaker, acidified to pH = 2 with diluted hydrochloric acid (1:1 v/v), and cooled to room temperature. The bottom of the beaker was scratched with a sharp glass rod to induce precipitation and stirred until the mixture solidified. After standing overnight in a refrigerator, the product was filtered off by suction, washed with ice water, and dried. The yield was 57%.

Nitrilotriacetic acid (L^4): 8.85 g monochloroacetic acid was put into the flask, stirred and heated until molten, and then an aqueous solution of NaOH was added (3.70 g in 10 mL). The solution was maintained at 80–90 °C for 5–6 h. Then, it was concentrated under vacuum, heated to 60 °C, and adjusted to a pH of 7–8 with saturated sodium carbonate solution. Under 80 °C, 4.0 g saturated solution of ammonium chloride was slowly added. Then the pH was adjusted to 9–10 with 1 M NaOH. After standing for 4 h at the room temperature, the mother liquors were then acidified (pH = 1) by the addition of HCl (6 M) and the product was removed by filtration, washed (cool water) and air-dried. The yield was 66%.

[Carboxymethyl(pyridin-2-ylmethyl)amino]acetic acid (L^5): Iminodiacetic acid (6.55 g, 50 mmol) was added to a solution of 4.0 g (100 mmol) of NaOH in 20 mL of H_2O and 60 mL of absolute ethanol. The solution was stirred vigorously. Solutions of 8.2 g (50 mmol) of 2-picolyl chloride-HCl in 17 mL of H_2O and 4.0 g (100 mmol) of NaOH in 7 mL of H_2O in separate addition funnels were then added over a period of 8 min. The reaction mixture was warmed to 70 °C and left to stir for 4 h. Then, an additional 4.0 g of NaOH was added, and the reaction mixture was stirred for another hour, after which the amber solution was evaporated, leaving a yellow solid. 50 mL of H_2O was added, and the solution was acidified with concentrated HCl to pH 1.5. The white solid obtained was recrystallized from methanol. The yield was 35.7%.

The five ligands were characterized by FTIR and ^{1}HNMR. The samples were characterized by FTIR using the potassium bromide compression method with a scanning range of 4000–500 cm^{-1} and 32 scans. The solvent choice for ^{1}H NMR characterization was water.

2.4. Properties of Scale under Uncontrolled Condition

In order to explore the effect of ligands on the size and morphology of coalesced particles, the coalesced particles without ligand control were first analyzed. After the simulated formation water had sat for 30 d, the scale in the water was filtered, dried in an oven, and then characterized by powder X-ray diffraction (XRD). The test conditions were as follows: operating voltage, 40 kV; operating current, 40 mA; scanning range, 5° to 40°; scanning speed, 2°/min. Scale particle size was analyzed with a S3500 laser particle size analyzer. The simulated formation water scale was treated ultrasonically for 20 min to test the particle size. The surface morphology of the particles was characterized by field emission scanning electron microscopy (SEM). A small amount of the sample was taken and pasted on a conductive adhesive and tested after gold spraying. The composition of the scale samples was determined by energy dispersive spectrometer (EDS). The sample was placed in an SEM sample chamber equipped with an EDS detector, which was used to analyze the chemical composition of the sample by converting its X-ray signal into energy spectrum data.

2.5. The Capturing of Ca^{2+} by L^1–L^5

In order to study the reaction of these ligands with Ca^{2+}, ESI/MS was used to characterize them. The atomizer flow rate was 3 L/min, the drying gas flow rate was 10 L/min, and the heater flow rate was 10 L/min.

2.6. $CaCO_3$ Particle Formation by Na_2CO_3 or $NaHCO_3$

There is a dynamic equilibrium between ligands, $CaCl_2$, and Na_2CO_3. This section mainly studies the reaction mechanism of ligands and Na_2CO_3 to generate $CaCO_3$ particles. The stability constant is calculated as follows:

$$K = \frac{(AXn)c}{(A)c * (X)c^n} \tag{1}$$

where $(AXn)c$ represents the concentration of the complex, $(A)c$ represents the concentration of the core substance, $(X)c$ represents the concentration of the ligand, and n represents the number of ligands.

To determine the presence of micro-nanoparticles, excess Na_2CO_3 solution was gradually dripped in and kept at room temperature for at least 4 d to produce a white precipitate. The $CaCO_3$ particles under the control of the ligands were characterized by XRD and SEM. The test method was the same as in Section 2.4.

2.7. $CaCO_3$ Nanoparticles Formation and Capturing Yield via L^1–L^5

In order to analyze the effects of L^1–L^5 on Ca^{2+} ion capture rate and $CaCO_3$ particle size in the presence of $NaHCO_3$ solution, the different ligands (L) were added, n_{Ca2+}/n_L was changed, and the concentration of free Ca^{2+} ions was determined by atomic absorption spectrometry after sitting for 72 h. Plasma gas flow was 12 L/min, nebulizer flow was 0.7 L/min, and auxiliary gas flow was 0.4 L/min. After 20 min of sonication, the was placed in a cuvette and the particle size of the $CaCO_3$ produced by the different ligands at different concentrations was measured using a Malvern laser particle size analyzer.

It was found that the ratio of n_{Ca2+}/n_{NaHCO3} also had an effect on the formation of $CaCO_3$, but there was no obvious rule among the different ligands. In this paper, the L^4 ligand was selected to set different proportions of n_{Ca2+}/n_{NaHCO3}. The concentration of free Ca^{2+} ions and the particle size of the $CaCO_3$ produced under different ratios of n_{Ca2+}/n_{NaHCO3} were determined by atomic absorption spectrometry and Malvern laser particle size analyzer, respectively, to determine the best ratio of n_{Ca2+}/n_{NaHCO3}.

Since the particle size of $CaCO_3$ particles is still large, it is necessary to further adjust the particle size by adding appropriate chemical reagents. According to previous studies, charged polymers such as polyaspartate (PASP) [31] and polyacrylic acid (PAA) [32] can stabilize amorphous calcium carbonate and may have mineralizing activity. In this paper, PASP was used to further control the size of $CaCO_3$ particles. The amount of added polyaspartic acid (PASP) was changed, and the particle size under different PASP dosages was tested by Marvin laser particle size analyzer to explore the effect of the addition amount on the particle size of $CaCO_3$.

2.8. The Flooding Experiment

A single core replacement experiment was used to simulate the oil flooding of agglomerated particles. The water used in the experiment was simulated formation water, with a salinity of 107,216 mg/L; its main components and concentration are listed in the Table 1. The simulated formation water refers to the formation water of an offshore oil field in China, which is highly mineralised and can reflect the effect of an agent well. The oil used in the experiment was simulated oil, which was obtained by mixing crude oil and aviation paraffin in a ratio of 1:1 (density: 0.8893 g/cm^3, viscosity: 17.5 mPa·s (50 °C)). An artificially homogeneous cylindrical core with a size of Ø2.5 × 10 cm was used for the experiment. The detailed experimental procedure was as follows:

1. Firstly, simulated formation water was placed in an intermediate container and injected into the core at a flow rate of 1 mL/min to establish the original water saturation. Core dry weight m_1 and core wet weight m_2 after water injection were recorded, respectively.
2. Simulated oil was then injected into the core to establish the original oil saturation. The simulated formation water was then injected into the core at a flow rate of 1 mL/min until no oil was produced. The injection volume and differential pressure were recorded, respectively, and the permeability was calculated via Darcy's equation to get the recovery rate of the water flooding.
3. The simulated formation water with and without additive were placed in an intermediate container, respectively. The water drive recovery rate was calculated by injecting the core at a flow rate of 1 mL/min when the pressure was stable and no oil was produced. The experiment was repeated for each solution to avoid error. The detailed information can be seen in Figure 2.

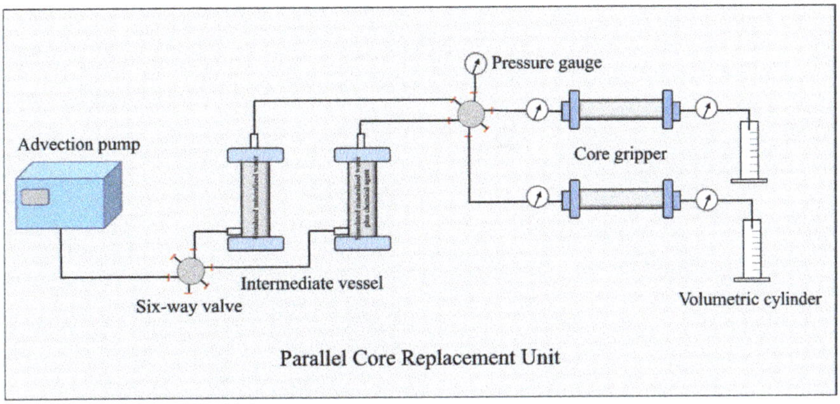

Figure 2. The core displacement device.

Table 1. The composition of simulated formation water (mg/L).

Na^+	K^+	Ca^{2+}	Mg^{2+}	Cl	SO_4^{2}	HCO_3^-	Total Mineralization
11,977	133	695	269	21,939	15	370	35,397

The relevant calculation formula is as follows:

$$V = \frac{m_2 - m_1}{\rho} \tag{2}$$

where V represents the effective pore volume in mL, m_1 represents the dry weight of the core in kg, m_2 represents the wet weight of the core in kg, and ρ represents the simulated formation water density in g/cm^3.

$$n = \frac{V}{V_1} \tag{3}$$

where n represents effective porosity, dimensionless; V represents effective pore volume in mL; and V_1 represents core volume in mL.

$$K = \frac{Q\mu l}{\Delta PA} \tag{4}$$

where K represents the permeability of the aqueous phase in units of 10^{-3} μm^2; Q represents the fluid flow rate through the core per unit time in units of cm^3/s; μ represents the viscosity of the fluid in units of mPa·s; L represents the length of the core in units of cm; ΔP represents the difference in pressure between the inlet and outlet in units of MPa; and A represents the cross-sectional area of the core in units of cm^2.

3. Results and Discussion

3.1. Preparation of Ion Regulator

N-Benzyl iminodiacetic acid (L^1): The elemental analysis of calculation result for $C_{11}H_{13}NO_4$ was C:H:N = 59.19%:5.87%:6.27% and the test results were C:H:N = 59.17%:5.89%:6.28%. The typical chemical shifts were δ 2.94 (s, 4H), 3.59 (s, 2H) and 7.26 (m, 5H).

2,2′,2″,2‴-((1,4-phenylenebis(methylene))bis(azanetriyl))tetraacetic acid (L^2): Figure 3 shows the FTIR result as υ(C=O) in 1732 and 1720 cm^{-1}. The elemental analysis of calculation result for $C_{16}H_{20}N_2O_8$ was C:H:N = 52.17%:5.47%:7.61% and the test results were C:H:N = 52.15%:5.49%:7.64. The typical chemical shifts were δ 3.56 (s, 4H) and 7.26 (s, 4H).

Trans-2,2′,2″,2‴-(cyclohexane-1,4-diylbis(azanetriyl))tetraacetic acid (L^3): Figure 3 showed the FTIR result as υ(C=O) 1729 and 1714 cm^{-1}. The elemental analysis of calculation result for $C_{11}H_{13}NO_4$ was C:H:N = 48.55%:6.40%:8.09% and the test results were C:H:N = 48.47%: 6.35%: 8.14%. The typical chemical shifts were δ 1.05 (q, 4H), 1.72 (q, 4H), 2.35 (s, 2H), 3.07 (s, 8H).

Nitrilotriacetic acid (L^4): Figure 3 shows the FTIR result as υ(C=O) 1740 and 1725 cm^{-1}. The elemental analysis of calculation result for $C_6H_9NO_6$ was C:H:N = 37.70%:4.75%:7.33% and the test results were C:H:N = 37.72%:4.71%:7.36%. The typical chemical shift was δ 3.01 (s, 6H).

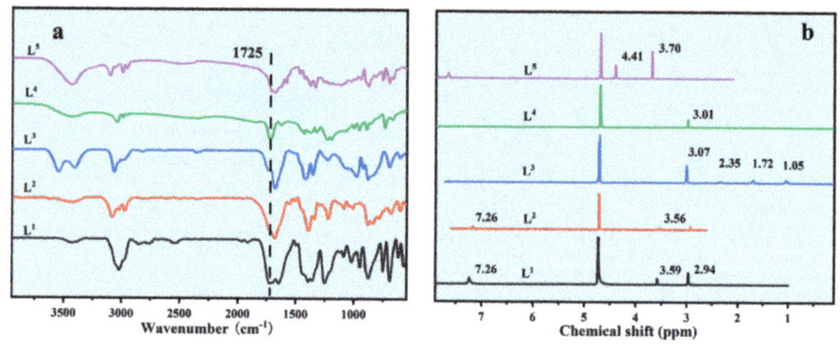

Figure 3. Spectra of compounds, (**a**): FTIR results of ligands; (**b**): ^{1}H NMR results of ligands.

[Carboxymethyl(pyridin-2-ylmethyl)amino]acetic acid (L^5): Figure 3 shows the FTIR as υ(C=O) 1727 and 1718 cm^{-1}. The elemental analysis of calculation result for $C_{10}H_{12}N_2O_4$ was C:H:N = 53.57%:5.39%:12.49% and the test results were C:H:N = 53.58%:5.37%:12.51%. The typical chemical shifts were δ 3.70 (s, 4 H), 4.41 (s, 2 H).

3.2. Properties of Scale under Uncontrolled Condition

First, the scale samples without ligand control were analyzed, as shown in Figure 4. This simulated real scale formation, and especially the size distribution. The XRD of the scale was consistent with the standard diffraction pattern of $CaCO_3$ (Figure 4a). The typical peaks were 21.00°, 24. 90°, 27.05°, 32.78°, 43.85°, 50.07°, 55.80°, 71.97°, and 73.59°. It was a typical calcite. The particle size of the scale was analyzed by S3500 laser particle size analyzer (Figure 4b). The median particle diameter (D50) was 56.02 μm. The mean volume diameter, mean number diameter, and mean area diameter were 78.08 μm, 26.64 μm, and 59.02 μm, respectively. It had a high particles size under the uncontrolled condition. The surface morphology of the particles was also characterized by SEM in Figure 5a. The surface of the particles was very rough and irregular under the uncontrolled condition. This indicated that the rough surface improved the specific surface of the particles, resulting in the micro-particles easily adsorbing onto the surface particles to form scale. In other words, it was also necessary to modify the surface of the particles to inhibit the adsorbing of the particles. Furthermore, the surface of composition was initially determined by EDS (Figure 5b–f). This revealed that it was mainly composed of O (38.3%), Ca (36.7%), C (14.3%), and Mg (8.2%). Combined with the results of EDS, the scale was mainly composed of $CaCO_3$ (~95.3%). Combining this with Figure 4a,b, it can be confirmed that, under normal conditions, the $CaCO_3$ precipitate formed easily and its particle size was big, which means it can block the pipeline easily and cannot be utilized directly. Hence, it is necessary to control the size of the particles and modify the surface of the particles such that they can flow with the formation water, thus realizing the sustainable utilization of Ca^{2+} in formation water.

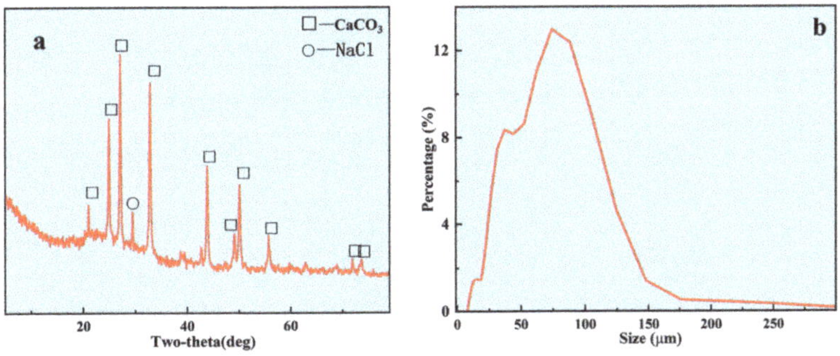

Figure 4. The properties of uncontrolled scale. (**a**): XRD of Simulated scale; (**b**): scale size analysis.

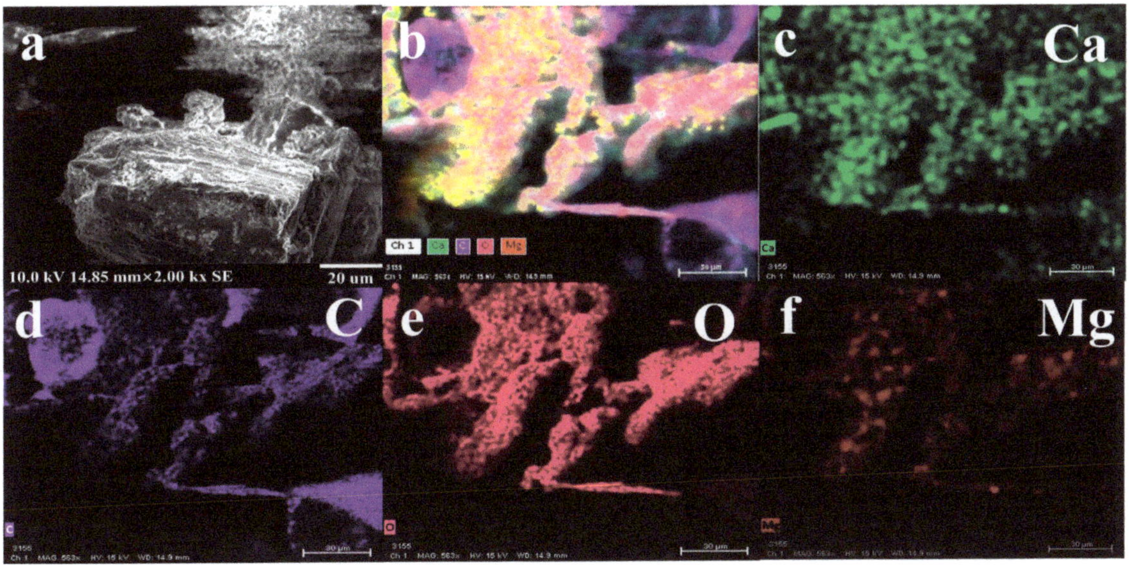

Figure 5. The SEM-EDS of particles, (**a**): SEM result; (**b**): EDS result; (**c–f**): C, Ca, O and Mg elements.

3.3. The Capturing of Ca^{2+} by L^1–L^5

All multi-dentate O-containing chelated ligands in this paper could form stable complexes with Ca^{2+} and Mg^{2+}. Therein, the reactions of these ligands with Ca^{2+} were investigated. L^1 was a typical tri-dentate ligand, which could easily chelate Ca^{2+} in H_2O. The ESI/MS results (Figure A1) showed a predominant peak appearing at m/z = 262, which was assigned to the species of $[L^1 + Ca^{2+} - H]^+$ (Figure 6). The result indicated that the mole ratio of L^1 and Ca^{2+} was 1:1. L^2 and L^3 were also used to react with Ca^{2+}. ESI/MS results showed two peaks appearing at the m/z = 423, and 445, respectively, which were assigned to $[L^2 + 2Ca^{2+} - 3H]^+$ and $[L^3 + 2Ca^{2+} - 3H]^+$. The mononuclear species were not found due to the steric effect. To tune the solubility in H_2O, the different cyclohexyl and phenyl groups in L^2 and L^3 were used. L^4 and L^5 were typical of tetra-dentate ligands and had a stronger 'chelated effect' [25] than those of L^1–L^3. ESI/MS results showed the predominant peaks appearing at m/z = 230 and 263, respectively, which were assigned to $[L^4 + Ca^{2+} - H]^+$ and $[L^4 + Ca^{2+} - H]^+$ (Figure 6). From the above analysis, it can be confirmed that Ca^{2+} can be effectively captured by L^1–L^5 because more carboxyl functional groups can chelate more Ca^{2+}.

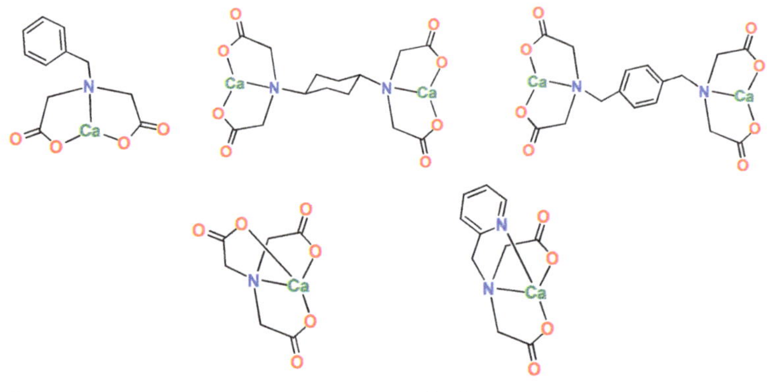

Figure 6. The possible coordination modes for L^1–L^5 with Ca^{2+} in H_2O.

3.4. $CaCO_3$ Particle Formation by Na_2CO_3 or $NaHCO_3$

Based on the ESI/MS investigation, these additives can quickly form stable complexes with Ca^{2+} when the mole ratio of ligand and Ca^{2+} is 1. The addition of excess Na_2CO_3 did not generate any obvious white precipitation, even when the mixture was left standing for over 3 days, with the presence of the previously described complexes, while the white $CaCO_3$ precipitation was immediately observed in the absence of these ligands. Although the white $CaCO_3$ precipitation could not be observed in the presence of these ligands, micro-nanoparticles were gradually formed. This result revealed the presence of a dynamic equilibrium between the ligand, $CaCl_2$, and Na_2CO_3. Taking L^1 as an example, L^1 can form a complex with Ca^{2+} with the stability constant K_1 (~1×10^9) (a). The large stability constant indicated that the Ca^{2+} in aqueous phase was nearly completely captured by L^1. When Na_2CO_3 was added, Ca^{2+} would combine with CO_3^{2-} to generate the micro-nanoparticle of $CaCO_3$ (K_{sp} ~ 2.8×10^{-9}) (b). Thus, the total reaction was (a) + (b) → (c). According to $K_1 \times K_2 = 1/K_{sp}$, $K_2 = 1/(K_{sp} \times K_1) = 0.36$. On the other hand, the formation rate (R) of $CaCO_3$ was significantly lowered due to the formation of a stable complex $[L{\cdot}Ca^{2+}]$. The $CaCO_3$ precipitation was not formed in the presence of L^1 mainly due to the very low concentration of Ca^{2+} in the reaction system.

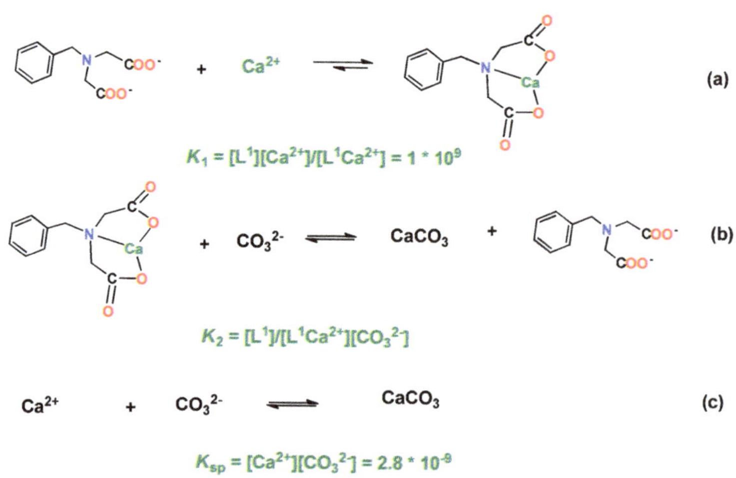

The XRD pattern was very consistent with that of the standard $CaCO_3$ sample. The typical peaks were 23°, 29°, 36°, 39°, 43°, 47.5°, and 48.5°. It was a typical calcite characteristic peak (Figure 7a). SEM showed that a spherical shape could be clearly observed

(Figure 7b,c). It also can be observed that the particles were relatively dispersed and showed a size of about 800 nm (Figure 7c), which was a smaller size compared to that of the uncontrolled $CaCO_3$ particles. Generally, high specific surface energy with more micro-nanoparticles coalescence to form micrometer-sized particles, and more smaller nanoparticles are expected to be formed under controlled conditions. This provided the possibility of enhanced oil recovery with micro-nanomaterials.

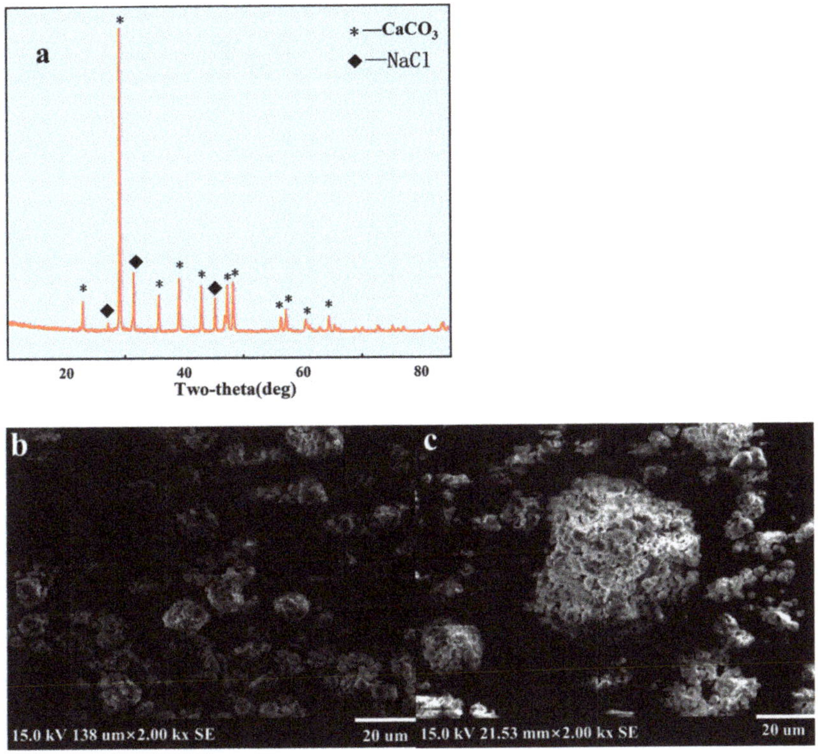

Figure 7. XRD and SEM for $CaCO_3$ particles. (**a**): XRD of nano-$CaCO_3$ under ligand control; (**b**,**c**): SEM of nano-$CaCO_3$ under ligand control.

The above experiments demonstrated that synthetic ligands could capture Ca^{2+} effectively in the aqueous phase when mole ratio of L and Ca^{2+} exceeded 1. Thus, when Na_2CO_3 was added, the $CaCO_3$ precipitate was not formed immediately. However, $CaCO_3$ precipitate rapidly formed when mole ratio of ligand and Ca^{2+} was lower than 1. To simulate the real situation in the oilfield, $NaHCO_3$ was used instead of Na_2CO_3 due to HCO_3^- needing to dissociate to CO_3^{2-}, resulting in a $CaCO_3$ precipitate that was relatively slow. Three possible equilibration reactions of $NaHCO_3$ occurred.

$$HCO_3^- + H_2O = H_2CO_3 + OH^- \tag{5}$$

$$HCO_3^- + H_2O = CO_3^{2-} + H_3O^+ \tag{6}$$

$$H_3O^+ + OH^- = 2H_2O \tag{7}$$

In a typical 0.1 M $NaHCO_3$ aqueous solution, the pH value was close to 8.34. c(HCO_3^-), c(CO_3^{2-}) and c(H_2CO_3) was close to 9.8×10^{-2} M, 9.8×10^{-4} M, and 9.8×10^{-4} M, respectively, ($K_{a1}(H_2CO_3) = 4.5 \times 10^{-7}$ and $K_{a2}(H_2CO_3) = 4.7 \times 10^{-11}$). Thus, $c(CO_3^{2-})$ was much lower than that of Na_2CO_3 and no obvious precipitate was found when $NaHCO_3$ was used instead of Na_2CO_3 in $CaCl_2$ solution.

3.5. $CaCO_3$ Nanoparticles Formation and Capturing Yield via L^1–L^5

The initial concentration of $c(Ca^{2+})$ was 460 mg/L. From Figure 6, it can be found that L^2 and L^3 could capture more Ca^{2+}, but the capturing yield was not higher in Table 2. Complexes formed with ligands stronger than L^4 are likely to release less Ca^{2+}. The Ca^{2+} in solution can be fixed via two possible pathways: One was the formation of $CaCO_3$, and another was the formation of stable complexes with ligands. As shown in Table 2, among the six ligands (L^1 to L^5), L^4 exhibits the max Ca^{2+} capturing yield. When $c(L^4)$ increased from 125 mg/L to 250 mg/L (n_{Ca2+}/n_{L4} 16.5:1 to 8.22:1), the free $c(Ca^{2+})$ in aqueous solution was significantly reduced, and the capturing rate increased from 73.0% to 87%. While the $c(L^4)$ further increased from 250 mg/L [(n_{Ca2+}/n_{L4}~8:1)] to 2000 mg/L [(n_{Ca2+}/n_{L4}~1:1)], the Ca^{2+} capturing yield did not increase significantly. Therefore, the optimal ratio of n_{Ca2+}/n_{L4} was ~8:1. The mean area diameter of particle size was near 1300 nm using L^4. It was lowest among the ligands. In other words, when L^4 was used, it had highest capturing yield and smallest particle size.

Table 2. Ca^{2+} capturing yield and particle size by adding different additives.

Ligands	c(L) (mg/L)	n_{Ca2+}/n_L	$c(Ca^{2+})$ (mg/L)	Capturing Yield (%)	Particle Size (nm)
L^1	125	38.4:1	150	67.4	1652
	250	19.2:1	144	68.7	1896
	500	9.6:1	142	69.0	1841
	1000	4.8:1	158	65.6	1653
	2000	2.4:1	165	64.1	1451
L^2	125	31.8:1	152	67.0	1668
	250	15.9:1	150	67.4	1539
	500	7.95:1	143	68.9	1519
	1000	3.98:1	163	64.5	1698
	2000	1.99:1	177	61.4	1843
L^3	125	60:1	163	64.5	1651
	250	30:1	161	65.0	1614
	500	15:1	157	65.8	1756
	1000	7.5:1	156	66.0	1687
	2000	3.74:1	165	64.1	1563
L^4	125	16.5:1	124	73.0	1362
	250	8.22:1	60	87.0	1467
	500	4.12:1	80	82.6	1324
	1000	2.06:1	74	83.8	1266
	2000	1.03:1	80	82.5	1385
L^5	125	38.7:1	157	65.8	1652
	250	19.4:1	150	67.4	1632
	500	9.7:1	143	69.0	1534
	1000	4.84:1	155	66.2	1654
	2000	2.42:1	160	65.4	1469

L^1 was a tridentate ligand and its complex with Ca^{2+} had relatively smaller stability constants than those of L^4 and L^5. As shown in Table 2, the capturing yield of Ca^{2+} for L^1 was in the range of 64.1–69.0%, which was essentially identical to those of the bis(tridentate) ligands L^2 and L^3 (61.4–69.0%). The particle sizes of $CaCO_3$ for L^1 (1451 nm~1896 nm) were also similar to those for L^2–L^3 (1380–1843 nm). These results indicated that no obvious 'cooperation effect' for these bis(tridentate) ligands existed. Actually, the tridentate units of L^1–L^3 at the same mass concentration were very close. These ligands, in theory, have similar stability constants. Based on the above results, Ca^{2+} in solutions was mainly bound via the formation of $CaCO_3$ rather than Ca^{2+} complexes.

Among these ligands, L^4 and L^5 are tetradentate ligands. As compared with L^1, L^5 has an additional pyridyl ring. The capturing yield of L^5 (65.4–69.0%) was slightly increased as compared with those of L^1/L^3. The particle sizes of $CaCO_3$ for L^5 (1469 nm to 1654 nm) were also smaller than those of L^1 and L^3. These results revealed that tetra-dentate ligand (L^5) exhibited a better performance than those of the tridentate ligands (L^1–L^3). In L^4, a carboxyl group was used instead of the pyridyl ring in L. The capturing yield of L^4 (73.0–87.0%) were higher than that of L^5. However, the capturing yield of Ca^{2+} did not increase with the increase of the stability constants of the ligands. As the Ca^{2+} needed to be captured and released in a dynamic equilibrium, the suitable stability constant of the ligands to Ca^{2+} was one of the important factors. It was also noteworthy that a complex with an extremely large stability constant was not beneficial to the formation of $CaCO_3$ nanoparticles. It was also noted that the particle sizes of $CaCO_3$ for L^4 (1362 to 1467 nm) were the smallest among these ligands. Thus, L^4 exhibited the best performance on the ion capturing yield and particle size. Although the size of the nanoparticles may be influenced by many factors including the temperature, stirring rate, liquid flow rate, solubility, etc., the stability of the complex that controls the concentration of Ca^{2+} released from the complex was one of the most important factors that then further controlled the formation rate and size of the $CaCO_3$ particles (Figure 8).

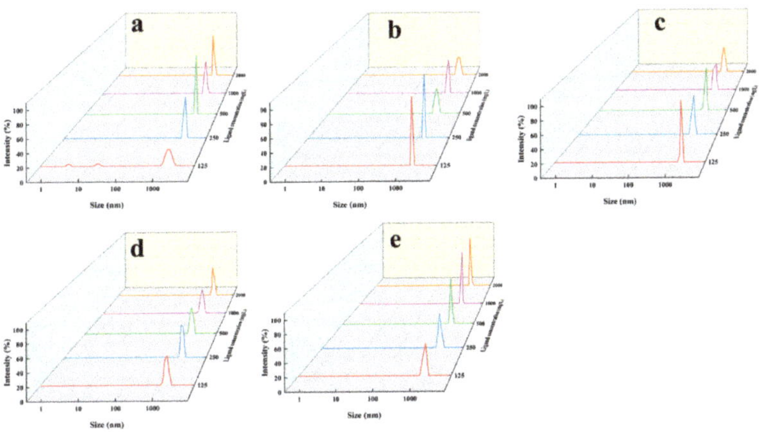

Figure 8. Particle size distribution plots, (**a**): different concentrations of L^1; (**b**): different concentrations of L^2; (**c**): different concentrations of L^3; (**d**): different concentrations of L^4; (**e**): different concentrations of L^5.

Different ratios of n_{Ca2+}/n_{NaHCO3} were also investigated. Among these, L^4 has been selected to investigate the best ratio of n_{Ca2+}/n_{NaHCO3} as it showed a max capturing yield when n_{Ca2+}/n_{L4} was 8:1. As shown in Table 3, it was clearly observed that the max capturing yield was 87.0% when the n_{Ca2+}/n_{NaHCO3} ratio was 0.363 under same conditions.

Table 3. Ca^{2+} capturing yield of L^4 by using different n_{Ca2+}/n_{NaHCO3} rations.

Entry	n_{Ca2+}/n_{L4}	n_{Ca2+}/n_{NaHCO3}	Capturing Yield (%)
1	8:1	0.726:1	72.3
2	8:1	0.363:1	87.0
3	8:1	0.181:1	81.2
4	8:1	0.09:1	78.4
5	8:1	0.045:1	61.2

In Figure 7, it can be found that smaller particles accumulated to form large particles. L^4 was chosen to demonstrate this. As shown in Table 4, the experiments were conducted with different concentrations of PASP under the optimal ratio (n_{Ca2+}/n_{L4} 8:1, n_{Ca2+}/n_{NaHCO3} 0.363). The results indicated that the PASP mainly affected the particle size of $CaCO_3$, while the capturing yield of Ca^{2+} remained almost unchanged. The particle size of $CaCO_3$ in the system without PASP addition was in the range of 1266–1467 nm, while after addition of PASP, the particle size of $CaCO_3$ decreased to the range of 764.8–1011 nm. The results showed that the optimal amount of PASP was 500 mg/L in the present system. The results also indicated that PASP probably could react with $CaCO_3$. As we all know, hydroxyl can be found on the surface of inorganic matter surfaces. Aspartic acid contains two carboxyl functional groups, which can react with the hydroxyl group on the surface of the $CaCO_3$ particle to form more hydrogen bonds. It would inhibit the particles accumulating to form large particles, leading to small particle formation. A high content of additives could form more hydrogen bonds and show relatively small particles of $CaCO_3$.

Table 4. Effect of addition of PASP on Ca^{2+} capturing yield and particle size.

$c(L^4)$ mg/L	c(PASP) mg/L	n_{Ca2+}/n	n_{Ca2+}/n_{NaHCO3}	Capturing Yield (%)	Particle Size (nm) without PASP	Particle Size (nm) with (PASP)
250	125	8:1	0.363:1	83.2	1362	997.6
250	250	8:1	0.363:1	84.0	1467	929.6
250	500	8:1	0.363:1	85.1	1324	764.8
250	1000	8:1	0.363:1	84.6	1266	1011
250	2000	8:1	0.363:1	85.2	1385	943.2

3.6. Controllable $CaCO_3$ Particles Improved Oil Recovery

To investigate the role of particles to enchained oil recovery (EOR), L^4 was selected for a simulated flooding experiment. As shown in Table 5, oil recovery reached 61.5% when the simulated formation water with L^4 and PASP was injected, while it was 55.2% when the simulated formation water without additive was injected. In other words, L^4 was an effective additive to improve EOR. It also demonstrated that the sustainability of Ca^{2+} in formation water was feasible. The results could be attributed to the interaction between L^4 and the Ca^{2+} present in the formation water, resulting in the micro- or nano-sized agglomerated particles formation during presence of PASP. Once these particles entered the reservoir, they would flow and deposit in the pores, effectively pushing oil droplets forward to improve EOR. On the other hand, the surface of nano $CaCO_3$ has more active functional groups like -OH, which would react with oil via hydrogen bond to change the interfacial tension [33,34]. It would benefit to improve EOR. Normally, it has a relatively high interfacial tension under high salinity condition. Hence, it needs to decrease the interfacial tension to improve EOR in formation water due to its high salinity. Normally, the rock surface properties would change from hydrophilic to lipophilic after long-term immersion in crude oil. It needed to change the wettability to enhance the oil film on the rock surface washing away. Tian et al. found that nanofluid changed the contact angle of oil film to improve EOR [34]. Therefore, micro- and nano-sized $CaCO_3$ particles had a excellent performance to improve EOR (Figure 9).

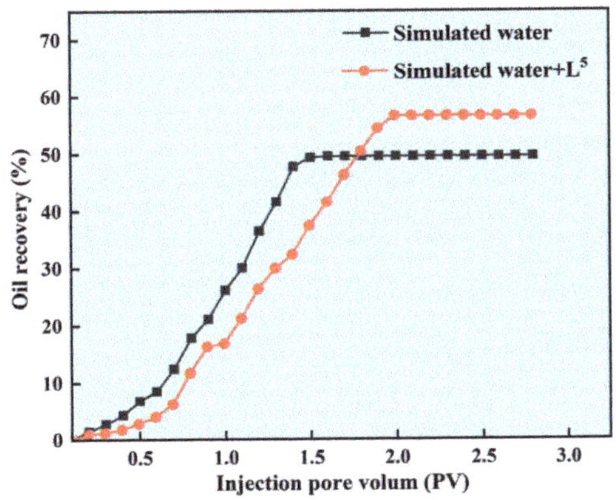

Figure 9. Plot of injection pore volume versus oil recovery.

Table 5. Experimental data of core displacement experiment.

Core Flooding Setup	Core Length/cm	Core Diameter/cm	Effective Porosity/%	Permeability /mD	Porosity/%	Oil Recovery/%
Simulated water	10 10	2.5 2.5	20.4 20.6	19.52 19.69	20.5 21.6	54.6 55.2
Simulated Water + L^4	10 10	2.5 2.5	20.9 21.5	19.89 20.01	21.3 22.5	60.3 61.5

The key process in this paper was controlled micro- and nano-sized $CaCO_3$ particle formation. Then, when the particles entered the reservoir pores, they exhibited an affinity for oil droplets, which facilitated the displacement of oil droplets within the reservoir. In addition, the presence of these particles could improve pore structure and reduce oil entrapment (Figure 10). Therefore, the efficiency of oil recovery was improved.

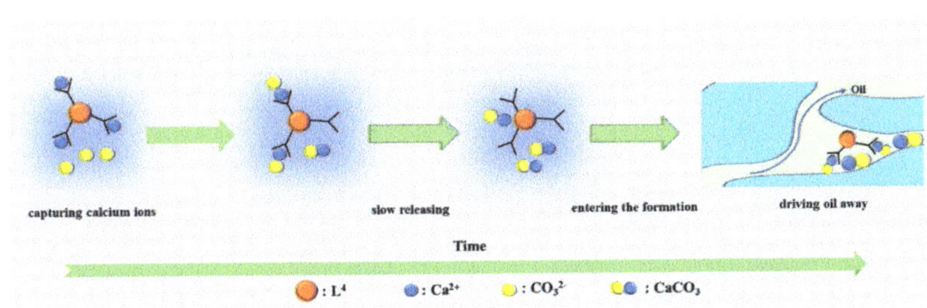

Figure 10. L^4 displacement mechanism diagram.

4. Conclusions

The paper reported the resource utilization of Ca^{2+} in formation water. The resource utilization of Ca^{2+} in formation water was investigated by organic ligands. The synthetic multi-dentate ligands can effectively capture Ca^{2+} in formation water. Then, Ca^{2+} slowly released to form $CaCO_3$ nanoparticles. Among the six ligands (L^1 to L^5), L^4 had highest capturing yield and smallest particle size. And it was clearly observed that the max capturing yield was 87.0% when n_{Ca2+}/n_{NaHCO3} ration was 0.363 under same conditions. The particle size can be further controlled by PASP. Therefore, adding the ligands studied in this paper to highly mineralized formation water can make calcium ions form calcium carbonate particles of controllable size, thus improving the recovery rate. The flooding test demonstrated it was an effective method, improving the EOR from 55.2% to 61.5%. The results in this study gives a new pathway to utilize the Ca^{2+} in formation water and provides a new method for scale treatment in petrochemical industry.

Author Contributions: Conceptualization, B.H. and S.H.; methodology, Z.H.; validation, J.Z. and Z.H.; formal analysis, H.Y.; investigation, Y.T.; data curation, B.W.; writing—original draft preparation, H.Y.; writing—review and editing, Y.Z.; project administration, S.H. All authors have read and agreed to the published version of the manuscript.

Funding: This research was funded by the Sixth Batch of Open Fund Projects of the State Key Laboratory of Offshore Oil and Gas Exploitation and grants from the National Natural Science Foundation of China (U22B6005).

Data Availability Statement: Data are contained within the article.

Conflicts of Interest: Authors Bo Huang, Shengzhen Hou, Zhao Hua, Jian Zhang, Yuejun Zhu were employed by the company CNOOC Research Institute Co., Ltd. The remaining authors declare that the research was conducted in the absence of any commercial or financial relationships that could be construed as a potential conflict of interest.

Appendix A

L^1

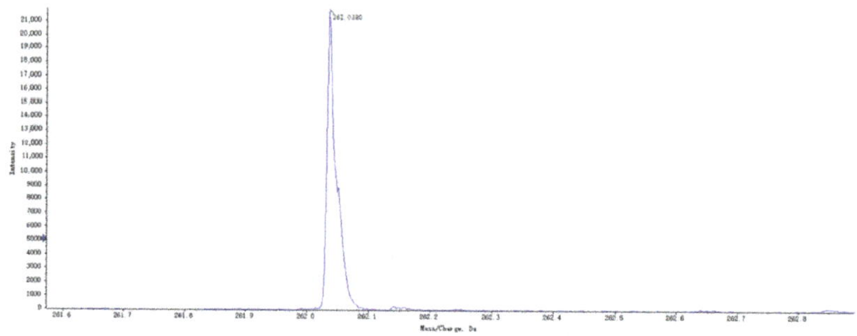

L^2

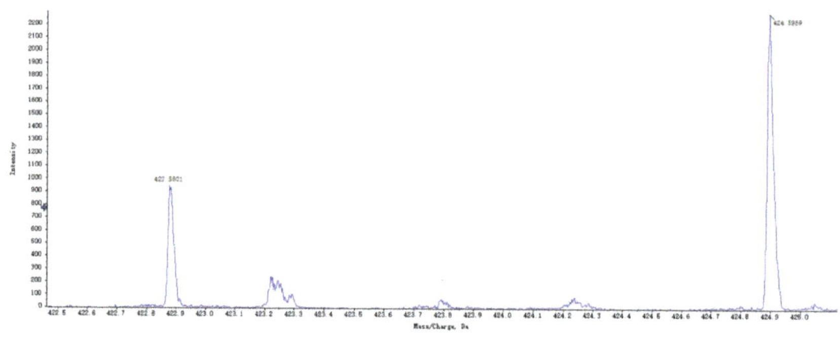

Figure A1. *Cont.*

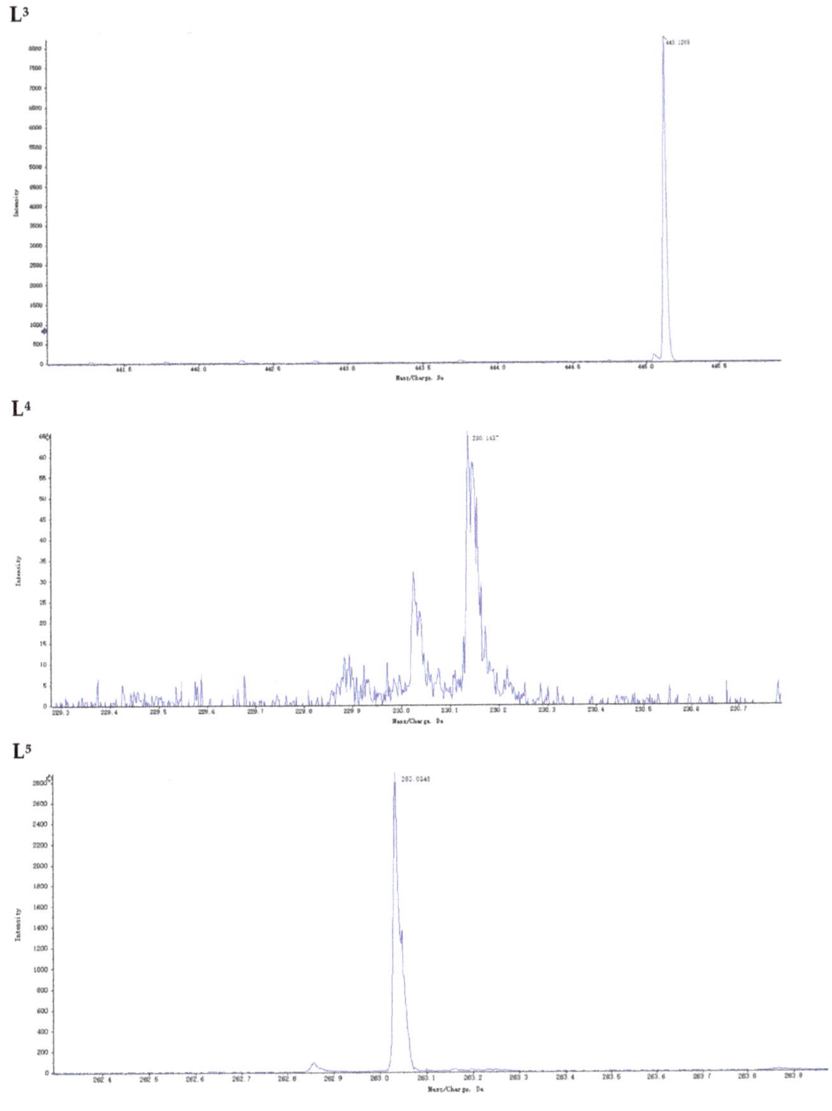

Figure A1. MS results of complexes of Ca^{2+}.

References

1. Suleimanov, B.A.; Abbasov, H.F.; Ismayilov, R.H. Enhanced oil recovery with nanofluid injection. *Pet. Sci. Technol.* **2023**, *41*, 1734–1751. [CrossRef]
2. Liang, T.; Hou, J.-R.; Qu, M.; Xi, J.-X.; Raj, I. Application of nanomaterial for enhanced oil recovery. *Pet. Sci.* **2022**, *19*, 882–899. [CrossRef]
3. Pourabdollah, K. Matrix acidizing: A fouling mitigation process in oil and gas wells. *Rev. Chem. Eng.* **2020**, *36*, 311–331. [CrossRef]
4. Mahmoodi, L.; Malayeri, M.R.; Tabrizi, F.F. Abatement of scale precipitation in oilfields using green scale inhibitors. *J. Pet. Sci. Eng.* **2022**, *208*, 109237. [CrossRef]
5. Sun, H.; Liu, Y.; Liu, D.; Li, S.; Li, X.; Chen, H.; Han, Z.; Li, L.; Li, X. Enhanced removal of scaling cations from oilfield produced water by carrier mineral floatation. *Water Sci. Technol.* **2021**, *84*, 3629–3640. [CrossRef]

6. Ituen, E.; Singh, A.; Li, R.; Yuanhua, L.; Guo, C. Nanostructure, surface and anticorrosion properties of phyto-fabricated copper nanocomposite in simulated oilfield descaling fluid. *Surf. Interfaces* **2020**, *19*, 100514. [CrossRef]
7. Kamal, M.S.; Hussein, I.; Mahmoud, M.; Sultan, A.S.; Saad, M.A.S. Oilfield scale formation and chemical removal: A review. *J. Pet. Sci. Eng.* **2018**, *171*, 127–139. [CrossRef]
8. Shafer-Peltier, K.; Kenner, C.; Albertson, E.; Chen, M.; Randtke, S.; Peltier, E. Removing scale-forming cations from produced waters. *Environ. Sci.* **2020**, *6*, 132–143. [CrossRef]
9. Zhang, N.; Wang, X.; Zhang, J.; He, X.; Kang, S.; Pu, J.; Fan, S.; Li, X. Experimental and statistical study on wellbore scaling mechanisms and characteristics for Huanjiang oilfield. *Geofluids* **2022**, *2022*, 9068440. [CrossRef]
10. Wang, H.; Dong, H.; Liu, X.; Zhang, P. Preparation and laboratory testing of polymeric scale Inhibitor colloidal materials for oilfield mineral scale control. *Polymers* **2022**, *14*, 4240. [CrossRef]
11. Samimi, A.; Bagheri, A.; Dokhani, S. Solousion for corrosion reducing gas pipe line with inspection for preventing fouling in oil exchangers. *Int. J. Basic Appl. Sci.* **2013**, *2*, 291–301.
12. Shen, C.; Zhao, Y.; Li, W.; Yang, Y.; Liu, R.; Morgen, D. Global profile of heavy metals and semimetals adsorption using drinking water treatment residual. *Chem. Eng. J.* **2019**, *372*, 1019–1027. [CrossRef]
13. Liu, Y.; Wang, J.; Xu, Y.; Wu, B. A deep desalination and anti-scaling electrodeionization (EDI) process for high purity water preparation. *Desalination* **2019**, *468*, 114075. [CrossRef]
14. Liu, Q.; Ren, J.; Lu, Y.; Zhang, X.; Roddick, F.A.; Fan, L.; Wang, Y.; Yu, H.; Yao, P. A review of the current in-situ fouling control strategies in MBR: Biological versus physicochemical. *J. Ind. Eng. Chem.* **2021**, *98*, 42–59. [CrossRef]
15. Malaeb, L.; Le-Clech, P.; Vrouwenvelder, J.S.; Ayoub, G.M.; Saikaly, P.E. Do biological-based strategies hold promise to biofouling control in MBRs? *Water Res.* **2013**, *47*, 5447–5463. [CrossRef]
16. Belattar, M.; Hadfi, A.; Ben-Aazza, S.; Karmal, I.; Mohareb, S.; El Housse, M.; Hafid, N.; Driouiche, A. Efficiency of one scale inhibitor on calcium carbonate precipitation from hot water sanitary: Effect of temperature and concentration. *Heliyon* **2021**, *7*, e06152. [CrossRef]
17. Zhang, P.; Kan, A.T.; Tomson, M.B. Oil Field Mineral Scale Control. In *Mineral Scales and Deposits;* Elsevier: Amsterdam, The Netherlands, 2015; pp. 603–617. [CrossRef]
18. Demadis, K.D.; Stavgianoudaki, N.; Grossmann, G.; Gruner, M.; Schwartz, J.L. Calcium-Phosphonate Interactions: Solution Behavior and Ca^{2+} Binding by 2-Hydroxyethylimino- bis (*methylenephosphonate*) Studied by Multinuclear NMR Spectroscopy. *Inorg. Chem.* **2009**, *48*, 4154–4164. [CrossRef] [PubMed]
19. Mady, M.F.; Rehman, A.; Kelland, M.A. Synthesis and Antiscaling Evaluation of Novel Hydroxybisphosphonates for Oilfield Applications. *ACS Omega* **2021**, *6*, 6488–6497. [CrossRef] [PubMed]
20. Jordan, M.M.; Williams, H.; Linares-Samaniego, S.; Frigo, D.M. New Insights on the Impact of High Temperature Conditions (176 °C) on Carbonate and Sulphate Scale Dissolver Performance. In Proceedings of the SPE International Oilfield Scale Conference and Exhibition, Aberdeen, UK, 14–15 May 2014. [CrossRef]
21. Zhou, Y.; Luo, Z.; Xu, M.; Zhao, T.; Ma, X.; Zhou, S.; Wen, B.; Yang, D. Preparation and Properties of Temperature-Sensitive P(NIPAM-AM) Nano–Microspheres in Enhanced Oil Recovery. *Energy Fuels* **2023**, *37*, 204–213. [CrossRef]
22. Wang, S.; Tang, Z.; Qu, J.; Wu, T.; Liu, Y.; Wang, J.; Liu, X.; Ju, Y.; Liu, F. Research on the mechanisms of polyacrylamide nanospheres with different size distributions in enhanced oil recovery. *RSC Adv.* **2021**, *11*, 5763–5772. [CrossRef]
23. Shagymgereyeva, S.; Sarsenbekuly, B.; Kang, W.; Yang, H.; Turtabayev, S. Advances of polymer microspheres and its applications for enhanced oil recovery. *Colloids. Surf. B* **2024**, *233*, 113622. [CrossRef]
24. Bu, X.; Jiang, K.; Wang, X.; Liu, X.; Tan, X.; Kong, Y.; Wang, L. Analysis of calcium carbonate scaling and antiscaling field experiment. *Geothermics* **2022**, *104*, 102433. [CrossRef]
25. Kumar, T.; Vishwanatham, S.; Kundu, S.S. A laboratory study on pteroyl-l-glutamic acid as a scale prevention inhibitor of calcium carbonate in aqueous solution of synthetic produced water. *J. Pet. Sci. Eng.* **2010**, *71*, 1–7. [CrossRef]
26. Duong, H.C.; Duke, M.; Gray, S.; Cooper, P.; Nghiem, L.D. Membrane scaling and prevention techniques during seawater desalination by air gap membrane distillation. *Desalination* **2016**, *397*, 92–100. [CrossRef]
27. Chen, T.; Neville, A.; Yuan, M. Assessing the effect of on scale formation–bulk precipitation and surface deposition. *J. Cryst. Growth* **2005**, *275*, e1341–e1347. [CrossRef]
28. Wang, Y.; Tian, J.; Han, H.; Sun, W.; Zhang, X. The enhanced flotation separation of magnesite and dolomite by introducing chelating reagent EDTA. *Colloids. Surf. A* **2024**, *682*, 132969. [CrossRef]
29. Mahmoud, M.; Attia, M.; Al-Hashim, H. EDTA chelating agent/seawater solution as enhanced oil recovery fluid for sandstone reservoirs. *J. Pet. Sci. Eng.* **2017**, *152*, 275–283. [CrossRef]
30. Sengupta, A.; Seitz, A.; Merz, K.M. Simulating the Chelate Effect. *J. Am. Chem. Soc.* **2018**, *140*, 15166–15169. [CrossRef]
31. Larsen, T.; Randhol, P.; Lioliou, M.; Jøsang, L.O.; Østvold, T. Kinetics of $CaCO_3$ Scale Formation During Core Flooding. In Proceedings of the SPE International Oilfield Scale Conference and Exhibition, Aberdeen, UK, 28–29 May 2008. [CrossRef]
32. Gower, L.B.; Odom, D.J. Deposition of calcium carbonate films by a polymer-induced liquid-precursor (PILP) process. *J. Cryst. Growth* **2000**, *210*, 719–734. [CrossRef]

33. Feng, S.J.; Zhao, J.F.; Gao, Y.L.; Yun, J.X. Polydopamine-$CaCO_3$ modified superhydrophilic nanocomposite membrane used for highly efficient separation of oil-in-water emulsions. *Colloids Surf. A* **2022**, *639*, 128355. [CrossRef]
34. Tian, K.P.; Pu, W.F.; Wang, Q.L.; Xie, M.K.; Wang, D.D.; Wang, M.M.; Liu, S. A Feasibility study on enhanced oil recovery of modified Janus nano calcium carbonate-assisted alkyl polyglycoside to form nanofluids in emulsification flooding. *Langmuir* **2024**, *40*, 4174–4185. [CrossRef] [PubMed]

nanomaterials

MDPI

Review

Adsorption of Different Pollutants by Using Microplastic with Different Influencing Factors and Mechanisms in Wastewater: A Review

Meng Zhao [1,†], Lei Huang [1,†], Samuel Raj Babu Arulmani [1], Jia Yan [1], Lirong Wu [1], Tao Wu [1], Hongguo Zhang [1,2,*] and Tangfu Xiao [1,3]

1 School of Environmental Science and Engineering, Guangzhou University, Guangzhou 510006, China; ymzguangzhou@163.com (M.Z.); huanglei@gzhu.edu.cn (L.H.); jarleumas@ymail.com (S.R.B.A.); jiayan@gzhu.edu.cn (J.Y.); a469949085@163.com (L.W.); wutaogzhu@163.com (T.W.); tfxiao@gzhu.edu.cn (T.X.)
2 Guangzhou University-Linköping University Research Center on Urban Sustainable Development, Guangzhou University, Guangzhou 510006, China
3 State Key Laboratory of Geohazard Prevention and Geoenvironment Protection, Chengdu University of Technology, Chengdu 610059, China
* Correspondence: hgzhang@gzhu.edu.cn
† These authors contributed equally to this work.

Abstract: The studies on microplastics are significant in the world. According to the literature, microplastics have greatly specific surface areas, indicating high adsorption capacities for highly toxic pollutants in aquatic and soil environments, and these could be used as adsorbents. The influencing factors of microplastic adsorption, classification of microplastics, and adsorption mechanisms using microplastics for adsorbing organic, inorganic, and mixed pollutants are summarized in the paper. Furthermore, the influence of pH, temperature, functional groups, aging, and other factors related to the adsorption performances of plastics are discussed in detail. We found that microplastics have greater advantages in efficient adsorption performance and cost-effectiveness. In this paper, the adsorptions of pollutants by microplastics and their performance is proposed, which provides significant guidance for future research in this field.

Keywords: organic pollutants; inorganic pollutants; adsorption

Citation: Zhao, M.; Huang, L.; Arulmani, S.R.B.; Yan, J.; Wu, L.; Wu, T.; Zhang, H.; Xiao, T. Adsorption of Different Pollutants by Using Microplastic with Different Influencing Factors and Mechanisms in Wastewater: A Review. *Nanomaterials* **2022**, *12*, 2256. https://doi.org/10.3390/nano12132256

Academic Editor: George Z. Kyzas

Received: 4 June 2022
Accepted: 26 June 2022
Published: 30 June 2022

1. Introduction

With the increasingly serious environmental pollution, sewage treatment has become a worldwide research hotspot, and the technology of microplastics adsorbing harmful substances to purify sewage has become a prominent choice [1]. Adsorption technology is developing rapidly, and the development of large-scale and sustainable hybrid technologies can overcome the shortcomings of traditional technologies. Microplastics have the advantages of large surface area and low cost [2], thus, leading to applications in the field of sewage treatment.

The American Association of Marine Conservation Scientists defines plastic as a synthetic water-insoluble polymer. According to the relevant literature, plastics are divided into thermosetting and thermoplastic plastics. The high abundance of microplastics (MPs) can be attributed to three factors: first, human activity, such as tourism waste, or waste nets from fishing can degrade to produce MPs [3]; second, industrial supplies, such as beverage cans, plastic bags, and plastic water cans.

Severe acute respiratory syndrome coronavirus 2 (SARS-CoV-2) is the virus responsible for the COVID-19 pandemic. Since it emerged, the disease has led to high mortality and morbidity [4]. In the process of testing for the SARS-CoV-2, a large of plastic products flowed into the environment. In addition, there are some modern products, such as the increased of using cosmetics [5] and textiles [3].

MPs are emerging pollutants that have been shown to provide toxic contaminants [6]. For example, the pathways by which the environment or its inherent toxic monomers and additives enter the aquatic food web were MPs [7]. MPs can enter agroecosystems with toxic effects [8] but also enter organisms [9] to cause cellular internalization and affect intracellular target organelles [10]. Moreover, there are many difficulties in fully solving nano plastics due to the limited methodologies for selectively recognizing nano plastics [11].

Traditional sewage treatment technology is complex in operation; has high costs, slow adsorption, limited capacity, and high regeneration costs; and may cause the risk of secondary pollution. Thus, finding a cheap and simple treatment method has become an urgent problem. Based on the principle of green environmental protection, waste utilization is the best way. Microplastics have been used to make electrospun micro/nanofibers (in the waste recycling) [12], PET nanoparticles (in the biological systems) [13], and antimicrobial coatings (in the healthcare sector) [14].

The unique structures of microplastics with benzene rings or carbon chains make them a good material choice for sewage treatment. As microplastics also contain rich functional groups, they have a strong affinity for pollutants under the action of chemical bonds, such as hydrogen bonds and van der Waals forces, and are suitable for sewage treatment [15]. Although pollutants are adsorbed on plastics and cause certain harm to the environment [16], we can achieve simultaneous removal of pollutants and microplastics through this method, which is a good water source treatment method [17].

Microplastics have attracted extensive attention due to their large specific surface area and stable properties [18]. Sewage treatment by microplastics mainly depends on adsorption, and the advantage of microplastics in the specific surface area greatly increases the contact area with pollutants. In addition, the resin carbon structure is stable and easy to separate. Therefore, after treating sewage, plastic can be separated from sewage by centrifugation or sedimentation.

Under the influence of the environment, light, weathering [19], breakage, corrosion, grafting, abrasion, single loading, aging [20], and other effects will not only hinder the treatment of sewage by microplastics but also increase the specific surface area of microplastics. When microplastics react with pollutants in the environment, modified microplastics can show strong adsorption performance for some contaminants under this influence [21]. This paper reviews the classification of microplastics and the adsorption mechanism of microplastics on organic and inorganic pollutants. In addition, the activities of microplastics in nature and their treatment methods and principles are briefly introduced. Hence, we provide new ideas for utilizing microplastics with contaminants.

2. Activity of Microplastics in the Environment

2.1. Activities of Microplastics in Soil

Although the activity of microplastics in the soil environment is limited, the amount of microplastics in the soil cannot be ignored. Most human activity is on land, which contributes to the accumulation of microplastics from landfills, fermentation and composting of organic fertilizers, and the use of tires. Under the influence of temperature, pH, and other factors, as shown in Figure 1, microplastics can still interact with pollutants in the soil environment [22]. Microplastics in the land mainly interacted with pollutants directly to realize the treatment of pollutants. Of course, the indirect effects of microplastics also affect soil properties [23], such as soil MPs, can provide sorption sites for soil microorganisms, establish unique microbial communities and influence the ecological role of soil microorganisms [24].

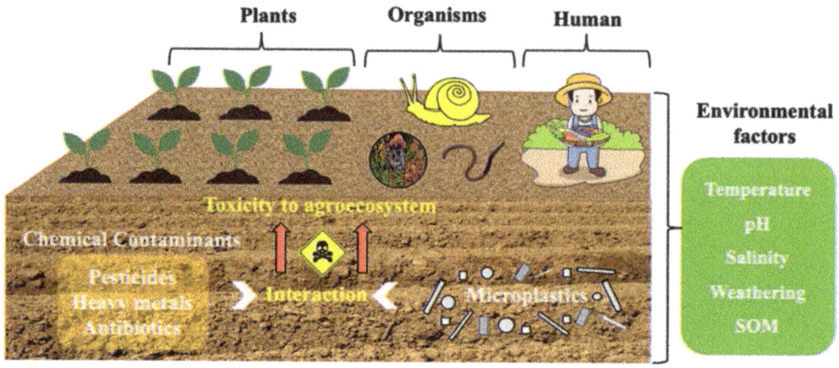

Figure 1. Interaction diagram between microplastics and pollutants in soil [22].

2.2. Activities of Microplastics in Water

At present, microplastics have been stored in large amounts and have frequent activities in the marine environment. The main phenomenon is the suspended atmosphere [25] as shown in Figure 2. Global marine plastic pollution is derived mainly from the input of vast amounts of land-based plastic waste [26]. Wastewater treatment plants (WWTPs) are known to be one of the main and most important sources of microplastic discharged into the environment [27].

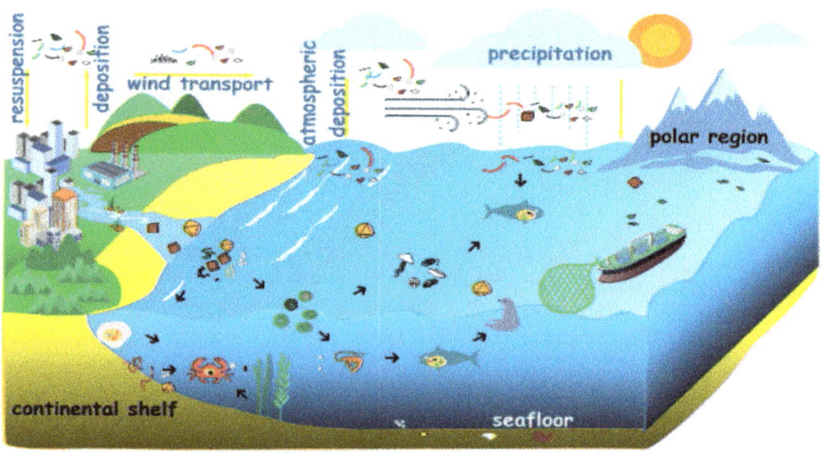

Figure 2. Origin and cycle process of microplastics [25].

The investigation found that a secondary wastewater treatment work (WWTW) (population equivalent 650,000) was releasing 65 million microplastics into the receiving water every day [28]. The abundance of MPs in the influents and effluents of domestic WWTPs was 18–890 and 6–26 $n \cdot L^{-1}$. No significant differences in MP abundance were demonstrated among effluents or wastewater from domestic, industrial, agricultural, and aquacultural sources, indicating they were all potential sources of MP pollution [29].

MPs release chemical additives during photodegradation [30] as shown in Figure 3. Microplastics deposited at the bottom of water, such as rivers, were easily swallowed and absorbed by aquatic animals and then circulated in the food chain superposition step by step. As a result, bioaccumulation was constantly amplified [31], such as microplastics used as anthracene carriers in medaka [32] and polyvinyl chloride (PVC) used as carriers of heavy metals [33].

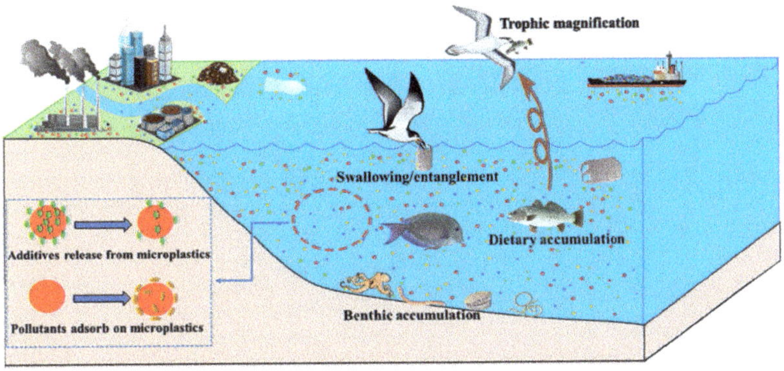

Figure 3. Photodegradation of microplastics and their activities in water environments [30].

Some contaminants also exhibited competitive behavior on the surface of microplastics, as DDT interferes with the adsorption of phenylalanine plastics [34]. To date, many kinds of microplastics have been found in the human body. Microplastics can easily combine with various additives and other pollutants in the water environment; however, the discovery that they can be used as carriers of harmful substances [35] brings hope to sewage treatment [36]. In waters, microplastics can not only interact with pollutants but also treat pollutants by winding, adsorption, and other methods. Similar laws inevitably exist in marine, river, lake, and other water environments.

3. Classification of Microplastics

Microplastic fragments are formed by the decomposition of larger plastic particles using photolysis, thermal oxidation, thermal degradation, and biodegradation as shown in Figure S1 [37]. Microplastics can be divided into three aspects: (i) polycarbonate polyurethane (PCU) fragments are found in foam plastics. (ii) Polypropylene (PP) and polyethylene (PE) + PP [38] films are found in boat paints. (iii) PET and PES are found in fishing nets and lines and textile production [39] (clothing, wool, and blankets) [40]. The classification of microplastics is determined by their physical and chemical properties and can be divided into five characteristics: their shape, polymer, color, source, and particle size [40] as shown in Table S1.

Shape: The main categories of microplastics by shape are microfibers or plastic fibers, followed by fragments, films, and particles. Microplastics are found in surface water and sediment, and beads make up only a small fraction of the types of microplastics [41].

Polymers: Polymers are identified as polypropylene, polyethylene, and polystyrene [42]. The most significant proportion of microplastic polymer particles is composed of polyethylene [42], followed by polyethylene terephthalate and polyacrylamide. The sediment is dominated by polypropylene. PE and polypropylene (PP) are the most abundant contaminants globally, followed by polymers, such as polyethylene terephthalate, polystyrene, and polyamide.

Grades: In the environment, microplastics are classified as primary plastics and secondary plastics according to their source. Primary microplastics include plastic particles manufactured as precursors for personal care products and other industrial uses [43]. Secondary microplastics are created through photolysis and hydraulic friction, which are found in plastic waste, such as fragments, scales, fibers, and particles, such as plastic bottles, plastic bags, and synthetic textile fibers, which are discarded manually. Second, microplastics are larger plastic fragments that are degraded by using (e.g., fibers released from washing clothing or textiles), waste management, or fragments of larger plastics in the natural environment (e.g., fibers released in water and water environments).

Microplastics in the environment have a large base, many types, and complex properties. In the natural environment, microplastics constantly change their forms and structures with the action of nature, and naturally, they have different properties, making their use more flexible in modern conditions [41].

4. Classification of Pollutants

There are many kinds of pollutants in complex forms. According to the literature [44], they can be divided into organic pollutants [2] and inorganic pollutants; however, some pollutants do not belong to the first two categories, and thus an additional pollutant is added. The primary organic pollutants are carbon chains and carbon rings. In addition, it contains acids, such as acetic acid, formic acid, and other carbon chains that have acid groups. There are also benzene rings such as phenanthrene, nitrobenzene, and naphthalene. Inorganic pollutants are mostly salts [45].

These substances in the water generally cannot maintain the original compound form, often through ionization hydrolysis and other effects in the form of ions free in the water, the most pollutants are metal ions. In addition to ions, there were a small number of metals that can maintain their original form in water, such as mercury and other metals with relatively stable physical and chemical properties. In addition to the common types of pollutants, some fibers, such as fiber carbon, fiber metal materials, etc., can also be classified as inorganic pollutants. Other pollutants are mixtures, such as oils and some medicines. As these substances can automatically decompose and recombine to form new substances under the action of temperature, it is impossible to obtain clear statistics on their content and composition. They are composite pollutants; therefore, they are classified separately.

5. Factors Affecting the Adsorption of Microplastics

There are many types of pollutants, and correspondingly, there are many types of microplastic adsorbents used to treat sewage. The literature review on the adsorption of plastics on various pollutants found that the factors influencing the adsorption of plastics on pollutants include many aspects. Figure 4 shows the adsorption of microplastics on contaminants and their influencing factors. Tables 1–3 briefly summarize the adsorption of various pollutants by different microplastics and their influencing factors: (i) the adsorption of organic pollutants by plastics as shown in Table 1; (ii) the adsorption of plastics to inorganic pollutants as shown in Table 2; and (iii) the adsorption of plastics to mixed pollutants as shown in Table 3.

The results were different under different influencing factors. For example, microplastics, heavy metals, and pollutants may form complexes, which exhibited different properties under the influence of atomic valence, atomic size, surface texture [46], and functional groups [47]. Of course, the adsorption effect was different for different adsorbents. Biochar may be an excellent adsorbent for repairing high-concentration NH_4^+ wastewater. The interaction between biochar and PE can change the surface properties of the adsorbent [48]. In addition, the same influencing factors, such as pH, temperature, and functional groups, also had different phenomena in various adsorption experiments [49].

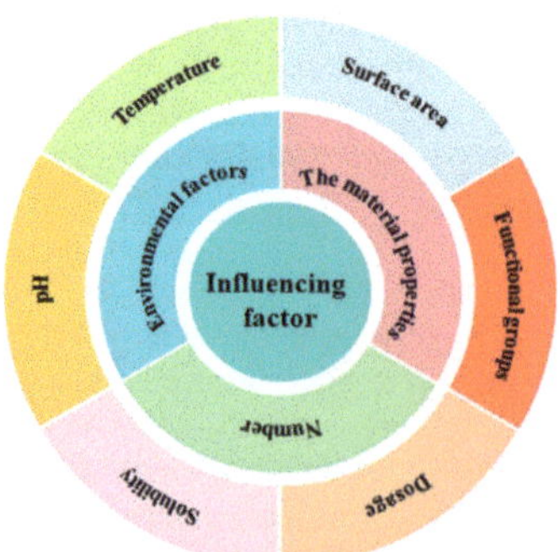

Figure 4. Different types of influencing factors in the process of adsorbing pollutants by microplastics.

5.1. Effect of pH

In the process of microplastic adsorption of pollutants, pH was the most significant influencing factor. It had the most crucial effect shown in the adsorption of organic pollutants by microplastics. One of the most important aspects was the effect of pH on the adsorption properties of microplastics, such as microplastics for sulfamethoxazole [41] and three triazole fungicides [50]. There were also some other organic pollutants or compounds adsorbed, and the amount of adsorption increased as the pH value increased.

Of course, the pH value also inhibited the adsorption capacity of microplastics, such as the adsorption capacity of 2,4-hexadiene (sorbic acid) [51], adsorption of perfluorooctane sulfonic acid on polyethylene and polystyrene [52], and adsorption of phenol on polymer resin decreased with increasing pH value [53].

In addition, the inhibition effect of pH value was also reflected in the adsorption of inorganic pollutants by microplastics. Under the influence of pH value, folate-polyaniline hybrid hydrogels tend to adsorb anionic dyes, and a low pH value was more conducive to improving the adsorption capacity of folate-polyaniline hybrid hydrogels [54]. Under basic conditions, iminodiacetic acid chelate resin can form a complex with metal ions [55]. However, alkaline conditions can sometimes produce fatal obstacles to the adsorption of plastics. For instance, the adsorption of bisphenol by polythioctylic acid-based coagulants can directly decrease the adsorption efficiency from 90% to 10% under alkaline conditions [55].

In the study of the adsorption of Cr(III) and Fe(III) by modified polyethyleneimine, the alkaline state would precipitate Cr(III) and Fe(III). In contrast, under acidic conditions, modified polyethyleneimine combines more easily with hydrogen ions, and the most favorable condition was a pH of 5 [56]. In addition, sulfamethoxazole mainly existed in the form of anions after ionization in an aqueous solution, and a suitable pH value was conducive to enhance the adsorption affinity of microplastics for organic pollutants [34]. When the pH was 5, the adsorption electrostatic repulsion of nano and microplastics were the lowest.

The effect of microplastic pH on pollutant adsorption did not follow the law of increasing or decreasing. In the case of inorganic pollutants, the impact of adsorption will be the highest under the influence of pH value. This was because pH can affect the group, hydrolysis, and ionization of substances in the solution and jeopardize the adsorption performance [31].

In general, microplastics had an optimal range for the adsorption of metal ion pollutants. The maximum adsorption value would appear in the adsorption process of styrene-DVB resin for homocarboxylic acid. For example, in a weakly acidic environment with a pH value of 4–6, poly (ethylimide) adsorbents were effective in decontamination [57]. The effect of pH on removing hexavalent chromium was carried out in the pH range of 2–9 [54]. When the pH was 5, the adsorption capacity of Cd and Ni ions on polystyrene nanofibers reached the highest value [58]. The most favorable adsorption conditions of polyacrylonitrile-2-aminothiazole resin for mercury, cadmium, lead, copper, zinc, and nickel are 2.50–6.50 [59].

When the pH was 5.95, D152 resin adsorbed cadmium(II). For the adsorption of cadmium and nickel ions on polystyrene nanofibers, the adsorption of Er(III) on acrylic grafted polytetrafluoroethylene fiber was the highest [60]. The positive charge of polystyrene microplastics was favorable for the attraction of HA and FA when the pH was less than 2.75. At pH 5, IDA-chelating resin adsorption of bivalent heavy metal ions follows the same principle. In the pH range of 5–9, the content of Cd adsorbed on MPs first increased and then decreased.

The adsorption effect of IRN77 cation exchange resin on Co(II), Cr(III), and Ni(II) was the best at the pH of the original solution [61], and the adsorption of polyacrylonitrile-2-amino-2-thiazoline resin for precious metal ions [62], adsorption treatment of Cu(II) and Pb(II) with amine and sulfur-chelating resin was similar [63]. In the adsorption of Rh(III),

Ru(IV), Pd(II), and Ir(IV) by polyacrylonitrile-thiosemicarbamide resin, a high pH was beneficial to the adsorption of Pd(II) [64].

When the pH was 5.8, the most suitable amine chelating resin for Cu(II) and Pb(II) absorption, if the pH was too low, amino groups were accessible to protons. When the pH was too high, metal ions easily precipitate by plugging the pore size. Moreover, it was possible to treat some metal ions in water when the pH was too high, such as the adsorption of copper by microplastics [65]. However, $Cu(OH)_2$ precipitation was generated instead of adsorption.

5.2. Effects of Particle Size

Microplastics with small particle sizes have a larger adsorption capacity because the specific surface area of plastics with small particle sizes is more prominent, which increases their contact area with pollutants, thus, significantly improving the adsorption performance of plastics. Weathering also increased the specific surface area of plastics, changed the particle size of microplastics, and produces oxygen groups, which would increase their polarity, charge, roughness [66], and porosity [49].

In addition, the same principle applies to crystallinity, with lower crystallinity polymers accumulating more contaminants. In the adsorption experiment of trinitrophenol and TCEP with polyethylene and polyvinyl chloride microplastics, plastic with a small particle size was better. The adsorption of phenanthrene on microplastics and medium plastics followed the same principle, and phenanthrene on microplastics had a more substantial adsorption capacity [67]. In conclusion, adsorbents with larger surface area had a larger adsorption capacity. Table S2 summarizes the surface area of different adsorbents.

5.3. Effects of Temperature

In the experiment of ion adsorption by microplastics, the temperature was a critical factor affecting the reaction result. In addition, hydrolysis and ionization can be promoted at high temperatures, and pollutants can be pyrolyzed at high temperatures to improve the sewage treatment capacity of microplastics. For the adsorption of mercury from coal and waste PVC, the adsorption efficiency of mercury increased from 600 to 800 °C. Similar to polystyrene sphere macromolecules with a positive charge, high-temperature polystyrene sphere macromolecules with intense movement can fully collided with anions and attracted each other to adsorb more anionic dyes. In addition, the adsorption of Cr(III) and Fe(III) on modified polyethyleneimine had similar results [68].

As the temperature increases, the adsorption performance of microplastics will also improve, such as the adsorption of cadmium(II) by D152 resin [60], adsorption of Er(III) on acrylic grafted polytetrafluoroethylene fiber [60], adsorption of metal ions, such as Cu(II), by cross-linked polystyrene supported low-generation diethanolamine dendrimers [69] and adsorption of precious metal ions on polyacrylonitrile-2-amino-2-thiazoline resin [62]. Moreover, the adsorption efficiency of polythioctylic acid-based coagulants for organic pollutants, such as bisphenol, decreased by approximately 10% when the temperature dropped from 330 to 277 K [55]. Although, according to various data, the adsorption effect of the grafted PTFE fiber on Cu(II) increased with increasing temperature [70], these spontaneous reactions and endothermic reactions under high-temperature conditions can have a good adsorption effect.

Subsequently, high temperatures were not always the supreme effect. For example, in the adsorption of TnBP and TCEP by polyethylene, the surface tension and van der Waals force of the organic light-emitting materials decreased at high temperatures, which hindered the adsorption performance. In addition, some exothermic reactions, such as the adsorption of As(III) by polytetrafluoroethylene (PTFE) microplastic particles, had an inhibiting effect on spontaneous adsorption [71].

5.4. Effects of Salinity

Salinity also played a specific role in the adsorption process of microplastics, such as the treatment of aromatic amines by polyethylene microplastics. With increasing salinity, the adsorption efficiency gradually increased [20]. Moreover, high salinity in water inhibited the release of carbon nanotubes and photodegradation, and the adsorption properties of microplastics were affected to a certain extent. In the experiment of removing heavy metal ions, coexisting sodium ions grabbed adsorption sites of heavy metals in a short time, thus, reducing the removal efficiency of heavy metals [72].

5.5. Effects of Functional Groups

Functional groups are also an essential factor that affects the adsorption of microplastics. The existence of functional groups endowed materials with different properties. As these functional groups showed unique properties, they exhibited excellent adsorption effects on pollutants. In addition to the functional groups of the resin itself, aging [73] and weathering yielded more surface area and produced all kinds of functional groups. This increased their polarity, charge, roughness, and porosity to a certain extent, which affected the adsorption capacity of microplastics [74]. It was proven that the adsorption capacity of PE for chlortetracycline and amoxicillin increased by 1.08–14.24 times after aging [75]. Furthermore, different functional groups may produce different results in acid–base, hydrogen bonding, complexation, etc.

Functional groups in the adsorption of microplastics generally promoted the experiment. For example, in the adsorption experiment of resin for phenol, the amine group on the resin, quaternary amine, primary amine, and secondary amine groups were conducive to the adsorption of resin for phenol, while the tertiary amine group is not [76]. The tert-butyl ester group in glass fiber-reinforced plastics also promoted the adsorption of polyacrylic acid [77]. When coal and waste polyvinyl chloride adsorbed mercury, the carbon chloride functional group could be converted to ionic chloride to realize the adsorption of mercury [78].

Calcium ion polymers can effectively adsorb organic micropollutants because of the π-rich network in porous polycalixarenes [79]. Similar to the adsorption of mercury by polyethyleneimine-modified porous cellulose carriers, the adsorption of metal pollutants by microplastics was free from the influence of pH value due to the action of the amine group [80]. Polyacrylonitrile-thiosemicarbazone resin contained triazolthione functional groups, Pd(II) adsorption required one active functional group, while Ru(IV) and Ir(IV) adsorption required two functional groups [64].

However, under exceptional circumstances, the functional groups inhibited the adsorption of microplastics. For example, alcohol groups in cellulose can be acetyl, which changed the structure of cellulose, and cellulose triacetate is more acetyl than cellulose acetate, which reduced its ability to adsorb 2,4-hexadiene (sorbic acid) [51]. Furthermore, with the increasing in the polarity of aromatic amines, the adsorption of polyethylene to aromatic amines was hindered [20]. In contrast to the adsorption experiment on polyethylene and nylon fibers, there was no obstruction of hydrophilic group amide, and polyethylene fiber had a decisive advantage in the adsorption capacity and adsorption speed; however, the other structure of the resin was unable to avoid the effect of inhibition [81].

In addition, in the adsorption experiment of Cu(II) by cross-linked polystyrene-supported low-generation diethanolamine dendritic polymer, too many functional groups caused a large area of blockage, and Cu(II) could not adsorb well on the resin [69]. To a certain extent, the adsorption efficiency of coal and waste PVC on mercury gradually increased with increasing PVC content. However, with the addition of abundant PVC, the adsorption efficiency of coal and waste PVC on mercury decreased because too much PVC easily blocked the pore size [78].

5.6. Effects of Ionic Strength

Ionic strength plays a small role in the adsorption experiments of microplastics; for example, the adsorption of pyrazystole and dioxystole could promote the experiment; however, for others, it had no similar effect [82]. In addition, due to the salting-out impact, a high salt concentration was more conducive to the adsorption of DPA molecules by microplastics [20]. Moreover, the presence of salt ions in the adsorption of triazole fungicides on polystyrene resulted in the highest adsorption capacity [50].

5.7. Effects of Surfactants

In resin adsorption experiments of pollutants, due to their hydrophobicity, the resin for pollutant adsorption efficiency was not high, and the solubility of a surfactant could change the resin. The adsorption performance of the resin, the most representative, was silica, which could adhere to the surface of organic matter or inorganic matter and be soluble in the inorganic or organic matter [83]. The adsorption of cationic methyl bromide on pure polyvinyl chloride was enhanced by mucopolysaccharides [84], which was a typical example of enhanced adsorption of ionic pollutants by surfactants. In addition to sewage treatment, surfactants were also widely used in facial, hair, and clothing cleaning work.

5.8. Effects of Solubility, Concentration and Dosage

The different solubilities of pollutants produced different adsorption results. For example, the results of the adsorption experiment on chloromethane by plastic indicated that the adsorption amount of trichloromethane by plastic was greater than the adsorption amount of carbon tetrachloride, which is the reason for the solubility [85], and the adsorption of lead and cadmium in intertidal sediments by facial scrub beads [86].

In the adsorption of 2,4,6-nitrotoluene by functionalized polystyrene nanospheres, the adsorption amount initially increased and then flattened because the high solubility of KH-570 obstructs the phosphorus–phosphorus accumulation interaction between the adsorbent and pollutant [87]. However, hydrophobic substances were more easily removed from an aqueous solution than hydrophilic substances, with different solubility and adsorption results [85].

In different experiments, the concentration altered the effects on the adsorption process of microplastics. As in the adsorption of three triazole fungicides by polystyrene microplastics, the concentration promoted the adsorption capacity of polystyrene [50]. Polystyrene can enhance the migration of nonpolar (pyrene) and weakly polar (2,2,4,4-tetrabromodiphenyl ether) compounds in saturated soils. However, in the adsorption of 5-sodium disulfoisphthalic acid by acrylate polymer YWB-7 resin, the methanol content had a hindrance effect on the performance of the adsorbent [88].

High levels of microplastics can absorb more pollutants. For example, with the increase in polystyrene nanofiber sorbent, the adsorption efficiency of cadmium and nickel ions can reach more than 90% [58]. The IRN-77 cation exchange resin for Co(II), Ni(II), and Cr(III), three kinds of ion adsorption efficiencies, is very high, and the increase in IRN-77 cation exchange resin dosage can reach 100% [89]. The same effect was observed with modified polyethyleneimine Cr(III) and Fe(III) adsorption [56].

The adsorption of Co(II), Cr(III), and Ni(II) on IRN-77 cation exchange resin followed the same principle [61]. The adsorption efficiency of phenanthrene and its mono hydroxyl derivatives increased with the amount of adsorbent [90]. In addition, some reagents can also affect the adsorption efficiency of microplastics. For example, in the adsorption of 5-sodium disulfoisylphthalic acid by acrylate polymer YWB-7 resin, the amount of methanol can seriously hinder the efficiency of the adsorbent [88].

On the other hand, the pore size of the adsorbent also followed the same principle, such as poly(ethylimide) adsorption of lead and mercury. The larger pore size of the adsorbent can provide better adsorption effect [57]. Thus, the difference in the amount of pollutants adsorbed by these plastics was due to the influence of solubility, concentration, and dosage on the adsorption performance of microplastics.

5.9. Effects of Adsorption Selectivity

Adsorption selectivity refers to the excellent adsorption performance of microplastics for some pollutants in the process of treating mixed contaminants. For example, polyaniline hybrid hydrogels showed high selective adsorption capacity for chromium(VI), eosin yellow, rose red, methyl orange, and other anionic pollutants but showed low adsorption capacity for mercury(II), lead(II), rhodamine B, bismuthite Brannia methyl blue, and neutral red [54].

On the other hand, polystyrene-deg-3-ap, polystyrene-TeG-3-AP 3-aminopyrine, and hydrophilic spacer arm chelating resin for Hg(II), Ag(I), Fe(III), Pb(II), Co(II), Cu(II), Ni(II), Cd(II), Zn(II) plasma, and Hg(II) showed a strong adsorption capacity [91]. In addition, phenanthrene had stronger adsorption on microplastics than did 1-nitronaphthalene. Polarity enhanced the adsorption of 1-nitronaphthalene on microplastics [92].

Table 1. Adsorption of organic pollutants.

Adsorbent	Size Range	Pollutants	Influencing Factors	Reference
plastic cellulose acetate	—	2,4-diallyl (sorbic acid)	pH	[51]
PE, PS, PVC	—	PFOS, FOSA	pH	[52]
polymer resin	—	phenol	pH	[53]
polylipoic acid ester-base coagulant	0.40–0.56 and 0.15–0.30 mm	organic pollutants (pops)	temperature, pH	[55]
HDPE, PS, LDPE, PVC, SSA	—	the Philippines, Nitrobenzene and naphthalene	particle size, crystallinity,	[67]
anion exchange resin	—	phenol	amino	[76]
glass fiber reinforced plastic	—	polyacrylic acid	tert-butyl ester groups	[77]
porous polycarboxarene	—	organic micropollutants	π	[79]
the original rusty water microplastic	0.2 mm	ions, organic pollutants	the surfactant	[84]
polyethylene, neoprene, polyvinyl chloride and polyurethane foam	12 mm	chlorinated methane	chlorinated methane solubility, concentration	[85]
functionalized polystyrene nano ball	—	2,4,6-trinitrotoluene (TNT)	the dosage of KH-570	[87]
acrylate polymer YWB-7 resin	0.4–0.6 mm	5–2 sulfo sodium isophthalic acid	methanol content	[88]
micro polyvinyl chloride (PVC) plastic	80–210 μm	Fe and single hydroxyl derivatives	adsorbent dosage	[90]
PA, PVC, PET	—	sulfanilamide	UV-irradiation, temperature	[93]
polyethylene and polyvinyl chloride (PVC) plastic	1–5, 0.425–1, 0.125–0.425 and 0.045–0.125 mm	san zhang butyl ester phosphate and phosphate (2-ethyl chloride)	particle size	[94]
polystyrene	80.4 ± 7.9 nm	organic pollutants (pops)	the concentration of	[95]
biological membrane reinforced plastic microfiber	2–3 mm	perfluorinated octane sulfonic acid (PFOS)	aging	[96]
PE, PS, PA, and PVC	152.53 ± 57.92, 168.55 ± 57.50, 109.44 ± 44.53, and 57.64 ± 26.50 μm	nonpolar organic compounds	adsorbent performance	[97]
PS and PP	3.5 mm in length +2.2 mm wide and 3–5 mm	fuel aromatics and ether	aging	[98]
polystyrene	29 μm	nonionic organic compounds	functional groups	[99]

Table 2. Adsorption of inorganic pollutants.

Adsorbent	Surface Area	Pollutants	Influencing Factors	Reference
polystyrene (MP_S)	—	nanometer oxide (CeNPs)	heavy metals (HM)	[46]
iminodiacetic acid chelating resin	0.40–0.56 and 0.15–0.30 mm	Sc(III), Y(III), La(III), Fe(III), Al(III), Ga(III), In(III)	pH	[55]
modification of polyethyleneimine	0.15 and 0.075 mm	Cr(III), Fe(III)	pH, adsorbent dosage, temperature	[56]
polyethylimine	—	Pb and Hg	the aperture	[57]
polystyrene nanofibers	—	Cd, Ni	pH, adsorbent dosage	[58]
polyacrylonitrile-2-amino thiazole resin	25.9 nm	Hg, Cd, Pb, Cu, Zn, and Ni	pH	[59]
acrylic acid grafted polytetrafluoroethylene fibers	—	Er(III)	pH, temperature, initial metal ion concentration	[60]
IRN77 cation exchange resin	<0.300 mm, >1.180 mm	Co(II), Cr(III), Ni(II)	dosage, pH, stirring time, and initial concentration	[61]
polyacrylonitrile-2-amino-2-thiazole moiety resin	—	precious metal ions	temperature, pH	[62]
amines and sulfur chelating resin	—	Zn (II), Cd (II), Hg (II)	pH	[63]
polyacrylonitrile-amino thiourea resin	25.1 nm	Rh(III), Ru(IV), Pd(II), Ir(IV)	pH, functional groups	[64]
methyl glycidyl ester of acrylic resin	0.07, 0.15 and 0.06 μm	Cu(II), Pb(II)	pH	[65]
cationic polystyrene balls	—	paper anion pollutants in water	temperature, distributed control system	[68]
crosslinked polystyrene diethanolamine load DiDai type of dendritic polymers	—	metal ions	temperature, content of functional groups	[69]
grafted polytetrafluoroethylene fibers	—	Cu(II)	temperature	[70]
coal and polyvinyl chloride (PVC) scrap	150–200 μm	Hg	temperature	[78]
porous cellulose modified polyethylene imine carrier	—	Hg	the adsorption selectivity	[80]
IRN-77 cation exchange resin	<0.300 mm, >1.180 mm	Co(II), Ni(II), Cr(III)	adsorbent dosage	[89]
3-aminopyridine hydrophilic spacer chelating resin	—	Hg(II), Ag(I), Fe(III), Pb(II), Co(II), Cu(II), Ni(II), Cd(II)	the adsorption capacity	[91]
IRC748 and NDC702	36.85 and 34.53 nm	Cu(II), Pb(II), Cd(II)	pH	[100]
new IDA-chelating resin	—	Cu(II), Pb(II), Cd(II)	pH	[101]
D152 resin	10.3 nm	Cd(II)	pH, temperature	[102]
polyacrylic acid-PVC composite adsorbent	—	cadmium pollution of wastewater	pH	[103]
PS	—	Cr(VI)	aging	[104]
doped polyaniline	—	anionic dye	doping	[105]
"X" shape of the cavity 2 d coordination polymer	—	oxygen anion pollutants	the adsorption selectivity	[106]
phytic acid doped polyaniline nanofibers	80–100 nm	water-borne Cu(II)	pH	[107]
new type of sulfur-containing polyamine chelating resin	—	precious metal	The solvent, temperature, and time	[108]
electrospinning fiber membrane	—	heavy metal	porosity, specific surface area	[109]
amination polyacrylonitrile fiber	—	Pb, Cu	pH	[110]

Table 3. Adsorption of mixed pollutants.

Adsorbent	Surface Area	Pollutants	Influencing Factors	Reference
porous super hydrophobic foam plastic	—	oily wastewater	a small amount of span 80 and silica particles	[83]
carbonized polypropylene	—	oil	NiO catalyst diameter	[111]
weak base anion exchange resin	—	benzene sulfonate	pH	[112]
polystyrene matrix	—	protein fiber connection	the concentration of	[113]
polystyrene	—	thrombin	cone methyl sulfonate and sulfanilamide essence	[114]
$La(OH)_3$@SA/PAM	—	methylene blue, crystal violet, and malachite green	ultraviolet light	[115]
metal ions impregnated polystyrene resin	—	antibiotics in water pollutants	pH	[116]
grafted polyethylene imine melamine formaldehyde	—	CO_2	temperature	[117]
polyvinyl chloride (PVC)/polystyrene fiber electrostatic spinning	70–300 μm	oil pollution	porosity	[118]
polyvinyl chloride (PVC) mesoporous membrane	45 nm	methylene blue	porosity	[119]
2-amino modified Chloromethylated polystyrene and GQ-08 resin	9.93 and 8.99 nm	glyphosate	pH	[120]
PTFE membrane	0.1 μm	crude oil	crude oil initial concentration, contact time, pH, ionic strength, temperature	[121]
polymer nanocomposites	—	harmful pollutants in the water or wastewater	pH	[122]
low poly beta cyclodextrin coupling polystyrene	—	puerarin	solubility	[123]

6. Adsorption Mechanism

Interlaced and complicated factors formed the adsorption mechanism of microplastics on pollutants. Figure 5 briefly depicts the mechanism of the pollutant adsorption process by microplastics. Based on the literature on the adsorption of plastics on various pollutants, the adsorption mechanism of plastics on pollutants include many aspects, and these are briefly summarized in Tables S3–S5: (i) The adsorption mechanism of plastic on organic pollutants as shown in Table S3; (ii) The adsorption mechanism of plastics on inorganic pollutants as shown in Table S4; and (iii) the adsorption mechanism of plastics on other pollutants as shown in Table S5. In addition, the adsorption affinity, hydrogen bonding, π–π interactions [124], and etc., as shown in Figure S2, had different results in different adsorption experiments, as follows.

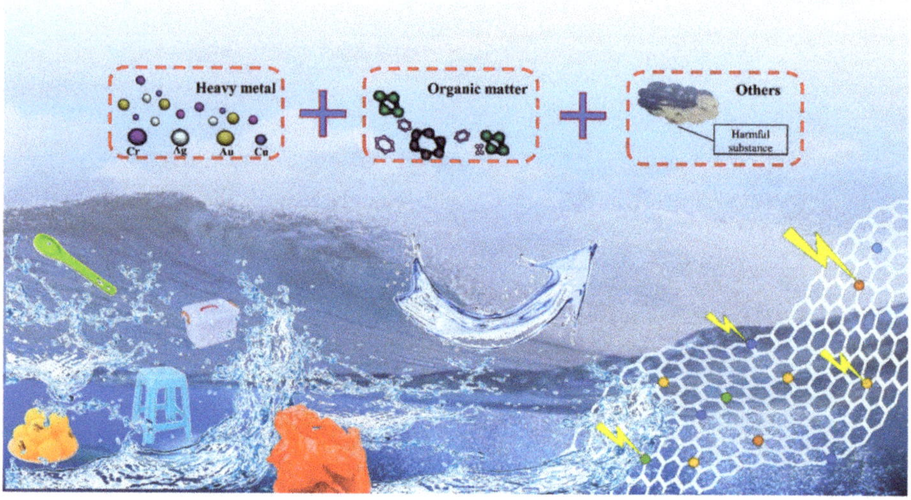

Figure 5. The adsorption machines in the process of adsorbing pollutants with microplastics.

6.1. Adsorption Affinity

The adsorption affinity of resin for pollutants was one of the main mechanisms for studying the properties of microplastics. In the adsorption process of microplastics, they often showed remarkable adsorption capacity for some contaminants. As in the process of chloromethane adsorption of plastic adsorbent, the low solubility of chloromethane was more readily adsorbed [85]. However, with the adsorption of styrene-DVB resin for homocarboxylic acids, carboxylic acids with high molecular weights were more easily adsorbed. The adsorption capacity of mercury was always higher than that of lead in the adsorption capacity of poly(ethylamine) to lead and mercury [57].

In the adsorption of mercury by a polyethyleneimine-modified porous cellulose carrier, the mercury amine complex was formed by a polyethyleneimine-modified porous cellulose carrier; thus, mercury had high stability, which made microplastics have a high mercury adsorption efficiency [80]. Additionally, the folate-polyaniline hybrid hydrogel can selectively adsorb anionic dyes and has a higher affinity for anionic pollutants, such as chromium (VI), eosin yellow, rose red, and methyl orange [54]. Likewise, polyacrylonitrile-2-aminothiazole resin also had a high adsorption efficiency for mercury-, cadmium-, lead-, copper-, zinc-, and nickel-polluted wastewater [59].

Under the condition of the coexistence of Hg(II), Ag(I), Fe(III), Pb(II), Co(II), Cu(II), Ni(II), Cd(II), and Zn(II) ions, 3-aminopyrine hydrophilic spacer arm chelating resin showed strong adsorption performance for Hg(II) [91]. Interestingly, polyvinylchloride microplastics had a higher affinity for TCEP in the adsorption of trin-butyl phosphate and tris (2-chloroethyl) phosphate [94]. When microplastics (MPs) were used to remove fungicides, the adsorption capacity of pyrrole cyclosporin was the highest, followed by microoxysporin and dioxysporin [82]. In general, polyethylene exhibited greater adsorption capacity than other types of plastics [125].

The role of functional groups, such as the adsorption effect of terlomycin on MPs, was in the order of PE (polyethylene) < PP (polypropylene) < PS (polystyrene) < PVC (polyvinyl chloride). In addition to its strength, the adsorption affinity sometimes needed the help of other substances. For example, silica played a vital role in treating oil-containing wastewater by porous waste plastic superhydrophobic foams, and silica had a strong affinity for both porous waste plastic superhydrophobic foams and oil-containing wastewater [125].

In the adsorption of iminodiacetic acid chelate resin for trivalent metal ions, Sc(III) > Ga(III) ≈ In(III) ≈ Fe(III) > Y(III) > La(III) > Al(III); however, in aqueous solution, Ga(III)

> Fe(III) > In(III) > Sc(III) > Al(III) > Y(III) > La(III) [55]. In the pH range of 2–8, metal ion-impregnated polystyrene resins had a strong adsorption capacity for antibiotic contaminants in water [116]. The adsorption of Cd and Ni ions on polystyrene nanofibers in an aqueous solution can achieve the best effect [58]. Adsorption was the main mechanism of oil pollution treatment for the porous structure of electrospun polyvinyl chloride/polystyrene fiber [118].

In addition, there was adsorption selectivity, as low concentrations of polystyrene enhance the migration of nonpolar (pyrene) and weakly polar (2,2,4,4-tetrabromodiphenyl ether) compounds but had little effect on the migration of three polar compounds (Bisphenol A, bisphenol F, and 4-nonylphenol) [95]. Moreover, when Co(II), Ni(II), and Cr(III) ions coexist in a solution, Cr(III) replaced them even if Co(II) and Ni(II) were adsorbed on the resin [89].

6.2. Chemical Bonds

In the process of resin treatment of microplastics, the role of various chemical bonds cannot be ignored, such as hydrogen bonds, van der Waals forces, π–π interactions, phosphorus–phosphorus accumulation interactions, carbon–chlorine bonds, etc.

Hydrogen bonds were an important factor affecting the mechanism of resin adsorption. For example, the hydrogen bond formed between the electron pair on nitrogen and the hydroxyl hydrogen of molecular phenol, the hydrogen bond connected to the nitrogen of amine, and the hydrogen bond formed between the hydrogen of the hydroxyl group and the free amine group played an essential role in the adsorption performance of microplastics. The adsorption of 5-sodium disulfoisphthalic acid by acrylate polymer YWB-7 resin was a similar principle [88].

Consequently, benzene rings and other functional groups inevitably appear in resins. Under the influence of the benzene ring and functional group, polystyrene had higher adsorption performance for echinophilin [82]. In the process of the adsorption test of polyethylene microplastics to organic luminescent materials, the characteristic band of the infrared spectrum showed no change [94]. Due to these benzene ring functional groups, resin adsorption containing pollutants will produce π–π interactions. For instance, polystyrene was adsorbed to echinophorin [82]. The adsorption of phenanthrene on plastic fibers (some plastics and phenanthrene contain benzene rings) [81], plastics and microplastics, phenylene, nitrobenzene, naphthalene, and other substances played a key role in the degradation of pollutants due to the presence of benzene ring functional groups.

Moreover, when coal and waste polyvinyl chloride adsorbed mercury, the carbon–chlorine bond can be changed into chloride ions, thus, realizing the chemisorption of mercury [78]. Phosphorus–phosphorus accumulation interactions played an important role in the interaction between adsorbents and trinitrotoluene [87]. In addition, the adsorption of Cd(II) by D152 resin was also investigated. The removal efficiency of Cu(II) was significantly improved by modifying PTFE fibers by radiation grafting to make them contain C=O and -OH functional groups [70].

6.3. Ion Exchange

As the experiment continued, the pH remained the same; however, the pollutant was gradually absorbed, indicating that the pollutant was removed by ion exchange. As with anion exchange resins that absorb phenol [76], ion exchange between carboxyl groups of proteins and functional groups of resins in the form of acids; phenol formaldehyde anion exchange resin can not only exchange ions in the form of acid but also absorb protons in the form of the free base under the action of amine functional group of benzene sulfonate (BS) removal [112].

In the adsorption of phenol by ion exchange resin, alkaline conditions were the most favorable [53]. Ion exchange of charged carboxyl groups of polyacrylic acid with cadmium(II) solution [103]. The iminodiacetic acid chelating resin had an ion exchange effect in the adsorption of Cu(II), Pb(II), and Cd(II) ions [101].

6.4. Hydrophobic Interaction

As hydrophobic organic matter, microplastics had abnormal adsorption capacity for hydrophobic organic matter [126], hydrophobic substances are more easily removed from aqueous solutions than hydrophilic substances [85]. For example, polythioctanoid-based coagulants preferred compounds with positive octanol-water partitioning coefficients [55]. The adsorption capacity of ectocystine on polyethylene was in the order of pyrazolesterim, dixoxystim, and pyrazolesterim. Between the functionalized polystyrene nanospheres and 2,4,6-trinitrotoluene, the adsorption capacity of polystyrene and KH570-polystyrene nanospheres for trinitrotoluene was the highest [87].

In the adsorption experiment of chlorinated methane by plastic, GAC > PLJ-PVC > NEO > PE [85]. Perfluorinated octane sulfonic acid (FOSA) was more easily adsorbed on polyethylene [52]. In the adsorption of three triazole fungicides by polystyrene microplastics, the resin with strong hydrophobicity had more substantial adsorption power [50]. Benzene derivatives were more hydrophobic than benzene monomers, therefore they were polyethylene rather than other polymer resins with higher adsorption properties [49].

In the adsorption of porous polycalixarene to organic micropollutants, CalP4 showed ultrafine adsorption [79]. After the adsorption of phenanthrene and its mono hydroxyl derivatives, the bands of PVC microplastics did not change significantly [90]. The adsorption efficiency of 5-sodium disulfoisylphthalic acid by acrylate polymer YWB-7 resin was higher in an aqueous solution without methanol [88].

6.5. Role of the Environment

In addition to the adsorption of pollutants by microplastics themselves, aging and other factors can enhance the adsorption performance of microplastics [127]. The adsorption effect of Cu(II) on metal ions was the best in the experiments of cross-linked polystyrene-supported low-generation diethanolamine dendrimers [69]. IDA-chelating resin had the best adsorption effect on Cu(II) and the worst adsorption effect on Cd(II) [100]. There were many kinds of microplastics, such as polyamide, rubber, and glass, and similar to the adsorption of metal pollutants, the adsorption effect was the best for polyamide and the worst for glass [30].

In the adsorption of Co(II), Cr(III), and Ni(II) by IRN77 cation exchange resin, Cr(III) was mainly adsorbed. Even though IRN77 cation exchange resin will adsorb Co(II) and Ni(II) at the beginning, it will desorb after some time, and the adsorption site is replaced by Cr(III) [61]. Among the adsorptions of rhodium(III), ruthenium(IV), iridium(IV), and palladium(II), the adsorption of polyacrylonitrile-ATAL was the highest for rhodium(III) [64]. Polyacrylonitrile-thiosemicarbazone resin had excellent adsorption selectivity for Rh(III), Ru(IV), Pd(II), and Ir(IV) [64]. In the adsorption of Cu(II) and Pb(II) by amine chelating resin, the absorption effect of copper was higher than that of lead under any conditions [63].

6.6. Electrostatic Interaction

Electrostatic interactions were due to the process of plastic adsorption, adsorption, and pollutants due to attraction or repelling of each other with charge. Glass-reinforced plastics, for example, were positively charged in acidic conditions and had a strong electrostatic attraction to acrylic polyamphoteric electrolytes [77]. At low pH, primary amines, secondary amines, and imines in folate-polyaniline hybrid hydrogels had groups containing holes, which can be protonated to form positively charged cations (e.g., $-NH_3^+$, $-NH_2^+$, $-NH^+$, etc.), these cationic groups had an electrostatic attraction with Cr(VI), Hg(II), and Pb(II), thus, improving the removal efficiency [54].

The positively charged cetyltrimethylammonium bromide can effectively adsorb anions in the presence of surfactants [84]. The adsorption of perfluorooctanesulfonamides by polyethylene microplastics was not affected by ionic strength; however, the influence of ionic strength was evident after ionization of pollutants, which was the effect of electrostatic force. Due to the pH value, the electrostatic interaction with pollutants affected the adsorption of three triazole fungicides by polystyrene microplastics [50].

6.7. Other Adsorption Mechanisms

In addition to the above common mechanisms of pollutant adsorption of several microplastics, there were also carboxyl functional groups in acrylic grafted PTFE fibers, which can bind with Er(III). Metal ions Er(III) have empty orbitals, and carboxyl groups provide their electron pairs, which belong to complexation. Under the action of surface-initiated free radicals, cationic substances can be connected on the surface of polystyrene spheres to obtain cationic microplastics and can then act as high-efficiency adsorption anionic pollutants in white papermaking water, which was grafted [68]. The properties of chelating groups and metal-ligand CO stability, such as amine and sulfur chelating resin for Cu(II) and Pb(II) adsorption [63]. The chelating mechanism of polyethylamine with lead and mercury are all examples of this [57].

7. Conclusions

7.1. Summary

The reasons and mechanisms of plastic adsorption for organic, inorganic, and mixed pollutants were discussed. There are many reasons affecting sewage treatment in using plastic. The influence of plastics includes many factors. As far as the most typical pH was concerned, it was divided into three aspects: (i) The existence of electrostatic repulsion at low pH reduced the adsorption efficiency. (ii) The adsorption efficiency was the highest when the pH was between 3 and 6. (iii) Precipitation at a high pH value. However, there were certain exceptions as mentioned earlier. In addition, the actions of the environment, illumination, weathering, crushing, etc. further increase the contact area between microplastics and pollutants.

The mechanism of sewage treatment by using microplastics was also complicated and interrelated. We found that the main mechanism of sewage treatment by using microplastics was mainly divided into: (i) Plastics naturally had other properties, such as showing a cationic effect, high-efficiency adsorption, and a porous structure, and they showed a unique and strong affinity for pollutants; (ii) microplastics with carboxylic acid, amino groups, and other functional groups had hydrogen bonds and electrostatic attraction with pollutants; and (iii) the pore-carbon structure benzene ring of microplastics could produce π–π interactions or phosphorus–phosphorus stacking interactions.

At present, microplastics have great potential in sewage treatment. In addition to the adsorption capacity (hydrophobic), adsorption efficiency, and other advantages, microplastics in sewage treatment have reusable benefits in comparison to other adsorbents. The excellent performance of microplastics in sewage treatment is still being investigated. In addition to sewage treatment, microplastics have been applied in industry. Microplastics will be even more valuable in the future.

7.2. Prospect

Adsorption technology is gradually applied to all aspects of environmental treatment, and it occupies a dominant position in sewage treatment. In this paper, the adsorption mechanism and influencing factors of different adsorbents for different pollutants were summarized as well as the activities of microplastics in soil and water environments. These have reference value for the study of microplastic adsorption. Although the influence of the water environment, soil fractionation, and speciation on the microplastic adsorption process are not completely clear, we hope that large-scale, sustainable, and low-cost adsorption technologies will be developed.

Supplementary Materials: The following supporting information can be downloaded at: https://www.mdpi.com/article/10.3390/nano12132256/s1, Figure S1: The process of forming large plastics into microplastics, Figure S2: Schematic representation of the six sorption mechanisms, Table S1: Classification of microplastics, Table S2: Surface area parameters of different adsorbents, Table S3: Adsorption mechanism of organic pollutants, Table S4: Adsorption mechanism of inorganic pollutants, and Table S5: Adsorption mechanism of other pollutants [128–134].

Author Contributions: M.Z.: Manuscript draft, drawing pictures, and modify paper. L.H.: modify paper and framework proposed. S.R.B.A.: Embellish the article. J.Y.: Investigation and writing—review and editing. L.W.: Supervision and writing—review and editing. T.W.: Supervision and writing—review and editing. H.Z.: Guidance, editing review, and funding acquisition. T.X.: Supervision and writing—review and editing. All authors have read and agreed to the published version of the manuscript.

Funding: The work was supported by the National Natural Science Foundation of China (nos. 51208122, 51778156, 51708142), Pearl River S&T Nova Program of Guangzhou (201806010191), Guangzhou Municipal Science and Technology Project (no. 201707010256), Talent Cultivation Program of Guangzhou University (RP2021014, YJ2021005), Guangzhou University-Hong Kong University of Science and Technology Joint Research Cooperation Fund (YH202102), and Youth Innovative Talents Project of Guangdong Province (2021KQNCX062).

Institutional Review Board Statement: Not applicable.

Informed Consent Statement: Not applicable.

Data Availability Statement: Not applicable.

Conflicts of Interest: The authors declare that they have no competing interest.

References

1. Liu, S.; Huang, J.; Zhang, W.; Shi, L.; Yi, K.; Yu, H.; Zhang, C.; Li, S.; Li, J. Microplastics as a vehicle of heavy metals in aquatic environments: A review of adsorption factors, mechanisms, and biological effects. *J. Environ. Manag.* **2022**, *302*, 113995. [CrossRef] [PubMed]
2. Costigan, E.; Collins, A.; Hatinoglu, M.D.; Bhagat, K.; Macrae, J.; Perreault, F.; Apul, O. Adsorption of Organic Pollutants by Microplastics: Overview of A Dissonant Literature. *J. Hazard. Mater. Adv.* **2022**, *6*, 100091. [CrossRef]
3. Cole, M.; Lindeque, P.; Halsband, C.; Galloway, T.S. Microplastics as contaminants in the marine environment: A review. *Mar. Pollut. Bull.* **2011**, *62*, 2588–2597. [CrossRef]
4. Kilavuz, S.; Kor, D.; Bulut, F.; Serbes, M.; Karagoz, D.; Altıntaş, D.; Bişgin, A.; Şeydaoğlu, G.; Mungan, H.Ö. Real-world patient data on immunity and COVID-19 status of patients with MPS, Gaucher, and Pompe diseases from Turkey. *Arch. Pediatr.* 2022, *In press*. [CrossRef]
5. Frias, J.; Sobral, P.; Ferreira, A.M. Organic pollutants in microplastics from two beaches of the Portuguese coast. *Mar. Pollut. Bull.* **2010**, *60*, 1988–1992. [CrossRef] [PubMed]
6. Prata, J.C. Microplastics in wastewater: State of the knowledge on sources, fate and solutions. *Mar. Pollut. Bull.* **2018**, *129*, 262–265. [CrossRef]
7. Migwi, F.K.; Ogunah, J.A.; Kiratu, J.M. Occurrence and spatial distribution of microplastics in the surface waters of Lake Naivasha, Kenya. *Environ. Toxicol. Chem.* **2020**, *39*, 765–774. [CrossRef]
8. Ma, J.; Chen, F.; Zhu, Y.; Li, X.; Yu, H.; Sun, Y. Joint effects of microplastics and ciprofloxacin on their toxicity and fates in wheat: A hydroponic study. *Chemosphere* **2022**, 135023. [CrossRef]
9. Yang, H.; He, Y.; Yan, Y.; Junaid, M.; Wang, J. Characteristics, Toxic Effects, and Analytical Methods of Microplastics in the Atmosphere. *Nanomaterials* **2021**, *11*, 2747. [CrossRef]
10. Lai, H.; Liu, X.; Qu, M. Nanoplastics and Human Health: Hazard Identification and Biointerface. *Nanomaterials* **2022**, *12*, 1298. [CrossRef]
11. Oh, S.; Hur, H.; Kim, Y.; Shin, S.; Woo, H.; Choi, J.; Lee, H.H. Peptide Specific Nanoplastic Detection Based on Sandwich Typed Localized Surface Plasmon Resonance. *Nanomaterials* **2021**, *11*, 2887. [CrossRef] [PubMed]
12. Li, X.; Peng, Y.; Deng, Y.; Ye, F.; Zhang, C.; Hu, X.; Liu, Y.; Zhang, D. Recycling and Reutilizing Polymer Waste via Electrospun Micro/Nanofibers: A Review. *Nanomaterials* **2022**, *12*, 1663. [CrossRef] [PubMed]
13. Lionetto, F.; Lionetto, M.G.; Mele, C.; Corcione, C.E.; Bagheri, S.; Udayan, G.; Maffezzoli, A. Autofluorescence of Model Polyethylene Terephthalate Nanoplastics for Cell Interaction Studies. *Nanomaterials* **2022**, *12*, 1560. [CrossRef]
14. Abramova, A.V.; Abramov, V.O.; Fedulov, I.S.; Baranchikov, A.E.; Kozlov, D.A.; Veselova, V.O.; Kameneva, S.V.; Ivanov, V.K.; Cravotto, G. Strong Antibacterial Properties of Cotton Fabrics Coated with Ceria Nanoparticles under High-Power Ultrasound. *Nanomaterials* **2021**, *11*, 2704. [CrossRef] [PubMed]
15. Camacho, M.; Herrera, A.; Gómez, M.; Acosta-Dacal, A.; Martínez, I.; Henríquez-Hernández, L.A.; Luzardo, O.P. Organic pollutants in marine plastic debris from Canary Islands beaches. *Sci. Total Environ.* **2019**, *662*, 22–31. [CrossRef] [PubMed]
16. Lionetto, F.; Esposito Corcione, C. An overview of the sorption studies of contaminants on poly (Ethylene Terephthalate) microplastics in the marine environment. *J. Mar. Sci. Eng.* **2021**, *9*, 445. [CrossRef]
17. Sun, J.; Dai, X.; Wang, Q.; van Loosdrecht, M.C.; Ni, B.-J. Microplastics in wastewater treatment plants: Detection, occurrence and removal. *Water Res.* **2019**, *152*, 21–37. [CrossRef]

18. Rodrigues, J.P.; Duarte, A.C.; Santos-Echeandía, J.; Rocha-Santos, T. Significance of interactions between microplastics and POPs in the marine environment: A critical overview. *TrAC Trends Anal. Chem.* **2019**, *111*, 252–260. [CrossRef]
19. Fred-Ahmadu, O.H.; Bhagwat, G.; Oluyoye, I.; Benson, N.U.; Ayejuyo, O.O.; Palanisami, T. Interaction of chemical contaminants with microplastics: Principles and perspectives. *Sci. Total Environ.* **2020**, *706*, 135978. [CrossRef]
20. Vockenberg, T.; Wichard, T.; Ueberschaar, N.; Franke, M.; Stelter, M.; Braeutigam, P. The sorption behaviour of amine micropollutants on polyethylene microplastics–impact of aging and interactions with green seaweed. *Environ. Sci. Proc. Impacts* **2020**, *22*, 1678–1687. [CrossRef]
21. Guo, X.; Pang, J.; Chen, S.; Jia, H. Sorption properties of tylosin on four different microplastics. *Chemosphere* **2018**, *209*, 240–245. [CrossRef] [PubMed]
22. Yin, J.; Huang, G.; Li, M.; An, C. Will the chemical contaminants in agricultural soil affect the ecotoxicity of microplastics? *ACS Agr. Sci. Tech.* **2021**, *1*, 3–4. [CrossRef]
23. Azeem, I.; Adeel, M.; Ahmad, M.A.; Shakoor, N.; Jiangcuo, G.D.; Azeem, K.; Ishfaq, M.; Shakoor, A.; Ayaz, M.; Xu, M.; et al. Uptake and Accumulation of Nano/Microplastics in Plants: A Critical Review. *Nanomaterials* **2021**, *11*, 2935. [CrossRef] [PubMed]
24. Chen, Y.; Wang, X.; Wang, X.; Cheng, T.; Fu, K.; Qin, Z.; Feng, K. Biofilm Structural and Functional Features on Microplastic Surfaces in Greenhouse Agricultural Soil. *Sustainability* **2022**, *14*, 7024. [CrossRef]
25. Liu, K.; Wu, T.; Wang, X.; Song, Z.; Zong, C.; Wei, N.; Li, D. Consistent transport of terrestrial microplastics to the ocean through atmosphere. *Environ. Sci. Technol.* **2019**, *53*, 10612–10619. [CrossRef]
26. Mai, L.; Sun, X.-F.; Xia, L.-L.; Bao, L.-J.; Liu, L.-Y.; Zeng, E.Y. Global riverine plastic outflows. *Environ. Sci. Technol.* **2020**, *54*, 10049–10056. [CrossRef]
27. Hamidian, A.H.; Ozumchelouei, E.J.; Feizi, F.; Wu, C.; Zhang, Y.; Yang, M. A review on the characteristics of microplastics in wastewater treatment plants: A source for toxic chemicals. *J. Clean. Prod.* **2021**, *295*, 126480. [CrossRef]
28. Murphy, F.; Ewins, C.; Carbonnier, F.; Quinn, B. Wastewater treatment works (WwTW) as a source of microplastics in the aquatic environment. *Environ. Sci. Technol.* **2016**, *50*, 5800–5808. [CrossRef]
29. Wang, F.; Wang, B.; Duan, L.; Zhang, Y.; Zhou, Y.; Sui, Q.; Xu, D.; Qu, H.; Yu, G. Occurrence and distribution of microplastics in domestic, industrial, agricultural and aquacultural wastewater sources: A case study in Changzhou, China. *Water Res.* **2020**, *182*, 115956. [CrossRef]
30. Chen, C.; Chen, L.; Yao, Y.; Artigas, F.; Huang, Q.; Zhang, W. Organotin release from polyvinyl chloride microplastics and concurrent photodegradation in water: Impacts from salinity, dissolved organic matter, and light exposure. *Environ. Sci. Technol.* **2019**, *53*, 10741–10752. [CrossRef]
31. Du, S.; Zhu, R.; Cai, Y.; Xu, N.; Yap, P.-S.; Zhang, Y.; He, Y.; Zhang, Y. Environmental fate and impacts of microplastics in aquatic ecosystems: A review. *RSC Adv.* **2021**, *11*, 15762–15784. [CrossRef] [PubMed]
32. Qiu, X.; Saovany, S.; Takai, Y.; Akasaka, A.; Inoue, Y.; Yakata, N.; Liu, Y.; Waseda, M.; Shimasaki, Y.; Oshima, Y. Quantifying the vector effects of polyethylene microplastics on the accumulation of anthracene to Japanese medaka (Oryzias latipes). *Aquat. Toxicol.* **2020**, *228*, 105643. [CrossRef] [PubMed]
33. Brennecke, D.; Duarte, B.; Paiva, F.; Caçador, I.; Canning-Clode, J. Microplastics as vector for heavy metal contamination from the marine environment. *Estuar. Coast. Shelf Sci.* **2016**, *178*, 189–195. [CrossRef]
34. Bakir, A.; Rowland, S.J.; Thompson, R.C. Competitive sorption of persistent organic pollutants onto microplastics in the marine environment. *Mar. Pollut. Bull.* **2012**, *64*, 2782–2789. [CrossRef] [PubMed]
35. Qu, H.; Ma, R.; Wang, B.; Zhang, Y.; Yin, L.; Yu, G.; Deng, S.; Huang, J.; Wang, Y. Effects of microplastics on the uptake, distribution and biotransformation of chiral antidepressant venlafaxine in aquatic ecosystem. *J. Hazard. Mater.* **2018**, *359*, 104–112. [CrossRef]
36. Zhang, H.; Pap, S.; Taggart, M.A.; Boyd, K.G.; James, N.A.; Gibb, S.W. A review of the potential utilisation of plastic waste as adsorbent for removal of hazardous priority contaminants from aqueous environments. *Environ. Pollut.* **2020**, *258*, 113698. [CrossRef] [PubMed]
37. Tang, Y.; Liu, Y.; Chen, Y.; Zhang, W.; Zhao, J.; He, S.; Yang, C.; Zhang, T.; Tang, C.; Zhang, C. A review: Research progress on microplastic pollutants in aquatic environments. *Sci. Total Environ.* **2021**, *766*, 142572. [CrossRef]
38. Hou, L.; Kumar, D.; Yoo, C.G.; Gitsov, I.; Majumder, E.L.-W. Conversion and removal strategies for microplastics in wastewater treatment plants and landfills. *Chem. Eng. J.* **2021**, *406*, 126715. [CrossRef]
39. Yang, Z.; Li, S.; Ma, S.; Liu, P.; Peng, D.; Ouyang, Z.; Guo, X. Characteristics and removal efficiency of microplastics in sewage treatment plant of Xi'an City, northwest China. *Sci. Total Environ.* **2021**, *771*, 145377. [CrossRef]
40. Wang, F.; Wong, C.S.; Chen, D.; Lu, X.; Wang, F.; Zeng, E.Y. Interaction of toxic chemicals with microplastics: A critical review. *Water Res.* **2018**, *139*, 208–219. [CrossRef]
41. Burns, E.E.; Boxall, A.B. Microplastics in the aquatic environment: Evidence for or against adverse impacts and major knowledge gaps. *Environ. Toxicol. Chem.* **2018**, *37*, 2776–2796. [CrossRef] [PubMed]
42. McCormick, A.R.; Hoellein, T.J.; London, M.G.; Hittie, J.; Scott, J.W.; Kelly, J.J. Microplastic in surface waters of urban rivers: Concentration, sources, and associated bacterial assemblages. *Ecosphere* **2016**, *7*, e01556. [CrossRef]
43. Godoy, V.; Martín-Lara, M.; Calero, M.; Blázquez, G. The relevance of interaction of chemicals/pollutants and microplastic samples as route for transporting contaminants. *Process Saf. Environ. Prot.* **2020**, *138*, 312–323. [CrossRef]

44. Ricardo, I.A.; Alberto, E.A.; Júnior, A.H.S.; Macuvele, D.L.P.; Padoin, N.; Soares, C.; Riella, H.G.; Starling, M.C.V.; Trovo, A.G. A critical review on microplastics, interaction with organic and inorganic pollutants, impacts and effectiveness of advanced oxidation processes applied for their removal from aqueous matrices. *Chem. Eng. J.* **2021**, *424*, 130282. [CrossRef]
45. Gao, X.; Hassan, I.; Peng, Y.; Huo, S.; Ling, L. Behaviors and influencing factors of the heavy metals adsorption onto microplastics: A review. *J. Clean. Prod.* **2021**, *319*, 128777. [CrossRef]
46. Ho, W.-K.; Leung, K.S.-Y. The crucial role of heavy metals on the interaction of engineered nanoparticles with polystyrene microplastics. *Water Res.* **2021**, *201*, 117317. [CrossRef]
47. Binda, G.; Spanu, D.; Monticelli, D.; Pozzi, A.; Bellasi, A.; Bettinetti, R.; Carnati, S.; Nizzetto, L. Unfolding the interaction between microplastics and (trace) elements in water: A critical review. *Water Res.* **2021**, *204*, 117637. [CrossRef]
48. Li, X.; Jiang, X.; Song, Y.; Chang, S.X. Coexistence of polyethylene microplastics and biochar increases ammonium sorption in an aqueous solution. *J. Hazard. Mater.* **2021**, *405*, 124260. [CrossRef]
49. Yu, F.; Yang, C.; Zhu, Z.; Bai, X.; Ma, J. Adsorption behavior of organic pollutants and metals on micro/nanoplastics in the aquatic environment. *Sci. Total Environ.* **2019**, *694*, 133643. [CrossRef]
50. Fang, S.; Yu, W.; Li, C.; Liu, Y.; Qiu, J.; Kong, F. Adsorption behavior of three triazole fungicides on polystyrene microplastics. *Sci. Total Environ.* **2019**, *691*, 1119–1126. [CrossRef]
51. Saski, W. Adsorption of sorbic acid by plastic cellulose acetates. *J. Pharm. Sci.* **1963**, *52*, 264–268. [CrossRef] [PubMed]
52. Wang, F.; Shih, K.M.; Li, X.Y. The partition behavior of perfluorooctanesulfonate (PFOS) and perfluorooctanesulfonamide (FOSA) on microplastics. *Chemosphere* **2015**, *119*, 841–847. [CrossRef] [PubMed]
53. Caetano, M.; Valderrama, C.; Farran, A.; Cortina, J.L. Phenol removal from aqueous solution by adsorption and ion exchange mechanisms onto polymeric resins. *J. Colloid Interface Sci.* **2009**, *338*, 402–409. [CrossRef] [PubMed]
54. Das, S.; Chakraborty, P.; Ghosh, R.; Paul, S.; Mondal, S.; Panja, A.; Nandi, A.K. Folic acid-polyaniline hybrid hydrogel for adsorption/reduction of chromium(VI) and selective adsorption of anionic dye from water. *ACS Sustain. Chem. Eng.* **2017**, *5*, 9325–9337. [CrossRef]
55. Yuchi, A.; Sato, T.; Morimoto, Y.; Mizuno, H.; Wada, H. Adsorption mechanism of trivalent metal ions on chelating resins containing iminodiacetic acid groups with reference to selectivity. *Anal. Chem.* **1997**, *69*, 2941–2944. [CrossRef]
56. He, Z.; Song, H.; Cui, Y.; Zhu, W.; Du, K.; Yao, S. Porous spherical cellulose carrier modified with polyethyleneimine and its adsorption for Cr(III) and Fe(III) from aqueous solutions. *Chin. J. Chem. Eng.* **2014**, *22*, 984–990. [CrossRef]
57. Delacour, M.L.; Gailliez, E.; Bacquet, M.; Morcellet, M. Poly (ethylenimine) coated onto silica gels: Adsorption capacity toward lead and mercury. *J. Appl. Polym. Sci.* **1999**, *73*, 899–906. [CrossRef]
58. Bahramzadeh, A.; Zahedi, P.; Abdouss, M. Acrylamide-plasma treated electrospun polystyrene nanofibrous adsorbents for cadmium and nickel ions removal from aqueous solutions. *J. Appl. Polym. Sci.* **2016**, *133*, 42944. [CrossRef]
59. Xiong, C.; Jia, Q.; Chen, X.; Wang, G.; Yao, C. Optimization of polyacrylonitrile-2-aminothiazole resin synthesis, characterization, and its adsorption performance and mechanism for removal of Hg(II) from aqueous solutions. *Ind. Eng. Chem. Res.* **2013**, *52*, 4978–4986. [CrossRef]
60. Xiong, C.; Yao, C. Preparation and application of acrylic acid grafted polytetrafluoroethylene fiber as a weak acid cation exchanger for adsorption of Er(III). *J. Hazard. Mater.* **2009**, *170*, 1125–1132. [CrossRef]
61. Rengaraj, S.; Yeon, K.-H.; Kang, S.-Y.; Lee, J.-U.; Kim, K.-W.; Moon, S.-H. Studies on adsorptive removal of Co(II), Cr(III) and Ni(II) by IRN77 cation-exchange resin. *J. Hazard. Mater.* **2002**, *92*, 185–198. [CrossRef]
62. Chen, Y.; Zhao, Y. Synthesis and characterization of polyacrylonitrile-2-amino-2-thiazoline resin and its sorption behaviors for noble metal ions. *React. Funct. Polym.* **2003**, *55*, 89–98. [CrossRef]
63. Atia, A.A.; Donia, A.M.; Yousif, A.M. Synthesis of amine and thio chelating resins and study of their interaction with zinc(II), cadmium(II) and mercury(II) ions in their aqueous solutions. *React. Funct. Polym.* **2003**, *56*, 75–82. [CrossRef]
64. Chen, Y.-Y.; Liang, C.; Chao, Y. Synthesis and characterization of polyacrylonitrile-thiosemicarbazide resin and its sorption behavior for Rh(III) Ru(IV) Pd(II) and Ir(IV) ions. *React. Funct. Polym.* **1998**, *36*, 51–58. [CrossRef]
65. Atia, A.A.; Donia, A.M.; Abou-El-Enein, S.A.; Yousif, A.M. Studies on uptake behaviour of copper(II) and lead(II) by amine chelating resins with different textural properties. *Sep. Purif. Technol.* **2003**, *33*, 295–301. [CrossRef]
66. Singh, N.; Khandelwal, N.; Tiwari, E.; Naskar, N.; Lahiri, S.; Lützenkirchen, J.; Darbha, G.K. Interaction of metal oxide nanoparticles with microplastics: Impact of weathering under riverine conditions. *Water Res.* **2021**, *189*, 116622. [CrossRef]
67. Wang, J.; Liu, X.; Liu, G. Sorption behaviors of phenanthrene, nitrobenzene, and naphthalene on mesoplastics and microplastics. *Environ. Sci. Pollut. Res. Int.* **2019**, *26*, 12563–12573. [CrossRef]
68. Xiao, H.; He, B.; Qian, L.; Li, J. Cationic polystyrene spheres for removal of anionic contaminants in white water of papermaking. *J. Appl. Polym. Sci.* **2015**, *132*, 41379. [CrossRef]
69. Sun, C.; Qu, R.; Ji, C.; Wang, C.; Sun, Y.; Yue, Z.; Cheng, G. Preparation and adsorption properties of crosslinked polystyrene-supported low-generation diethanolamine-typed dendrimer for metal ions. *Talanta* **2006**, *70*, 14–19. [CrossRef]
70. Li, J.-H.; Miao, X.-X.; Chen, X.-Y.; Lu, L.; Yang, Y.; Fu, Y.-Q.; Xiong, C.-H. Application and characterization of grafted polytetrafluoroethylene fiber for enhanced adsorption of Cu(II) in aqueous solutions. *J. Cent. South Univ.* **2016**, *23*, 2513–2519. [CrossRef]
71. Dong, Y.; Gao, M.; Song, Z.; Qiu, W. Adsorption mechanism of As(III) on polytetrafluoroethylene particles of different size. *Environ. Pollut.* **2019**, *254*, 112950. [CrossRef] [PubMed]

72. Uwayid, R.; Guyes, E.N.; Shocron, A.N.; Gilron, J.; Elimelech, M.; Suss, M.E. Perfect divalent cation selectivity with capacitive deionization. *Water Res.* **2022**, *210*, 117959. [CrossRef] [PubMed]
73. Zhong, L.-B.; Liu, Q.; Wu, P.; Niu, Q.; Zhang, H.; Zheng, Y. Facile on-site aqueous pollutant monitoring using a flexible, ultra-light and robust SERS substrate: Interface self-assembly of Au@Ag nanocubes on polyvinyl chloride template. *Environ. Sci. Technol.* **2018**, *52*, 5812–5820. Available online: https://cpfd.cnki.com.cn/Article/CPFDTOTAL-ZGHY201806002474.htm (accessed on 16 April 2018). [CrossRef] [PubMed]
74. Liu, N.; Yu, F.; Wang, Y.; Ma, J. Effects of environmental aging on the adsorption behavior of antibiotics from aqueous solutions in microplastic-graphene coexisting systems. *Sci. Total Environ.* **2022**, *806*, 150956. [CrossRef] [PubMed]
75. Duan, J.; Bolan, N.; Li, Y.; Ding, S.; Atugoda, T.; Vithanage, M.; Sarkar, B.; Tsang, D.C.; Kirkham, M. Weathering of microplastics and interaction with other coexisting constituents in terrestrial and aquatic environments. *Water Res.* **2021**, *196*, 117011. [CrossRef]
76. Pollio, F.X.; Kunin, R. Sorption of phenols by anion-exchange resin. *Environ. Sci. Technol.* **1967**, *1*, 160–163. [CrossRef]
77. Terabayashi, T.; Maruyama, T.; Shimizu, Y.; Komatsu, M.; Takahashi, T. Surface modification with acrylic polyampholytes 1, adsorption of acrylic polyampholytes on fiberglass reinforced plastics, and characterization of the polymer adsorbed surfaces. *J. Appl. Polym. Sci.* **2006**, *101*, 4454–4461. [CrossRef]
78. Xu, Y.; Zeng, X.; Zhang, B.; Zhu, X.; Zhou, M.; Zou, R.; Sun, P.; Luo, G.; Yao, H. Experiment and kinetic study of elemental mercury adsorption over a novel chlorinated sorbent derived from coal and waste polyvinyl chloride. *Energy Fuels* **2016**, *30*, 10635–10642. [CrossRef]
79. Shetty, D.; Jahovic, I.; Raya, J.; Asfari, Z.; Olsen, J.-C.; Trabolsi, A. Porous polycalix [4] arenes for fast and efficient removal of organic micropollutants from water. *ACS Appl. Mater. Interfaces* **2018**, *10*, 2976–2981. [CrossRef]
80. Navarro, R.R.; Sumi, K.; Fujii, N.; Matsumura, M. Mercury removal from wastewater using porous cellulose carrier modified with polyethyleneimine. *Water Res.* **1996**, *30*, 2488–2494. [CrossRef]
81. Wang, Z.; Chen, M.; Zhang, L.; Wang, K.; Yu, X.; Zheng, Z.; Zheng, R. Sorption behaviors of phenanthrene on the microplastics identified in a mariculture farm in Xiangshan Bay, southeastern China. *Sci. Total Environ.* **2018**, *628*, 1617–1626. [CrossRef] [PubMed]
82. Hai, N.; Liu, X.; Li, Y.; Kong, F.; Zhang, Y.; Fang, S. Effects of microplastics on the adsorption and bioavailability of three strobilurin fungicides. *ACS Omega* **2020**, *5*, 30679–30686. [CrossRef] [PubMed]
83. Yu, C.; Lin, W.; Jiang, J.; Jing, Z.; Hong, P.; Li, Y. Preparation of a porous superhydrophobic foam from waste plastic and its application for oil spill cleanup. *RSC Adv.* **2019**, *9*, 37759–37767. [CrossRef]
84. Xia, Y.; Zhou, J.-J.; Gong, Y.-Y.; Li, Z.-J.; Zeng, E.Y. Strong influence of surfactants on virgin hydrophobic microplastics adsorbing ionic organic pollutants. *Environ. Pollut.* **2020**, *265*, 115061. [CrossRef]
85. Cook, F.C., Jr.; Hartz, K.E. Adsorption of chlorinated methanes from aqueous solution on selected plastic adsorbents. *J. Am. Water Works Assoc.* **1983**, *75*, 423–426. [CrossRef]
86. Boucher, C.; Morin, M.; Bendell, L. The influence of cosmetic microbeads on the sorptive behavior of cadmium and lead within intertidal sediments: A laboratory study. *Reg. Stud. Mar. Sci.* **2016**, *3*, 1–7. [CrossRef]
87. Zhang, Y.; Wang, X.; Lv, F.; Chu, P.K.; Ye, Z.; Zhou, F.; Zhang, R.; Wei, F. Adsorption behavior and mechanism of 2, 4, 6-trinitrotoluene by functionalized polystyrene nanospheres. *J. Appl. Polym. Sci.* **2013**, *128*, 3720–3725. [CrossRef]
88. Yang, W.; Li, A.; Zhang, Q.; Fei, Z.; Liu, F. Adsorption of 5-sodiosulfoisophthalic acids from aqueous solutions onto acrylic ester polymer YWB-7 resin. *Sep. Purif. Technol.* **2005**, *46*, 161–167. [CrossRef]
89. Kang, S.-Y.; Lee, J.-U.; Moon, S.-H.; Kim, K.-W. Competitive adsorption characteristics of Co^{2+}, Ni^{2+}, and Cr^{3+} by IRN-77 cation exchange resin in synthesized wastewater. *Chemosphere* **2004**, *56*, 141–147. [CrossRef]
90. Bao, Z.-Z.; Chen, Z.-F.; Zhong, Y.; Wang, G.; Qi, Z.; Cai, Z. Adsorption of phenanthrene and its monohydroxy derivatives on polyvinyl chloride microplastics in aqueous solution: Model fitting and mechanism analysis. *Sci. Total Environ.* **2021**, *764*, 142889. [CrossRef]
91. Ji, C.; Song, S.; Wang, C.; Sun, C.; Qu, R.; Wang, C.; Chen, H. Preparation and adsorption properties of chelating resins containing 3-aminopyridine and hydrophilic spacer arm for Hg(II). *Chem. Eng. J.* **2010**, *165*, 573–580. [CrossRef]
92. Xu, J.; Zhang, K.; Wang, L.; Yao, Y.; Sun, H. Strong but reversible sorption on polar microplastics enhanced earthworm bioaccumulation of associated organic compounds. *J. Hazard. Mater.* **2022**, *423*, 127079. [CrossRef] [PubMed]
93. Fu, J.; Li, Y.; Peng, L.; Gao, W.; Wang, G. Distinct Chemical Adsorption Behaviors of Sulfanilamide as a Model Antibiotic onto Weathered Microplastics in Complex Systems. *Colloids Surf. A* **2022**, 129337. [CrossRef]
94. Chen, S.; Tan, Z.; Qi, Y.; Ouyang, C. Sorption of tri-n-butyl phosphate and tris (2-chloroethyl) phosphate on polyethylene and polyvinyl chloride microplastics in seawater. *Mar. Pollut. Bull.* **2019**, *149*, 110490. [CrossRef]
95. Liu, J.; Ma, Y.; Zhu, D.; Xia, T.; Qi, Y.; Yao, Y.; Guo, X.; Ji, R.; Chen, W. Polystyrene nanoplastics-enhanced contaminant transport: Role of irreversible adsorption in glassy polymeric domain. *Environ. Sci. Technol.* **2018**, *52*, 2677–2685. [CrossRef]
96. Bhagwat, G.; Tran, T.K.A.; Lamb, D.; Senathirajah, K.; Grainge, I.; O'Connor, W.; Juhasz, A.; Palanisami, T. Biofilms Enhance the Adsorption of Toxic Contaminants on Plastic Microfibers under Environmentally Relevant Conditions. *Environ. Sci. Technol.* **2021**, *55*, 8877–8887. [CrossRef]
97. Hüffer, T.; Hofmann, T. Sorption of non-polar organic compounds by micro-sized plastic particles in aqueous solution. *Environ. Pollut.* **2016**, *214*, 194–201. [CrossRef] [PubMed]

98. Müller, A.; Becker, R.; Dorgerloh, U.; Simon, F.-G.; Braun, U. The effect of polymer aging on the uptake of fuel aromatics and ethers by microplastics. *Environ. Pollut.* **2018**, *240*, 639–646. [CrossRef]
99. Uber, T.H.; Hüffer, T.; Planitz, S.; Schmidt, T.C. Sorption of non-ionic organic compounds by polystyrene in water. *Sci. Total Environ.* **2019**, *682*, 348–355. [CrossRef]
100. Ling, P.; Liu, F.; Li, L.; Jing, X.; Yin, B.; Chen, K.; Li, A. Adsorption of divalent heavy metal ions onto IDA-chelating resins: Simulation of physicochemical structures and elucidation of interaction mechanisms. *Talanta* **2010**, *81*, 424–432. [CrossRef]
101. Li, L.; Liu, F.; Jing, X.; Ling, P.; Li, A. Displacement mechanism of binary competitive adsorption for aqueous divalent metal ions onto a novel IDA-chelating resin: Isotherm and kinetic modeling. *Water Res.* **2011**, *45*, 1177–1188. [CrossRef] [PubMed]
102. Xiong, C.; Yao, C. Study on the adsorption of cadmium(II) from aqueous solution by D152 resin. *J. Hazard. Mater.* **2009**, *166*, 815–820. [CrossRef] [PubMed]
103. Park, S.W.; Bediako, J.K.; Song, M.-H.; Choi, J.-W.; Lee, H.-C.; Yun, Y.-S. Facile fabrication of polyacrylic acid-polyvinyl chloride composite adsorbents for the treatment of cadmium-contaminated wastewater. *J. Environ. Eng.* **2018**, *6*, 2401–2408. [CrossRef]
104. Li, Y.; Zhang, Y.; Su, F.; Wang, Y.; Peng, L.; Liu, D. Adsorption behaviour of microplastics on the heavy metal Cr (VI) before and after ageing. *Chemosphere* **2022**, *302*, 134865. [CrossRef] [PubMed]
105. Mahanta, D.; Madras, G.; Radhakrishnan, S.; Patil, S. Adsorption and Desorption Kinetics of Anionic Dyes on Doped Polyaniline. *J. Phys. Chem. B* **2009**, *113*, 2293–2299. [CrossRef] [PubMed]
106. Nath, K.; Maity, K.; Biradha, K. Two-dimensional coordination polymers with "X"-Shaped cavities as adsorbents of oxoanion pollutants and toxic dyes. *Cryst. Growth Des.* **2017**, *17*, 4437–4444. [CrossRef]
107. Kim, H.J.; Im, S.; Kim, J.C.; Hong, W.G.; Shin, K.; Jeong, H.Y.; Hong, Y.J. Phytic acid doped polyaniline nanofibers for enhanced aqueous copper(II) adsorption capability. *ACS Sustain. Chem. Eng.* **2017**, *5*, 6654–6664. [CrossRef]
108. Qu, R.; Wang, C.; Ji, C.; Sun, C.; Sun, X.; Cheng, G. Preparation, characterization, and metal binding behavior of novel chelating resins containing sulfur and polyamine. *J. Appl. Polym. Sci.* **2005**, *95*, 1558–1565. [CrossRef]
109. Huang, Y.; Miao, Y.E.; Liu, T. Electrospun fibrous membranes for efficient heavy metal removal. *J. Appl. Polym. Sci.* **2014**, *131*, 40864. [CrossRef]
110. Deng, S.; Bai, R.; Chen, J.P. Aminated polyacrylonitrile fibers for lead and copper removal. *Langmuir* **2003**, *19*, 5058–5064. [CrossRef]
111. Gong, J.; Liu, J.; Chen, X.; Jiang, Z.; Wen, X.; Mijowska, E.; Tang, T. Striking influence of NiO catalyst diameter on the carbonization of polypropylene into carbon nanomaterials and their high performance in the adsorption of oils. *RSC Adv.* **2014**, *4*, 33806–33814. [CrossRef]
112. Hinrichs, R.; Snoeyink, V. Sorption of benzenesulfonates by weak base anion exchange resins. *Water Res.* **1976**, *10*, 79–87. [CrossRef]
113. Lhoest, J.B.; Detrait, E.; Van Den Bosch De Aguilar, P.; Bertrand, P. Fibronectin adsorption, conformation, and orientation on polystyrene substrates studied by radiolabeling, XPS, and ToF SIMS. *J. Biomed. Mater. Res.* **1998**, *41*, 95–103. [CrossRef]
114. Boisson, C.; Jozefonvicz, J.; Brash, J.L. Adsorption of thrombin from buffer and modified plasma to polystyrene resins containing sulphonate and sulphamide arginyl methyl ester groups. *Biomaterials* **1988**, *9*, 47–52. [CrossRef]
115. Qian, D.; Bai, L.; Wang, Y.-S.; Song, F.; Wang, X.-L.; Wang, Y.-Z. A bifunctional alginate-based composite hydrogel with synergistic pollutant adsorption and photocatalytic degradation performance. *Ind. Eng. Chem. Res.* **2019**, *58*, 13133–13144. [CrossRef]
116. Chao, Y.; Zhu, W.; Ye, Z.; Wu, P.; Wei, N.; Wu, X.; Li, H. Preparation of metal ions impregnated polystyrene resins for adsorption of antibiotics contaminants in aquatic environment. *J. Appl. Polym. Sci.* **2015**, *132*, 41803. [CrossRef]
117. Yin, F.; Peng, P.; Mo, W.; Chen, S.; Xu, T. The preparation of a porous melamine–formaldehyde adsorbent grafted with polyethyleneimine and its CO_2 adsorption behavior. *New J. Chem.* **2017**, *41*, 5297–5304. [CrossRef]
118. Zhu, H.; Qiu, S.; Jiang, W.; Wu, D.; Zhang, C. Evaluation of electrospun polyvinyl chloride/polystyrene fibers as sorbent materials for oil spill cleanup. *Environ. Sci. Technol.* **2011**, *45*, 4527–4531. [CrossRef]
119. Guo, N.N.; Zhang, Q.G.; Li, H.M.; Wu, X.M.; Liu, Q.L.; Zhu, A.M. Facile fabrication, structure, and applications of polyvinyl chloride mesoporous membranes. *Ind. Eng. Chem. Res.* **2014**, *53*, 20068–20073. [CrossRef]
120. Xiao, G.; Wen, R. Comparative adsorption of glyphosate from aqueous solution by 2-aminopyridine modified polystyrene resin, D301 resin and 330 resin: Influencing factors, salinity resistance and mechanism. *Fluid Phase Equilib.* **2016**, *411*, 1–6. [CrossRef]
121. Zhang, B.; Yu, S.; Zhu, Y.; Shen, Y.; Gao, X.; Shi, W.; Tay, J.H. Adsorption mechanisms of crude oil onto polytetrafluoroethylene membrane: Kinetics and isotherm, and strategies for adsorption fouling control. *Sep. Purif. Technol.* **2020**, *235*, 116212. [CrossRef]
122. Soetaredjo, F.E.; Ismadji, S.; Foe, K.; Woworuntu, G.L. Removal of hazardous contaminants from water or wastewater using polymer nanocomposites materials. *Nanotechnol. Sustain. Water Resour.* **2018**, *2*, 103–139. [CrossRef]
123. Yang, L.; Zhang, H.; Tan, T.; Rahman, A.U. Thermodynamic and NMR investigations on the adsorption mechanism of puerarin with oligo-β-cyclodextrin-coupled polystyrene-based matrix. *J. Chem. Technol. Biotechnol.* **2009**, *84*, 611–617. [CrossRef]
124. Torres, F.G.; Dioses-Salinas, D.C.; Pizarro-Ortega, C.I.; De-la-Torre, G.E. Sorption of chemical contaminants on degradable and non-degradable microplastics: Recent progress and research trends. *Sci. Total Environ.* **2021**, *757*, 143875. [CrossRef]
125. Alimi, O.S.; Farner Budarz, J.; Hernandez, L.M.; Tufenkji, N. Microplastics and nanoplastics in aquatic environments: Aggregation, deposition, and enhanced contaminant transport. *Environ. Sci. Technol.* **2018**, *52*, 1704–1724. [CrossRef]
126. Zhang, Y.; Goss, G.G. Potentiation of polycyclic aromatic hydrocarbon uptake in zebrafish embryos by nanoplastics. *Environ. Sci. Nano* **2020**, *7*, 1730–1741. [CrossRef]

127. Lang, M.; Yu, X.; Liu, J.; Xia, T.; Wang, T.; Jia, H.; Guo, X. Fenton aging significantly affects the heavy metal adsorption capacity of polystyrene microplastics. *Sci. Total Environ.* **2020**, *722*, 137762. [CrossRef]
128. Qin, Y.; Wang, Z.; Li, W.; Chang, X.; Yang, J.; Yang, F. Microplastics in the sediment of lake Ulansuhai of Yellow river basin, China. *Water Environ. Res.* **2020**, *92*, 829–839. [CrossRef]
129. Godoy, V.; Blázquez, G.; Calero, M.; Quesada, L.; Martín-Lara, M. The potential of microplastics as carriers of metals. *Environ. Pollut.* **2019**, *255*, 113363. [CrossRef]
130. Abdurahman, A.; Cui, K.; Wu, J.; Li, S.; Gao, R.; Dai, J.; Liang, W.; Zeng, F. Adsorption of dissolved organic matter (DOM) on polystyrene microplastics in aquatic environments: Kinetic, isotherm and site energy distribution analysis. *Ecotoxicol. Environ. Saf.* **2020**, *198*, 110658. [CrossRef]
131. Yu, H.; Yang, B.; Waigi, M.G.; Peng, F.; Hu, X. The effects of functional groups on the sorption of naphthalene on microplastics. *Chemosphere* **2020**, *261*, 127592. [CrossRef] [PubMed]
132. Li, H.; Xiong, J.; Xiao, T.; Long, J.; Wang, Q.; Li, K.; Liu, X.; Zhang, G.; Zhang, H. Biochar derived from watermelon rinds as regenerable adsorbent for efficient removal of thallium(I) from wastewater. *Process Saf. Environ. Prot.* **2019**, *127*, 257–266. [CrossRef]
133. Dong, Y.; Gao, M.; Qiu, W.; Song, Z. Adsorption of arsenite to polystyrene microplastics in the presence of humus. *Environ. Sci. Proc. Impacts* **2020**, *22*, 2388–2397. [CrossRef]
134. Liu, P.; Lu, K.; Li, J.; Wu, X.; Qian, L.; Wang, M.; Gao, S. Effect of aging on adsorption behavior of polystyrene microplastics for pharmaceuticals: Adsorption mechanism and role of aging intermediates. *J. Hazard. Mater.* **2020**, *384*, 121193. [CrossRef] [PubMed]

nanomaterials

MDPI

Review

Chitosan Nanoparticles as Potential Nano-Sorbent for Removal of Toxic Environmental Pollutants

Asmaa Benettayeb [1,*], Fatima Zohra Seihoub [1], Preeti Pal [2], Soumya Ghosh [3], Muhammad Usman [4,*], Chin Hua Chia [5], Muhammad Usman [6] and Mika Sillanpää [7,8,9,10,11]

1 Laboratoire de Génie Chimique et Catalyse Hétérogène, Département de Génie Chimique, Université de Sciences et de la Technologie-Mohamed Boudiaf, USTO-MB, BP 1505 EL-M'NAOUAR, Oran 31000, Algeria
2 Accelerated Cleaning Systems India Private Limited, Sundervan Complex, Andheri West, Mumbai 400053, India
3 Department of Genetics, Faculty of Natural and Agricultural Sciences, University of the Free State, Bloemfontein 9301, South Africa
4 School of Civil Engineering, Hamburg University of Technology, Am Schwarzenberg-Campus 3, 20173 Hamburg, Germany
5 Department of Applied Physics, Faculty of Science and Technology, Universiti Kebangsaan Malaysia, Bangi 43600, Selangor, Malaysia
6 PEIE Research Chair for the Development of Industrial Estates and Free Zones, Center for Environmental Studies and Research, Sultan Qaboos University, Al-Khoud, Muscat 123, Oman
7 Department of Chemical Engineering, School of Mining, Metallurgy and Chemical Engineering, University of Johannesburg, Doornfontein 2028, South Africa
8 School of Chemical and Metallurgical Engineering, University of the Witwatersrand, Johannesburg 2050, South Africa
9 Chemistry Department, College of Science, King Saud University, Riyadh 11451, Saudi Arabia
10 School of Resources and Environment, University of Electronic Science and Technology of China (UESTC), No. 2006, Xiyuan Ave., West High-Tech Zone, Chengdu 611731, China
11 Faculty of Science and Technology, School of Applied Physics, University Kebangsaan Malaysia, Bangi 43600, Selangor, Malaysia
* Correspondence: asmaa.benettayeb@univ-usto.dz (A.B.); muhammad.usman@daad-alumni.de (M.U.)

Citation: Benettayeb, A.; Seihoub, F.Z.; Pal, P.; Ghosh, S.; Usman, M.; Chia, C.H.; Usman, M.; Sillanpää, M. Chitosan Nanoparticles as Potential Nano-Sorbent for Removal of Toxic Environmental Pollutants. *Nanomaterials* **2023**, *13*, 447. https://doi.org/10.3390/nano13030447

Academic Editors: Lei Huang, Junye Cheng, Hongguo Zhang and Zhenxing Wang

Received: 10 December 2022
Revised: 13 January 2023
Accepted: 16 January 2023
Published: 21 January 2023

Abstract: Adsorption is the most widely used technique for advanced wastewater treatment. The preparation and application of natural renewable and environmentally friendly materials makes this process easier and more profitable. Chitosan is often used as an effective biomaterial in the adsorption world because of its numerous functional applications. Chitosan is one of the most suitable and functionally flexible adsorbents because it contains hydroxyl (-OH) and amine (-NH_2) groups. The adsorption capacity and selectivity of chitosan can be further improved by introducing additional functions into its basic structure. Owing to its unique surface properties and adsorption ability of chitosan, the development and application of chitosan nanomaterials has gained significant attention. Here, recent research on chitosan nanoparticles is critically reviewed by comparing various methods for their synthesis with particular emphasis on the role of experimental conditions, limitations, and applications in water and wastewater treatment. The recovery of pollutants using magnetic nanoparticles is an important treatment process that has contributed to additional development and sustainable growth. The application of such nanoparticles in the recovery metals, which demonstrates a "close loop technology" in the current scenarios, is also presented in this review.

Keywords: nano-sorbent; adsorption; chitosan nanoparticles; pollutant and metal recovery; functionally flexible adsorbent; water treatment

1. Introduction

The diversity and complexity of pollutants greatly affect the efficiency of wastewater treatment [1]. To overcome this limitation, extensive research has focused on the development of biosorbents [2,3] and their variable applications with the help of nanotechnology [3,4]. Nanotechnology exploits the properties of any material at the nanoscale. The materials of this new technology are termed nanoparticles (NPs) [5]. Nanotechnology is an ideal solution to ensure high quality water. It can be considered a powerful 21st century tool for protecting the environment and improving environmental quality [6]. The application of nanotechnology in water purification and environmental sanitation has potential, as conventional methods do not always provide cost-effective solutions for the removal of common pollutants. Conventional technologies have a limited lifetime, generate hazardous and toxic environmental wastes, and are non-renewable. For several years, NPs have been the subject of numerous research publications, and patents, because of their large surface area, high resistance to heat and chemicals, and high adsorption capacity for the removal of organic and inorganic contaminants [7–11].

Adsorption technology is one of the most reliable strategies in wastewater treatment, and the use of a variety of nanosized adsorbents enables preferential surface adsorption [12–16]. The increase in surface area can increase the sorption capacity towards pollutants on the surface of NPs [6,17–20]. In addition to wastewater treatment, NPs are used as antimicrobial agents [21], as catalysts [22], in biomedicine, energy conversion [23], agriculture, electronics, and optoelectronics industries [24]. According to Vakili et al. (2014), nano-chitosan is one such nanomaterial that is a natural substance with excellent physicochemical properties and is harmless to humans [25]. Therefore, chitosan biopolymer has become the environmentally friendly substance of choice. Several modifications have been carried out on the alginate for the introduction of the amine functional group (-NH_2) on its surface [26–28], or other biosorbents by introducing other actives functions [29–34], this active site is found naturally in chitosan. Chitosan is rich in amino (-NH_2) and hydroxyl (-OH) groups, which gives it a powerful adsorption capacity and reactivity to most pollutants [35–38]. Thus, chitosan is an excellent natural adsorbent that can be modified to increase its efficiency and improve its basic properties [6,17,39–41].

The main problem with using chitosan in its natural form is its low adsorption capacity, which can be improved by physical or chemical modification. Hence, researchers have developed more effective chitosan-based adsorbents. Chitosan NPs are among the best nano-adsorbents due to their large surface area, high adsorption capacity, and environmental friendliness. Chitosan is abundant in nature, reusable, and can be easily modified with various chemical and biological agents so that it can be regenerated and reused over several cycles. Chitosan NPs can be categorised as nano adsorbents that meet the essential criteria for use in wastewater treatment. Chitosan NPs can be chemically inert, and their morphology resists various complex conditions.

The preparation of chitosan NPs cannot only improve the surface area and adsorption capacity, but the presence of functional groups also makes it selective [42–44]. Since chitosan is biodegradable, it does not cause additional environmental pollution. Apart from its ecological nature, it also has antibacterial properties that enhance its use as an adsorbent for water treatment. According to Saxena et al. (2020), it needs chemical modification using chemical cross-linking to increase its stability over time [45].

In the present review, the basic properties of chitosan NPs and chitosan magnetic NPs, including chitosan nanocomposites, and other types based on chitosan NPs are discussed. Some effective preparation methods, such as covalent crosslinking, ionic gelation, change in pH, and other methods are also discussed. The characteristics of chitosan and operational conditions are compared in terms of their efficiency as adsorbents for the removal of various pollutants, such as uranium, and rare metals. The parameters affecting the adsorption mechanism, and advantages and limitations of using chitosan NPs in fields of sorption/biosorption are presented.

2. Unique Properties of Chitosan Nanomaterials and Magnetic Chitosan

Chitosan (poly [β-(1-4)-2-amino-2-deoxy-D-glucopyranose]) is a linear cationic biopolymer with high molecular weight. The primary source of usable chitosan is the deacetylation of chitin obtained from the shells of crustaceans (crabs, lobsters, shrimp, and crayfish) [46], which are an abundant natural resource. The natural material is commercially obtained from the deacetylation of chitin by thermochemical treatment [35]. Natural chitosan is non-toxic, non-allergenic, biodegradable, biocompatible, inexpensive, hydrophilic, biologically active, and can form fibers and films [47–50].

Figure 1 summarizes the chemical process to obtain chitosan from chitin. Chitosan is a semi-crystalline cationic polysaccharide that attracts positively charged molecules and enhances bonding due to the presence of the –NH_2 group. In addition, the –OH group is also present in the structure and helps to increase the adsorption capacity [51].

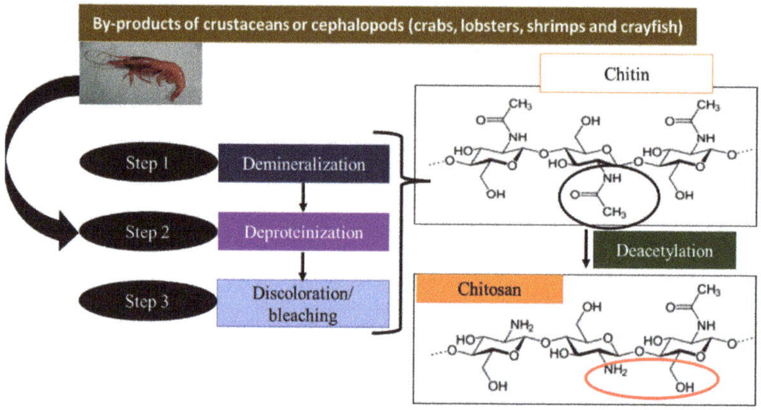

Figure 1. Different stages of the chitosan manufacturing process.

Nano-sized particles have characteristics that cannot be achieved with solid, normal-sized material. For example, the electronic and optical properties of metallic materials can be modified by controlling their size below the Bohr radius (usually between 1–10 nm). The interest in NPs is due to their ability to act as an effective bridge between solid materials and atomic structures. Solid materials exhibit constant physical properties, regardless of their size and mass. However, NPs have properties that depend on their size due to the large proportion of atoms on their surfaces relative to their volume, resulting in a large specific surface area. In view of this, the electronic, optical, and magnetic properties of materials change as their size decreases towards the nanoscale. Therefore, controlling the size of NPs is of particular interest because it can influence their properties. The exceptional physicochemical properties of nanomaterials are due to three main reasons:

(i) The size of the nanomaterial is comparable to the Bohr radius of the excitons. This dramatically alters the optical, luminescent, and redox properties of nanomaterials compared to their bulk counterparts.
(ii) The surface area atom largely determined the thermodynamic properties of solids and also determines the melting temperature and structural transitions of nanomaterials.
(iii) When the particle size is decreased, the net internal cohesive forces increase. Thus, reducing the particle size increases the surface area to bulk volume ratio, i.e., particle size is inversely proportional to the surface area to bulk volume ratio [52].

In general, nanotechnology is used to produce materials with specific properties and a high degree of reproducibility. In this regard, researchers are currently focusing on the synthesis of new nanostructured materials capable of cleaning the environment; they know that nanomaterials are entirely or partially composed of nano-objects, which gives them improved or specific properties in nanometric dimensions. In the family of nanomaterials, there are three types, namely NPs, nano-fibers, and nano-films. NPs are elements with a nanometric size between 1–100 nm [53] and are used daily in products such as cosmetics, paints, electronics, and computers. NPs can be in the form of powders, suspensions, solutions, or gels from which other physical forms, such as nano-beads are formed.

In recent years, researchers have focused on the use of NPs, particularly magnetic NPs, as effective adsorbents for the treatment of pollutants present in wastewater [54,55].

Magnetic NPs have been used as adsorbents for water treatment. These adsorbents have remarkable properties such as nanometric size, high specific surface area per volume ratio, and resistance to internal diffusion leading to a high adsorption capacity [56,57], biocompatibility, biodegradability and low toxicity [58], low cost of fabrication, green sources, magnetic intensity, so, for these reasons that chitosan magnetic received considerable attention.

In addition, iron oxide NPs have the advantage of being superpara-magnetic [59] and the powder can be easily recovered using an external magnetic field. Magnetic separation technology is easy to use and preferable to avoid slow separation techniques such as filtration, centrifugation, and precipitation. Some important characteristics of nano-adsorbents compared to classical ones are summarized in Figure 2.

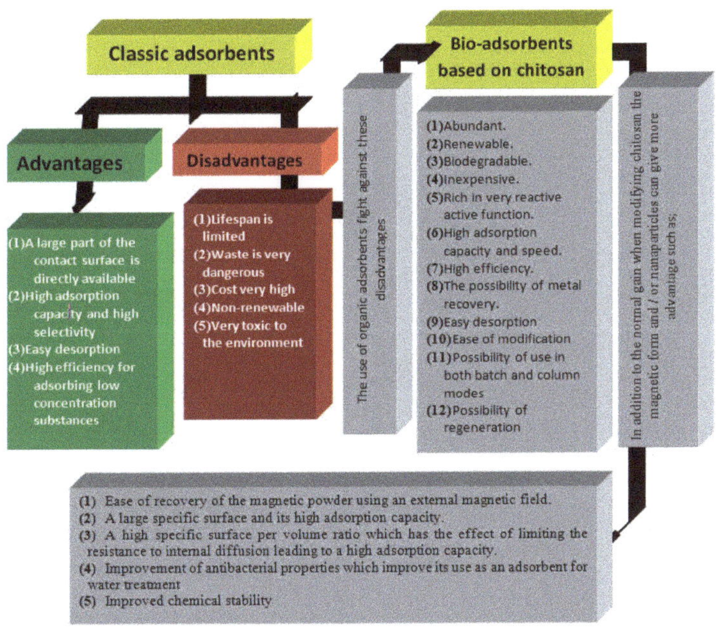

Figure 2. Advantages and disadvantages of NPs and use of chitosan NPs to overcome the disadvantages of traditional NPs in wastewater treatment.

3. Standard Methods of Chitosan Nanoparticles Synthesis

3.1. Synthesis of Chitosan NPs

Research shows that chitin and chitosan can be easily transformed into gels, membranes, nanofibers, beads, and micro-nanoparticles. Generally, there are four ways of modifying chitosan to obtain NPs, which are summarized in Figure 3.

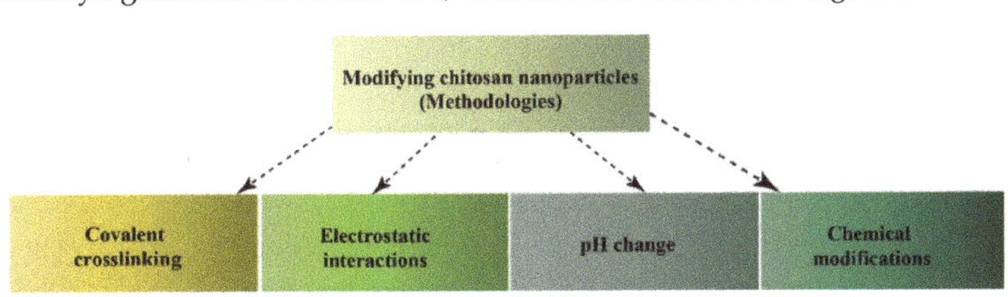

Figure 3. Methods to modify chitosan NPs.

3.1.1. Covalent Cross-Linking Methodologies

Different cross-linking agents can be used for the fabrication of chitosan cross-linking especially glutaraldehyde because it can be easily used in various physicochemical reactions to prepare functional materials from amino polysaccharides, proteins, natural and synthetic polymers due to the -NH_2 and -OH groups. Ohya et al. (1994) obtained chitosan NPs with a diameter between 250 and 300 nm by emulsifying the chitosan solution in toluene and then using glutaraldehyde for covalent cross-linking [60]. Banerjee et al. (2002) obtained chitosan NPs with a reverse micelle method using a surfactant (sodium bis (ethylhexyl) sulfosuccinate) in hexane. The aqueous core of the micelles contained a chitosan solution, which was cross-linked with glutaraldehyde [61]. The researchers confirmed that the particle size depended on the rate of cross-linking with glutaraldehyde. The obtained NPs between 30 to 100 nm when cross-linked with 10% and 100% of the amine, respectively [61].

3.1.2. Ionic Gelation by Electrostatic Interactions, Encapsulation, or Adsorption

The ionic gelation for chitosan nanoparticle synthesis with the cross-linking of tripolyphosphate is another method that is widely method due to its simplicity and ease of preparation [62,63]. Synthesis of chitosan NPs uses cross-linking agents with opposite charges, causing an electrostatic attraction, encapsulation, or adsorption has also been applied (Figure 4).

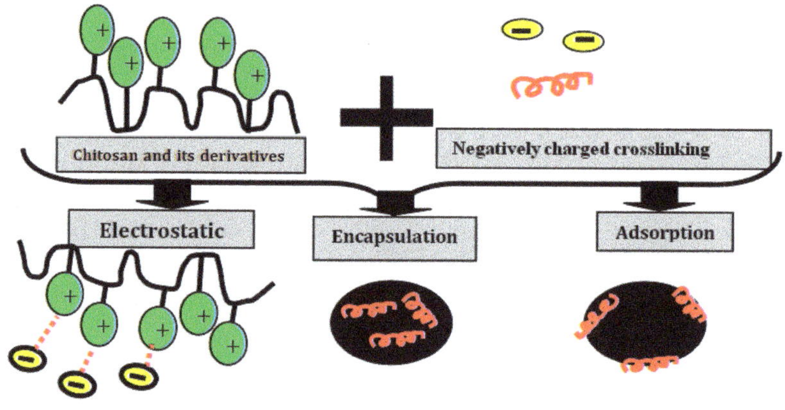

Figure 4. Schematic representation of different techniques for obtaining NPs based on chitosan and a cross-linking agent carrying an opposite charge.

These researchers [62,63] confirmed the formation of NPs of different sizes by adding a tripolyphosphate solution to an acidic chitosan solution. They know that other anionic spaces can play the same role as tripolyphosphate.

In the work of Ali et al. (2018), chitosan NPs were prepared according to the method of Qi and Xu, (2004) by dissolving 0.5 g of chitosan in 100 mL of acetic acid (1% (v/v)) and adjusting the pH of the solution to a 4.6–4.8 using 1N NaOH. Chitosan NPs formed spontaneously after adding 3 mL of chitosan solution under vigorous magnetic stirring to 1.0 mL of an aqueous solution of sodium tripolyphosphate (0.25% w/v), with a chitosan/tripolyphosphate ratio of 3/1 at room temperature [64,65]. In this type of modification, the formation of chitosan NPs is attributed to the aid of electrostatic attraction between the positively charged amine group of chitosan and the anionic cross-linking agent such as tripolyphosphate. The synthesis can be carried out by adding polyethylene glycol as a cross-linking agent to the solution of chitosan in the acetic acid-containing stabilizer. The size and surface charge of NPs depend on the ratio of chitosan to the stabilizer. Generally, the size of the chitosan NPs obtained by this method ranges from 40–100 nm [64,66].

3.1.3. Change in pH

The microemulsion method is one of the newer techniques for preparing inorganic NPs. Microemulsions are isotropic, macroscopically homogeneous, and thermodynamically stable solutions containing at least three components, a polar phase (usually water), a non-polar phase (usually oil), and a surfactant [67].

Brunel et al. (2008) developed a technique to obtain chitosan nanogels without chemical cross-linking. The principle is to apply an aqueous gelation process to a water/oil reverse microemulsion. In this method of manufacturing chitosan nanogels, an acid solution of chitosan is dispersed in an organic phase containing a surfactant and then the nanodroplets are gelled by modifying the pH of the medium [68]. The chitosan nano-hydrogels are then washed to remove the organic phase and redispersed in an acid buffer. During emulsification, the product formation depends on the diffusion rate of the solvent in the dispersed phase. The ratio of stabilizer and oil-polymer in an aqueous solution offers greater solvent diffusion into the external phase [69].

3.1.4. Chemical Modification of Chitosan

Chemical modification of chitosan with hydrophobic groups is used to obtain an amphiphilic polymer with self-assembly properties in solution. The grafting of these groups occurs generally via the free amine functions of chitosan. Chitosan, which has become amphiphilic, then organizes into sub-micronic core-shell structures in an aqueous medium. The hydrophobic segments are located in the core of the particle, while the hydrophilic segments form the crown. The size of the NPs depends on the grafted molecule and the rate of substitution [70,71]. Furthermore, according to Morales et al. (2013) [72], the size of NPs increases with increasing the amount of cross-linking agents. For example, in the work of Galhoum et al. (2015), the size of the crystallites of magnetic particles varies between 11 and 13 nm when cross-linked with glutaraldehyde [73]. When tripolyphosphate is used as a crosslinking agent with a chitosan/TPP ratio of (3:1), we can obtain chitosan NPs of small size [65].

3.2. Synthesis of Chitosan-Magnetic NPs

Several methods have been used to prepare high-quality magnetic NPs of different sizes. Co-precipitation (Figure 5), thermal decomposition, hydrothermal and solvothermal methods, microemulsion, and sol-gel methods are used for the preparation of magnetic chitosan NPs. The most commonly used manufacturing methods are co-precipitation and microemulsion. The ease of preparation of chitosan and the fact that NPs intermix during co-precipitation is an important advantage in the use of magnetic composites. The functional groups on the surface and iron oxide react easily with chitosan and its derivatives [74]. Moreover, there is a link between co-precipitation and iron oxide-chitosan blends, so we can obtain magnetic NPs when we use either method, but, in the case of chitosan, we need to merge both two methods to obtain magnetic NPs from chitosan.

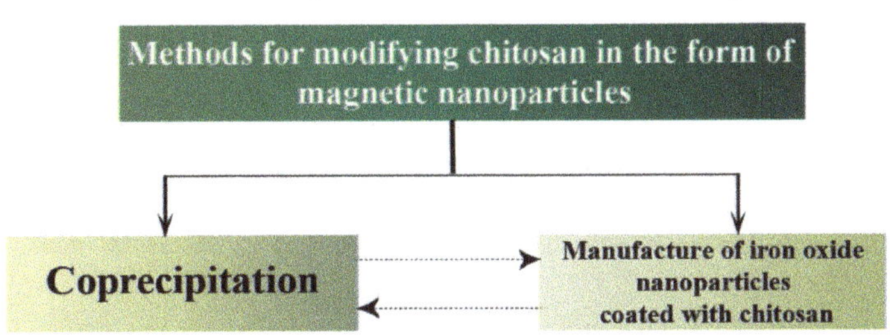

Figure 5. A well-known method of modifying chitosan to form magnetic NPs.

From several methods of manufacturing magnetic chitosan, we can conclude that magnetic chitosan can be prepared in a one-step or two-step process. In the one-step process, ferric and ferrous salts (Fe_3O_4) or (γ-Fe_2O_3) are dissolved in a chitosan solution by raising the pH of the solution (above 9). Then, magnetite formation and magnetite/chitosan precipitation take place simultaneously in one pot [75,76]. In the two-step method, the magnetic particles and the chitosan solution are prepared separately, then the magnetic particles are dispersed in the chitosan solution, followed by a precipitation step and crosslinking to finally form the magnetic chitosan.

3.2.1. Chemical Co-Precipitation

Chemical co-precipitation of Fe(II) and Fe(III) in alkaline solution (changing pH) is widely used for the preparation of magnetite (Fe_3O_4) or (γ-Fe_2O_3) NPs due to its simplicity, reproducibility, energy efficiency, and the possibility of large-scale preparation.

This technique has been used by several researchers such as Namdeo and Bajpai (2008) and Gregorio-Jauregui et al. (2012) [35,77]. NPs were prepared with sizes ranging from 5 to 100 nm using co-precipitation. Chemical co-precipitation is based on the reaction of aqueous solutions of Fe(II)/Fe(III) salts, usually in a 1/2 molar ratio with a base such as ammonia, potassium hydroxide, or sodium hydroxide under an inert atmosphere at 40–50 °C. The size and shape of the magnetite particles are generally controlled by the synthesis conditions such as temperature, pH, ionic strength, Fe(II) and Fe(III) concentrations, and the nature and concentration of the base.

3.2.2. Iron Oxide-Chitosan Blends

Most of the literature dealing with magnetic NPs for the decontamination of wastewater by removing heavy metals relates to magnetic NPs encapsulated in polymers, especially chitosan. The chitosan biopolymer is grafted onto the surface of the magnetic cores, or the magnetic powder is encapsulated in chitosan using different syntheses to prepare Fe_3O_4 NPs modified by chitosan (magnetic core and multi-core in chitosan-based adsorption material [78]). Several methods, such as co-precipitation [79], cross-linking [80], and covalent bonding using coupling agents can be used to prepare new materials. The resulting magnetic composites are mostly micrometric in size, have low stability, and aggregate.

There are several ways to manufacture magnetic chitosan NPs; the preferred and simplest method is proposed by Galhoum et al. (2015a) and Galhoum et al. (2015c). The method consists of the synthesis of magnetic NPs with the deposition of a biopolymer on magnetic cores. Magnetic NPs are synthesized by a hydrothermal process in a single vessel involving co-precipitation under thermal conditions with Fe(III) and Fe(II) salts in the presence of chitosan. The magnetic/chitosan composite material is cross-linked with epichlorohydrin and modified by the grafting/functionalization of amino acids such as alanine, cysteine, and serine [73,81–84]. The crystal size of magnetic NPs in the work of [85] has been found to be close to 13.5 nm, while in [81], the nanometric size of diethylenetriamine-functionalized chitosan magnetic nano-based particles is in the range of 30–50 nm.

According to Liu et al. (2015), magnetic chitosan nanoparticles (MCNPs) with a diameter of about 10 nm, can be prepared by co-precipitation in the following steps. Dissolve 0.5 g of chitosan in 200 mL of acetic acid (0.5% v/v) with continuous stirring. Add 4.7 g $FeCl_3 \cdot H_2O$ and 2.4 g $FeSO_4 \cdot 7H_2O$, which was dissolved in 22 mL of distilled water. Mix with the chitosan solution for 20 min at a stable stirring speed of 1000 rpm. Add dropwise 40 mL of ammonia (NH_4OH) to the reaction system and 6 mL during continuous stirring at 1000 rpm for 3 h. Finally, the resulting magnetic chitosan NPs are separated by a magnetic field [85].

A second preparation method is reported by Zhou et al. (2014) where chitosan and magnetic NPs are bound to glutaraldehyde using cyclohexane, 10 mg/mL chitosan, 2% acetic acid, and Fe_3O_4 in water. The materials are mixed in a beaker at 1800 rpm. After that, 20 mL NaOH at a concentration of 50 mol/L is added quickly to the solution, and the mixture is placed in a water bath at 60 °C for 2 h [86].

Chitosan magnetic NPs are collected with a magnet and rinsed with ethanol, then washed with deionized water three times. Finally, the NPs are modified by adding different concentrations of glutaraldehyde with constant stirring at 150 rpm for 60 min at room temperature to obtain chitosan magnetic NPs. A diagrammatic representation of the process is shown in Figure 6, and the interaction between the chitosan chain and magnetic particles Fe_3O_4 is shown in Figure 7. Ionic liquids also play an important role in the improvement of chitosan derivatives [87–89].

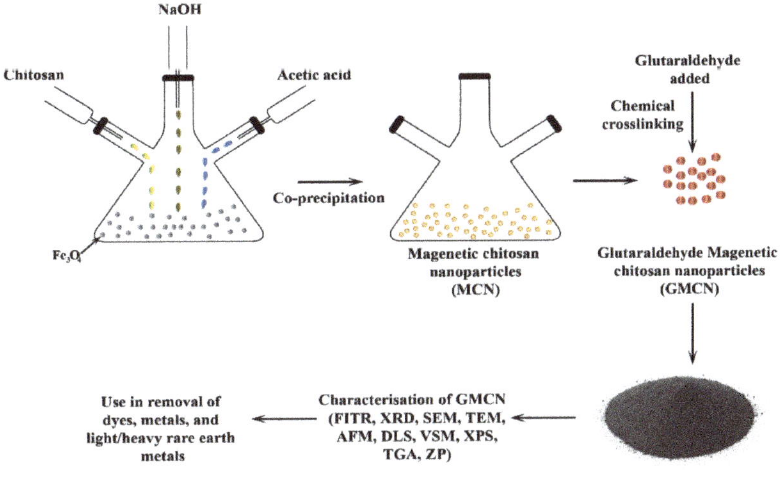

Figure 6. Reaction method of magnetic chitosan NPs cross-linked by glutaraldehyde.

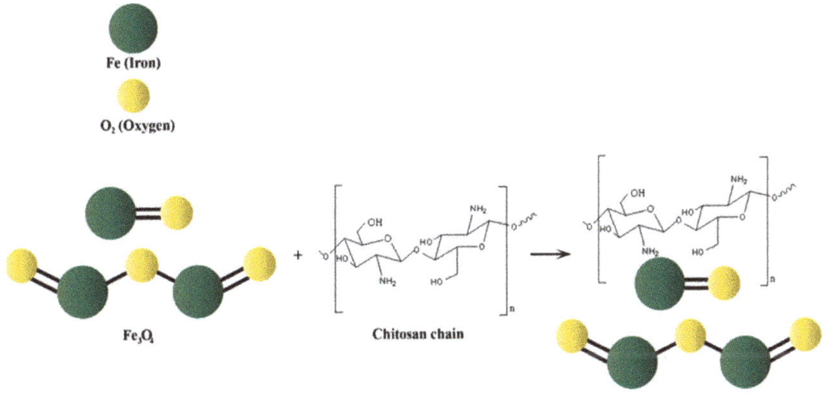

Figure 7. Interaction between the chitosan chain and the magnetic particles.

4. Characteristics of Chitosan NPs and Their Applications

According to Dasgupta et al. (2015), the physicochemical properties of NPs depend on their surface characteristics. Unlike chemical compounds, where the characterization is usually confined to chemical composition and purity, nanomaterials demand comprehensive characterization. Therefore, an extensive and complete characterization, including size distribution, shape, surface area, surface chemistry, crystallinity, porosity, agglomeration state, surface charge, and solubility is recommended for nanomaterials [90]. There are several methods of characterization to confirm the formation of NPs and to examine the efficiency of the nanomaterials with the target pollutant. The following table (Table 1) summarizes some types of analyzes (FTIR/SEM/TEM/XRD etc.) and their respective functions for the analysis of chitosan NPs.

Table 1. Methods commonly used to characterize chitosan NPs.

Transmission Electron Microscopy (TEM)	Fourier-Transform Infrared Spectroscopy (FTIR)	X-ray Diffraction (XRD)
used to examine the size, shape, and morphology of the NPs	used to detect or identify the functions of chitosan NPs, and to confirm the presence of iron in the structure	used to examine the crystallinity of chitosan and chitosan NPs
Atomic force microscopy (AFM)	Dynamic light scattering device (DLS)	Scanning electron microscopy and transmission electron microscopySEM and SEM-EDX
used to visualize the rough nature of chitosan NPs, which facilitate the adsorption of heavy metals	used to determine the size of NPs. To confirm the synthesis of magnetic NPs, it is necessary to use a vibrating sample magnetometer, which characterizes the magnetic properties of chitosan.	analyses the shape and type of surface used to determine the morphology of chitosan NPs

4.1. Application of Chitosan NPs in Wastewater Treatment

Different protocols can be involved during the manufacture of chitosan magnetic NPs to produce, for example, chitosan-magnetite nanocomposites. These are synthesized by chemical co-precipitation of Fe(II) and Fe(III) ions by NaOH in the presence of chitosan and have been evaluated for the adsorption of several pollutants. Chitosan magnetic NP-based adsorbents adsorb differently and the best performance is achieved during grafting functional groups such as amino, sulfur, carboxyl, and alkyl groups onto the backbone of chitosan. Therefore, there are several modifications that increase the percentage of active sites ($-NH_2$, –COOH, –SH, and –OH) responsible for the fixation of heavy metals, which consequently increase adsorption/sorption efficiency, different types of interaction between chitosan NPs and pollutants such as ion exchange, electrostatic attraction, and others according to the results of isotherms and kinetics, all this information gives us a clear idea about the mechanisms.

New chitosan magnetic NPs combine the metal/dye-binding potential of chitosan with an enormous surface area, dispersed character, and easy regeneration when a magnetic field of suitable strength is applied. Chitosan NPs exhibit good adsorption for dyes and heavy metal ions. It adsorbs radioactive heavy metals and rare earth metals effectively, with additional modifications improving adsorption capacity and efficiency.

4.1.1. Heavy Metal Removal

The term heavy metal defines any metallic element with an atomic weight between 63.5 and 200.6 and a density greater than 5.0 [91]. Current research suggests that chitosan NPs with large specific surface areas and the inclusion of $-NH_2$ and –OH groups, could effectively remove metal ions [17]. In addition, in the adsorption process, the magnetic separation advancements are effective, fast, and cost-effective compared to other separation methods since magnetic chitosan can be easily removed using an external magnetic field from the media [92].

Magnetic chitosan NP materials prepared to remove Hg(II), Cu(II), and Ni(II), were also examined for removal of Cu(II), Pb(II), and Cd(II) [93] and Ni(II) and Co(II) [94]. Table 2 summarizes research on NPs and chitosan magnetic NPs in the treatment of heavy metal ions. For instance, the preparation of chitosan NPs by ionic gelation was reported by [95,96]. Ali et al. (2018) utilized chitosan NPs for Fe(II) and Mn(II) removal with an adsorption capacity of 116.2 mg/g and 74.1 mg/g, respectively, with a dosage of 0.5–15 g/L [65]. Basaad et al. (2016) reported the adsorption of Fe(II) and Mn(II) with an effectiveness of 99.94% and 80.85%, respectively [97].

The preparation of magnetic modified chitosan using co-precipitation and cross-linking by glutaraldehyde is reported in the work of [98] that was used for the removal of Zn(II) [85], while magnetic chitosan NPs were used for the removal of As(V) and

As(III) [98]. The initial Zn(II) concentration was taken as 1300 mg/L, and a magnetically modified chitosan dose of 300 g/L was applied, which removed 99% of the metal ions with the adsorption capacity of 32.16 mg/g, under these experimental conditions we have a high efficiency and a low adsorption capacity. The adsorption mechanism followed Langmuir-Freundlich isotherm means that we have a multilayer adsorption and not a specific adsorption [98]. Magnetic chitosan nanoparticles (MCNPs) have shown good results with a removal capacity of 95% (≈144.75 mg/g) for both As(V) and As(III) at a dose of 1.0 g/L and a concentration of 2 at nearly 11 mg/L within just 15 min and followed by Sips (Langmuir-Freundlich) isotherm [85]. In the work of Liu et al., the dominant mechanism, in this case, is the electrostatic attraction between the positive surface charges of the protonated chitosan amine functions and the negative charges of the arsenate ions. knowing that according to the authors, the MCNP is rich in –N of about 1.93 mmol/g (N/C ratio = 0.252) and the amine groups participate in the hydrogen bonding of the chitosan, and are known by their strong affinity against heavy metals.

Table 2. Chitosan-based nanomaterials for heavy metal removal.

Biosorbent	Modification Method	Characterization	Metal	Ad. Efficiency (mg/g)	Models	Parameters					References
						pH	Time and Speed	Temperature	SD	C_0 (mg/L)	
Chitosan nanoparticles (CNPs)	Ionic gelation	FT-IR XRD TEM SEM	Fe(II) Mn(II)	116.2 74.1	Langmuir	2.5–7	120 min 180 tr/min	Room temperature	0.5–15 g/L	10–120	[65]
	Ionic gelation	FT-IR XRD SEM AFM TGA DSD	Cr (VI)	1395.32	Langmuir Freundlich Temkin	3–9	60min	Room temperature	1 g/100 mL	200	[96]
	Ionic gelation	TEM	Pb(II) Cd(II) Cr(V)	398 358 323.6	Langmuir	4	/	/	/	/	[99]
Magnetic CS—Gluta. (CMMC)	Co-precipitation and cross-linking	SEM TEM FTIR	Zn(II)	32.2	Langmuir Freundlich	5.0	30 min 100 rpm	25 °C	0.3 g/mL	50 and 1300	[98]
Magnetic chitosan particle (MCNPs)	Co-precipitation and cross-linking	UV	As(V) As(III)	65.5 60.2	Langmuir Freundlich	6.8	15 min 150 tr/min	Room temperature	1 g/L	0.2–50	[85]
Chitosan magnetic beads	Magnetite powder added to chitosan solution to form magnetic beads	SEM	Cd(II)	580	/	6.5	100 rpm	25 °C	0.15 g of dry chitosan	/	[100]
Thiourea-modified magnetic chitosan microsphere composite (TMCS)	Co-precipitation	XRD FTIR VSM Zeta Potential analyzer	Hg(II) Cu(II) Ni(II)	625.2 66.7 15.3	Langmuir Freundlich Temkin	5.0 5.0 5.0	8 h 150 rpm	28 °C	0.3 g of dry TMCS	/	[101]

Table 2. *Cont.*

Biosorbent	Modification method	Characterization	Metal	Adsorption efficiency or capacity	Adsorption isotherm	Parameters					References
						pH	Stirring time and speed	Temp.	SD	C_0 (mg/L)	
CTS NPs	Ion gelation method	/	As(V)	90%	Sips	3.5	3 h 200 rpm	24 °C	0.025 g/50 mL	100	[102]
Magnetic chitosan nanoparticles (MCNPs)		DLS dynamic light scattering	Cd(II)	358 mg/g	Langmuir	4.6	12 h 300 tr/min	28 °C	1 g/L	0.5	[103]
Cross-linked chitosan magnetic modified with methionine-glutaraldehyde	New nano biosorbent based on cross-linking of chitosan and magnetic NPs (Chi-MG/Fe_3O_4)	XRD FTIR SEM	Cu(II) Pb(II) Cd(II)	172.4 mg/g 175.4 mg/g 163.9 mg/g	Langmuir	5.5	40 min 150 rpm	Room temperature	/	/	[93]
Composite chitosan—magnetite microparticles	Magnetic material was produced by oxidation of ferrous ions incorporated into a chitosan-Fe(II) complex	/	Ni(II) Co(II)	833.34 mg/g 588.24 mg/g	Langmuir	5.5	24 h 200 rpm	25 °C	0.1 g	340	[94]
Chitosan NPs	Ion gelation method using tripolyphosphate	FT-IR TEM	Fe(II) Mn)II) Zn(II) Cu(II)	99.94% 80.85% 90.49% 95.93%	/	7	30 min 100 tr/min	Room temperature	2 g/L	20 ppm	[97]
Magnetite Fe_3O_4 NPs coated with chitosan	Iron oxide NPs were synthesized using co-precipitation	TEM, FTIR XRD VSM	Cd(II) Cr(VI)	99% 98%	Langmuir Fre-undlich	6.5 2.5	10 min 400 rpm	22 °C	/	10, 20, 40 and 80 mg/L	[104]

Magnetic chitosan NPs Seyedi et al., (2013) and chitosan magnetic beads [100], used for Cd(II) removal, followed the Langmuir adsorption isotherm. Magnetic chitosan NPs, chitosan NPs, and Chi-MG/Fe_3O_4 have the adsorption capacity for cadmium removal of 580 mg/g, 358 mg/g, and 163.9 mg/g, respectively [93,99,100]. Sivakami et al. (2013) and Tang et al. (2007) reported the removal of hazardous Cr(VI). The initial concentration of Cr(VI) was 200 mg/L with an adsorbent dose of 10 g/L, kept for 60 min. With the optimum time of one hour, chitosan NPs prepared by ionic gelation removed 61–75% of the metal ions from an aqueous solution with an adsorption capacity of 1395.32 mg/g, which was considered to be very good [95,96]. Qi & Xu, (2004) prepared chitosan NPs by ionotropic gelation of chitosan and tripolyphosphatecross–linking to use for the removal of Pb(II) and it is used in the work of Sivakami et al. (2013) to eliminate the Cr(VI) to Cr(III) ions, the adsorption capacity was 398 mg/g and 323.6 mg/g for Pb(II) and Cr(V), respectively [64,96]. According to the authors [96], the high efficiency of materials against Cr(VI) is due to the electrostatic attraction between two oppositely charged ions, since the chromium(VI) ions in the solutions are present in the form of dichromate ions which are negatively charged and chitosan NPs having an overall positive surface charge. Zhou et al. (2009) reported Hg(II) removal with an adsorption capacity of 625.2 mg/g using chitosan composites designated as thiourea-modified magnetic chitosan microsphere composites (TMCS) [101], that is characterized by its high percentage of nitrogen (-N) and sulfur (-S). These composites showed and adsorption capacity of 66.7 mg/g and 15.3 mg/g for Cu(II) and Ni(II) respectively. The amine and sulfur groups present on the surface of TMCS are responsible for metal ion binding through chelation mechanisms. Amine and sulfur sites are the main reactive groups for metal ions though hydroxyl groups in the C-3 position [105,106]. Ample research has also been reported for the removal of heavy metals such as As(V) [103], Cu(II) [94,98], and Co(II) [95] as summarized in Table 2.

Table 3. Chitosan-based nanomaterials for dye removal.

Biosorbent	Modification Method	Characterization Method	Dyes	Adsorption Efficiency or Capacity	Adsorption Isotherm	Parameters					References
						pH	Stirring Time and Speed	Temperature	SD	C Initial (mg/L)	
Chitosan NPs	Ionic gelation	TEM	Rougeacide73 -Orange acide12 -Rouge acide18 -Noir acide26 -Direct bleu78	2.1 mmol/L 1.8 mmol/L 1.4 mmol/L 34.5 mg/g 52.6 mg/g	Langmuir	4	/	/	/	/	[99]
Glutaraldehyde cross-linked magnetic chitosan NPs	Glutaraldehyde-bonded tripolyphosphate	FT-IR TEM	FD&C blue1 D&C jaune5	475.6 mg/g 292.1 mg/g	Langmuir	3.0	60 min 150 tr/min	25 °C	50 mg/50 mL	50–1500	[86]
Magnetic β-cyclodextrin–chitosan/graphene oxide as nanoadsorbent	Co-precipitation method and chemical cross-linking	FTIR SEM TEM XRD	Methylene Bleu	84.3 mg/g	Langmuir Freundlich	11.0	180 rpm	25 °C	0.01 g and 100 mL	/	[107]
MCNCs	Glutaraldehyde cross-linking	/	Crystal violet Acid Red	333.3 mg/g (72%)	Langmuir Freundlich Temkin	7.0	140 min	25 °C	1.0 g	77	[108]
MCNCs	Reduction precipitation method	XRD	Acid Red	90.1 mg/g	Redlich–Peterson Langmuir Freundlich	3.0	/	/	1.0 g/L	/	[109]
Chitosan-coated magnetic mesoporous silica NPs	Chemical co-precipitation and chemical modification	FTIR XRD TEM TGA	Methylene Blue	43 mg/g	Langmuir Freundlich	7.0	90 min	25 °C	0.02 g	20	[110]
Magnetic chitosan NPs	Chemical co-precipitation	SEM, TEM FTIR, XRD VSM	RR-141 RY-14	99.5% (98.8 mg/g) 92.7% (89.7 mg/g)	Freundlich	4.0–10.00	30 min 120 rpm	25 °C	/	/	[111]

4.1.2. Dye Removal

Dyes are chemical compounds that impart color when attached to the surface of fabrics, papers, and cosmetics. Dyes are used in the manufacture of ink, paint, and laboratories. The annual production of commercial dyes used in industry is estimated to be over 7×10^5 tonnes per year. An average of 1×10^2 tonnes per year is discharged into water bodies as waste. Such an amount of dye effluent in water can cause serious harm to all living things [45]. Prolonged exposure to toxic dyes leads to skin irritation, respiratory disorders, and even cancer [9]. For these reasons, researchers are interested in eliminating toxic molecules using materials such as NPs and magnetic NPs of chitosan, as shown by [99,112]. Table 3 shows some examples of chitosan NPs and experimental conditions for the elimination of toxic dyes. The main dyes used in laboratories, and textile and color industries are methylene blue, crystal violet, and acid red. Chitosan composites can successfully remove these dyes from wastewater along with other dyes such as rougeacide73, orange acide12, rouge acide18, noir acide26, and direct blue 78 [99].

Zhou et al. (2014) successfully prepared glutaraldehyde cross-linked magnetic chitosan NPs to remove FD&C blue1, and D&C jaune5 dyes with the adsorption capacity of 475.61 mg/g and 292.07 mg/g, respectively, within 1.0 h at a dosage of 1.0 g/L. The adsorption processes were spontaneous and exothermic, thus following the Langmuir isotherm, which suggested monolayer adsorption on the surface and confirmed low cytotoxicity [86].

Methylene blue was successfully adsorbed from an aqueous solution by various type of magnetic sorbents [107], including, magnetic β-cyclodextrin–chitosan/graphene oxide (as nano adsorbent). The study concluded that, with the least dose of 0.1 g/L, the composite of cyclodextrin-chitosan-graphene obtained an adsorption capacity of 84.3 mg/g, which is higher than that of chitosan-coated magnetic mesoporous silica NPs (43.03 mg/g) (Li et al. 2015) [85]. Magnetic chitosan nanocomposite and glutaraldehyde cross-linked magnetic chitosan nanocomposites were prepared for the removal of crystal violet and acid red, respectively [108,109]. Adsorption of crystal violet was possible with an adsorption efficiency of 72% and adsorption capacity of 333.3 mg/g at SD of 1g/L, $C_0 = 77$ mg/L, and contact time of 140 min, while the adsorption capacity for acid red was 90.06 mg/g. The interaction between chitosan and CV is due to the binding between the $-NH_2$ and –OH groups of chitosan and amino cationic groups of dye CV. Furthermore, according to the results of kinetics data obtained by researchers [108] the adsorption of CV on MC followed the intra-particle diffusion model, which implies CV adsorption on MCNC was a chemical process and electrostatic interaction, at the same time (a mixed mechanism).

4.1.3. Uranium and Light/Heavy Rare Earth Metal Removal

Uranium belongs to the family of actinoid or actinide elements, which contains 15 consecutive chemical elements in the periodic table. These elements, especially uranium, are important due to the common property of radioactivity. The importance of uranium is its use in atomic weapons for their explosive power and in nuclear plants for the production of electrical power. Studies on the adsorption of traces of uranium U(VI) by using different nanomaterials such as magnetic chitosan NPs have increased research interest. Chitosan magnetic NPs are not only limited to heavy metal or dye removal but also extended to the removal of rare earth metals, Some studies are summarized in Table 4. In the study of Galhoum et al. (2015), materials based on magnetic chitosan NPs were prepared to extract La(III), Nd(III), and Yb(III) ions [84]. The authors reported La(III), Nd(III), and Yb(III) removal with adsorption capacities of 16.2 mg/g, 14.6 mg/g, and 12.9 mg/g, respectively, using cysteine-functionalized chitosan magnetic NPs.

There are limited studies on the removal of rare earth elements using chitosan composites. For instance, alanine-functionalized chitosan magnetic nano-based particles and serine-functionalized chitosan magnetic NPs were used by Galhoum et al. (2015d) for the scavenging of U(VI). The authors reported the successful removal of U(VI) with adsorption capacities of 85.3 mg/g and 116.5 mg/g of the prepared composites [73], respectively. Later, Galhoum et al. (2017) reported various NPs such as alanine-chitosan magnetic

NPs, cysteine-chitosan magnetic NPs, serine-chitosan magnetic NPs, and ethylenediamine-tetraacetic acid- chitosan magnetic NPs for U(VI) removal [82]. It was observed that ethylenediamine-tetraacetic acid-chitosan magnetic NPs had the best results with an adsorption capacity of 185.0 mg of U(VI)/g, which was the same as diethylenetriamine-chitosan magnetic nano-based particles [113], Alanine-chitosan magnetic NPs had the least adsorption capacity (88.5 mg/g). The adsorption process onto composites followed the Langmuir isotherm model. Tetra-ethylenepentamine-modified chitosan magnetic resins obtained an adsorption capacity of 397 mg/g with an adsorbent dose of 0.1 g, an initial concentration of uranium of 6 mmol/L and an optimum time of 30 min [114].

Table 4. Chitosan-based nanomaterials for uranium and rare earth metals removal.

Biosorbent	Modification Method	Characterization Method	Metal	Adsorption Efficiency or/and Capacity	Adsorption Isotherm	Parameters					References
						pH	Stirring and Speed	Temperature	SD	C Initial	
Alanine-functionalized chitosan magnetic nano-based particles	Co-precipitation followed by hydrothermal treatment and after grafting	FTIR XDR TEM VSM	U(VI)	85.3 mg/g	Langmuir Dubinin-Radushkevich	3.6	50 min 200 rpm	25 °C	0.05 g and 50 mL	110 mg/L	[73]
Serine-functionalized chitosan magnetic nano-based particles		FTIR XDR TEM VSM	U(VI)	116.5 mg/g	Langmuir Dubinin-Radushkevich	3.6	50 min 200 rpm	25 °C	0.05 g and 50 mL	110 mg/L	[73]
EDTA-chitosan magnetic nanoparticles (CMNPs) Alanine-CMNPs Cysteine-CMNPs Serine-CMNPs	Grafting of –N functions by combined polymer precipitation and hydrothermal treatment	VSM XDR TEM	U(VI)	185 mg/g 88.5 mg/g 101.0 mg/g 120.5 mg/g 185 mg/g	Langmuir	4.0	45–60 min 200 rpm	25 °C	1 g/L	110 mg/L	[82]
TEPA modified chitosan magnetic resin	Cross-linking	SEM EDX	U(VI)	397 mg/g	Langmuir Freundlich Dubinin–Radushkevich	4.0	30 min 200 rpm	25 °C	0.1 g	6 mmol/L	[113]
DETA-chitosan magnetic nano-based particles	Co-precipitation method	FT-IR XRD AFM TEM VSM	U(VI)	185 mg/g	Langmuir Dubinin–Radushkevich (D–R)	4.0	40 min 200 rpm	25 °C	0.02 g	110 mg/L	[114]
Cysteine-functionalized chitosan magnetic nano-based particles	Co-precipitation and cross-linking	FTIR XRD TEM VSM	La(III) Nd(III) Yb(III)	16.2 mg/g 14.6 mg/g 12.9 mg/g	Langmuir	5.0	4 h 300 rpm	27 °C	0.05 g and 100 mL	100 mg/L	[84]
Alanine and serine functionalized magnetic nano-based particles	Co-precipitation	/	Nd(III) Yb(III)	9 mg/g 18 mg/g	Langmuir	5.0	4 h	/	/	/	[115]

4.2. Parameters Affecting the Chitosan Biosorbent Performance

There are various factors that affect the adsorption process, such as pH, temperature, contact time, initial concentration of pollutant, and the amount of adsorbent. These factors must be optimized to obtain maximum adsorption capacity.

The pH of a solution strongly affects the adsorption capacity of chitosan nanoparticles, such as increasing pH will increase adsorption. Magnetic chitosan nano-composites contain functional groups, such as $-NH_2$, –NH, –COOH, –OH, and C=S. At higher H^+ concentrations or lower pH, metal cations have to compete with H^+ to be adsorbed onto the surface of magnetic chitosan nano-composites. However, beyond a certain pH range, a further increase leads to precipitation [45]. The pH also affects the surface of the adsorbent by charging the molecule itself or attached functional groups, which further enhances the attraction between the adsorbate and the adsorbent. Contact time is an important parameter in determining the maximum adsorption capacity at a fixed pollutant concentration; it is the minimum time required for a particular concentration of dye or metal to interact with the adsorbent [45]. After reaching T_{eq}, the adsorption becomes constant or slow, which shows that all the adsorbent sites are occupied and that no vacant site is there to adsorb more pollutants. Among the advantages of using NPs in adsorption are the reduction in equilibrium time and the rapid adsorption rate. Generally, the equilibrium time of heavy metal adsorption onto magnetic chitosan nano-composites is lower than 1 h, except for certain types of metal and most dyes.

It is necessary to measure the adsorption capacity by varying the mass of the adsorbent. The percentage of removal of heavy metals and dyes increases with increasing doses of adsorbent. These increases are attributed to the increase in the number of active sites and the surface area of chitosan NPs. Sometimes, equilibrium is achieved if a certain adsorbent mass is exceeded. In other cases, a decrease in the adsorption capacity at an extra amount of adsorbent is observed. Researchers explain this phenomenon by aggregation of particles, which may also have a negative impact on the process because of the decrease in the surface area of the adsorbent. In practical applications, a minimum amount of adsorbent that is capable of satisfying the needs should be employed [116,117].

Similar to contact time and dosage, the initial pollutant concentration directly relates to the active sites during adsorption and helps determine the adsorption capacity of the adsorbent. Pollutant removal efficiency decreases with an increasing initial concentration of these pollutants because active sites become occupied and saturated over time. According to the results obtained by researchers, when the initial concentration of adsorbate increases, the adsorption first increases, and then at some point, it becomes constant, which shows that almost all the sites on the surface of the adsorbent are filled, and no vacant sites are available for adsorption [118–120].

Temperature is an important factor that affects the efficiency of adsorption. Studying the effect of temperature helps decide whether the process is exothermic or endothermic. When adsorption increases with increasing temperature, the process is an endothermic process. The increase in temperature increases the mobility of the pollutant, so the adsorption becomes faster due to the easy access of the adsorbate to the adsorbent. For example, Seyed Massoud Seyedi (2013) showed that the elimination of Cd(II) took place at 28 °C for chitosan and nanochitosan [103]. Furthermore, according to Table 4, researchers prefer to use chitosan NPs at an ambient temperature of 25 °C.

5. Predominant Mechanisms When Using Chitosan NPs

The adsorption mechanisms depend on the pH of the solution, pH_{pzc}, chemical composition of chitosan and composite structure, chemical nature of the metal ions, nature of their charges (cations or anions) as well as the number of atoms that compose them (monoatomic or polyatomic) [121].

Knowing that the pH affects the surface of the adsorbent by charging the molecule itself or the functional groups attached, further enhances the attraction between the adsorbate and the adsorbent [45]. The adsorption of methyl orange (MO) on magnetic chitosan beads

(MagCH) was tested in the work of L. Obeid et al., (2013) they noticed that the adsorption is strongly linked by the pH value of the medium The results indicated a better efficiency of 99% (≈779 mg/g) in the pH range of 3 to 5 due to electrostatic attraction between the negative functional groups of the dye and the positively charged magnetic beads [122].

In the work of Massoudinejad et al. (2019) [108], the pH_{pzc} of the Fe_3O_4 NPs was about 7.1, confirming the coating of chitosan, and it is evident that the MC nanocomposites were positively charged at pH < 6.5, which is consistent with the work of Chang and Chen (2005) who found the value of 6.7 [123]. While in the work of Zhou et al. (2009), the isoelectric point (IEP) of the TMCS was 5.92. This revealed that the TMCS microspheres were positively charged at pH < 5.92 and negatively charged at pH > 5.92. So, TMCS is positively charged with a pI of 5.92. Therefore, in acidic solutions, it is protonated and possesses electrostatic properties. Thus, it is also possible to adsorb metal ions through anion exchange mechanisms.

All types of nano-adsorbents have functional groups (active sites), such as carboxyl, hydroxyl, and amine groups, which bind to heavy metals either by ion exchange (where –COOH groups are involved) or by a complexation mechanism (where –COOH groups, –OH, –SH, and $-NH_2$ may be involved). Metal ions generally bind to the magnetic chitosan nano-composites via the available functional groups, and the mechanism is proposed according to the active sites available on the surface of these materials. Each site gives several possibilities of interaction; therefore, the mechanism can be considered by the researcher, taking into consideration the nature of the site and the pollutant.

The binding mechanisms are influenced by both the type of ion/dyes and the functional groups of the active sites. The amine groups ($-NH_2$) were the most reactive and mainly responsible for fixing metals rather than the hydroxyl groups carried by the C-3 carbon [123]. The strong attraction of the free electron pair to the nitrogen atom was greater than that of the oxygen atom, and it contributed to the formation of a metal complex by sharing its free electron pair with a metal ion [124]. In particular, the existence of the functional groups containing oxygen, such as carboxylic, phenolic, or lactone functions, results in an acidic character, while the presence of pyrones, chromene-type functions induces a basic character [125]. Depending on their nature and concentration, these surface functional groups can influence an adsorbent's adsorption capacity and the hydrophilic/hydrophobic character.

The three main mechanisms for fixing pollutants on chitosan mentioned in the literature are (i) complexation and/or chelation, (ii) ion-exchange/electrostatic interaction with anions, and (iii) formation of ternary complexes. Adsorption mechanisms are much more complex with magnetic NPs than with normal chitosan.

There are several models for modeling and analyzing the experimental data, such as the Freundlich, Langmuir, Sips, and Dubinin-Radushkevich isotherm models. According to the best-adapted model of isotherms and kinetics, we can understand the sorption mechanisms and can give an idea that is not confirmed on the nature/type of adsorption, so the successful fitting of the kinetic or isotherm model alone does not validate any evidence about chemisorption or physisorption.

For example, the Dubinin-Radushkevich model discriminates between physical and chemical mechanisms [83]. The application of the Sips or/and Langmuir model is linked directly to the nature of the surface, the homogeneity or heterogeneity of the surface (identical/or different active site), monolayer (in the case of chemical and specific adsorption, which must be carried out in a specific active site) or multilayer (in the case of physical adsorption) or also mixed sorption according to the results obtained. Despite the complexity of this mechanism, we can say that chemisorption occurs via the hydroxyl and/or amine groups of chitosan or additional functional groups present (thiol, amino, carboxyl, etc.).

6. Conclusions and Future Perspectives

A wide range of treatment technologies is developing daily for water purification to meet the current demands. Advances in nanoscale science and engineering offer new opportunities to develop more cost-effective and environmentally acceptable water treatment technology. Nano means reducing the size of any material to the nanoscale, and in doing so, the properties of that material change dramatically. Recent research has indicated that chitosan nanomaterials, due to their unique physicochemical properties, structure, and surface characteristics, are useful tools for the removal of organic and inorganic pollutants. These materials can remove metal ions at low concentrations, with high selectivity and adsorption capacity. Nowadays, nanotechnology has become an advanced technology that gives scientists a tool to prepare effective adsorbents to remove contaminants from water. Nanoscale adsorbents have high performance due to their large specific surface area and quantum size effect, which could cause them to exhibit higher capacities for pollutants, particularly metal ions. We conclude that there are different methods for the development of chitosan NPs. Among these methods, covalent cross-linking, electrostatic interactions, and microemulsion methods are widely accepted. At the same time, there are different methods of analysis to confirm the success of the chemical modification, such as FT-IR, XRD, SEM, TEM, AFM, DLS, and VSM.

In addition, the use of NPs and magnetic NPs of chitosan offers several advantages in the field of adsorption compared to the initial powder, because chitosan is already known for its great affinity with respect to most pollutants and this form gives more advantages and reinforces its stability and efficiency when used in the field of wastewater treatment (rapid separation, rapid kinetics, stability, easy and total desorption, large specific surface).

Further improvements should be made in the direction of developing materials with greater stability and the ability to simultaneously remove multiple contaminants under complex conditions. In addition, there is a need to synthesize inexpensive, efficient, and recyclable adsorbents for their practical applications. These techniques constitute reproducible processes with excellent size control, but the use of a chemical cross-linker, such as glutaraldehyde, and some organic solvents, can be detrimental to the final biocompatibility of the NPs. This review is a compilation of recent chitosan-nanomaterial composites reported recently for the removal of heavy metals, dyes, rare earth metals, or radioactive materials. Comparison of different composites, their preparation method, and their working conditions are summarized in this paper, which allows the reader to apply the modifications to previously available methods. Chitosan is a wonder material, which can be utilized in different forms. One can take the opportunity to read this review and further research the unlocked potential of this marine waste-extracted chitosan. In the present era of fast development, where environmental pollution is a huge global problem, chitosan gives us the opportunity to prepare a plethora of composites for different applications.

Author Contributions: Conceptualization, A.B., F.Z.S., P.P. and M.U. (Muhammad Usman 1); validation, S.G., C.H.C., M.U. (Muhammad Usman 1) and M.S.; formal analysis, A.B.; investigation, A.B. and F.Z.S.; resources, M.U. (Muhammad Usman 1); data curation, A.B.; writing—original draft preparation, A.B., F.Z.S., P.P., S.G. and M.U. (Muhammad Usman 1); writing—review and editing, A.B., S.G., C.H.C., M.U. (Muhammad Usman 1), M.U. (Muhammad Usman 2) and M.S.; visualization, A.B.; supervision, M.S. All authors have read and agreed to the published version of the manuscript.

Funding: This research received no external funding.

Data Availability Statement: Not applicable.

Acknowledgments: The authors would like to thank all the institutes for providing the necessary facilities to carry out this review. This work is supported by DGRSDT and the ministry of higher education and scientific research, Algeria, and le Laboratoire de Génie Chimique et de catalyse hétérogène, university USTO-MB, Algeria, which is under construction although he was born in 1986 by Pr. KESSAS Rachid.

Conflicts of Interest: The authors declare no conflict of interest.

References

1. Usman, M.; Waseem, M.; Mani, N.; Andiego, G. Optimization of soil aquifer treatment by chemical oxidation with hydrogen peroxide addition. *Pollution* **2018**, *4*, 369–379.
2. Gao, J.; Yuan, Y.; Yu, Q.; Yan, B.; Qian, Y.; Wen, J.; Ma, C.; Jiang, S.; Wang, X.; Wang, N. Bio-inspired antibacterial cellulose paper-poly(amidoxime) composite hydrogel for highly efficient uranium(vi) capture from seawater. *Chem. Commun.* **2020**, *56*, 3935–3938. [CrossRef] [PubMed]
3. Jjagwe, J.; Olupot, P.W.; Menya, E.; Kalibbala, H.M. Synthesis and Application of Granular Activated Carbon from Biomass Waste Materials for Water Treatment: A Review. *J. Bioresour. Bioprod.* **2021**, *6*, 292–322. [CrossRef]
4. Obey, G.; Adelaide, M.; Ramaraj, R. Biochar derived from non-customized matamba fruit shell as an adsorbent for wastewater treatment. *J. Bioresour. Bioprod.* **2022**, *7*, 109–115. [CrossRef]
5. Ahila, K.G.; Vasanthy, M.; Thamaraiselvi, C. Green Synthesis of Magnetic Iron Nanoparticle Using Moringa oleifera Lam Seeds and Its Application in Textile Effluent Treatment. In *Utilization and Management of Bioresources*; Springer: Singapore, 2018; pp. 315–324.
6. Hamza, M.F.; Wei, Y.; Benettayeb, A.; Wang, X.; Guibal, E. Efficient removal of uranium, cadmium and mercury from aqueous solutions using grafted hydrazide-micro-magnetite chitosan derivative. *J. Mater. Sci.* **2019**, *55*, 4193–4212. [CrossRef]
7. Zhang, W.X. Nanoscale iron particles for environmental remediation: An overview. *J. Nanopart. Res.* **2003**, *5*, 323–332. [CrossRef]
8. Ju-Nam, Y.; Lead, J.R. Manufactured nanoparticles: An overview of their chemistry, interactions and potential environmental implications. *Sci. Total Environ.* **2008**, *400*, 396–414. [CrossRef] [PubMed]
9. Osagie, C.; Othmani, A.; Ghosh, S.; Malloum, A.; Kashitarash Esfahani, Z.; Ahmadi, S. Dyes adsorption from aqueous media through the nanotechnology: A review. *J. Mater. Res. Technol.* **2021**, *14*, 2195–2218. [CrossRef]
10. Rashtbari, Y.; Sher, F.; Afshin, S.; Hamzezadeh bahrami, A.; Ahmadi, S.; Azhar, O.; Rastegar, A.; Ghosh, S.; Poureshgh, Y. Green synthesis of zero-valent iron nanoparticles and loading effect on activated carbon for furfural adsorption. *Chemosphere* **2022**, *287*, 132114. [CrossRef]
11. Igwegbe, C.A.; Ighalo, J.O.; Ghosh, S.; Ahmadi, S.; Ugonabo, V.I. Pistachio (Pistacia vera) waste as adsorbent for wastewater treatment: A review. *Biomass Convers. Biorefin.* **2021**. [CrossRef]
12. Thirunavukkarasu, A.; Nithya, R.; Sivashankar, R. A review on the role of nanomaterials in the removal of organic pollutants from wastewater. *Rev. Environ. Sci. Biotechnol.* **2020**, *19*, 751–778. [CrossRef]
13. Kurniawan, T.A.; Sillanpää, M.E.T.; Sillanpää, M. Nanoadsorbents for Remediation of Aquatic Environment: Local and Practical Solutions for Global Water Pollution Problems. *Crit. Rev. Environ. Sci. Technol.* **2012**, *42*, 1233–1295. [CrossRef]
14. Ghosh, S.; Malloum, A.; Bornman, C.; Othmani, A.; Osagie, C.; Esfahani, Z.K.; Khanday, W.A.; Ahmadi, S.; Dehghani, M.H. Novel green adsorbents for removal of aniline from industrial effluents: A review. *J. Mol. Liq.* **2022**, *345*, 118167. [CrossRef]
15. Usman, M.; Belkasmi, A.I.; Katsoyiannis, I.A.; Ernst, M. Pre-deposited dynamic membrane adsorber formed of microscale conventional iron oxide-based adsorbents to remove arsenic from water: Application study and mathematical modeling. *J. Chem. Technol. Biotechnol.* **2021**, *96*, 1504–1514. [CrossRef]
16. Usman, M.; Katsoyiannis, I.; Mitrakas, M.; Zouboulis, A.; Ernst, M. Performance evaluation of small sized powdered ferric hydroxide as arsenic adsorbent. *Water* **2018**, *10*, 957. [CrossRef]
17. Benettayeb, A.; Morsli, A.; Elwakeel, K.Z.; Hamza, M.F.; Guibal, E. Recovery of Heavy Metal Ions Using Magnetic Glycine—Modified Chitosan—Application to Aqueous Solutions and Tailing Leachate. *Appl. Sci.* **2021**, *11*, 8377. [CrossRef]
18. Ghosh, S.; Malloum, A.; Igwegbe, C.A.; Ighalo, J.O.; Ahmadi, S.; Dehghani, M.H.; Othmani, A.; Gökkuş, Ö.; Mubarak, N.M. New generation adsorbents for the removal of fluoride from water and wastewater: A review. *J. Mol. Liq.* **2022**, *346*, 118257. [CrossRef]
19. Benettayeb, A.; Ghosh, S.; Usman, M.; Seihoub, F.Z. Some Well-Known Alginate and Chitosan Modifications Used in Adsorption: A Review. *Water J.* **2022**, *14*, 1353. [CrossRef]
20. Benettayeb, A.; Usman, M.; Tinashe, C.C.; Haddou, B. A critical review with emphasis on recent pieces of evidence of Moringa oleifera biosorption in water and wastewater treatment. *Environ. Sci. Pollut. Res.* **2022**, *29*, 48185–48209. [CrossRef]
21. Choi, O.; Deng, K.K.; Kim, N.J.; Ross, L.; Surampalli, R.Y.; Hu, Z. The inhibitory effects of silver nanoparticles, silver ions, and silver chloride colloids on microbial growth. *Water Res.* **2008**, *42*, 3066–3074. [CrossRef]
22. Shiju, N.R.; Guliants, V.V. Recent developments in catalysis using nanostructured materials. *Appl. Catal. A Gen.* **2009**, *356*, 1–17. [CrossRef]
23. Kim, K.; Jung, B.; Kim, J.; Kim, W. Effects of embedding non-absorbing nanoparticles in organic photovoltaics on power conversion efficiency. *Sol. Energy Mater. Sol. Cells* **2010**, *94*, 1835–1839. [CrossRef]
24. Phillips, J.; Bowen, W.; Cagin, E.; Wang, W. Electronic and Optoelectronic Devices Based on Semiconducting Zinc Oxide. In *Comprehensive Semiconductor Science and Technology*; Elsevier: Amsterdam, The Netherlands, 2011; Volume 1–6, pp. 101–127; ISBN 9780444531537.
25. Vakili, M.; Rafatullah, M.; Salamatinia, B.; Abdullah, A.Z.; Ibrahim, M.H.; Tan, K.B.; Gholami, Z.; Amouzgar, P. Application of chitosan and its derivatives as adsorbents for dye removal from water and wastewater: A review. *Carbohydr. Polym.* **2014**, *113*, 115–130. [CrossRef]
26. Benettayeb, A.; Guibal, E.; Morsli, A.; Kessas, R. Chemical modification of alginate for enhanced sorption of Cd(II), Cu(II) and Pb(II). *Chem. Eng. J.* **2017**, *316*, 704–714. [CrossRef]

27. Benettayeb, A.; Guibal, E.; Bhatnagar, A.; Morsli, A.; Kessas, R. Effective removal of nickel (II) and zinc (II) in mono-compound and binary systems from aqueous solutions by application of alginate-based materials. *Int. J. Environ. Anal. Chem.* **2021**, 1–22. [CrossRef]
28. Benettayeb, A.; Morsli, A.; Guibal, E.; Kessas, R. New derivatives of urea-grafted alginate for improving the sorption of mercury ions in aqueous solutions. *Mater. Res. Express* **2021**, *8*, 035303. [CrossRef]
29. Benettayeb, A.; Haddou, B. New biosorbents based on the seeds, leaves and husks powder of Moringa oleifera for the effective removal of various toxic pollutants. *Int. J. Environ. Anal. Chem.* **2021**, 1–26. [CrossRef]
30. Hamza, M.F.; Fouda, A.; Wei, Y.; El Aassy, I.E.; Alotaibi, S.H.; Guibal, E.; Mashaal, N.M. Functionalized biobased composite for metal decontamination–Insight on uranium and application to water samples collected from wells in mining areas (Sinai, Egypt). *Chem. Eng. J.* **2022**, *431*, 133967. [CrossRef]
31. Hamza, M.F.; Khalafalla, M.S.; Wei, Y.; Hamad, N.A. Effect of bi-functionalization silica micro beads on uranium adsorption from synthetic and washing pregnant uranyl solutions. *J. Radioanal. Nucl. Chem.* **2021**, *330*, 191–206. [CrossRef]
32. Wang, J.; Sun, Y.; Zhao, X.; Chen, L.; Peng, S.; Ma, C.; Duan, G.; Liu, Z.; Wang, H.; Yuan, Y.; et al. A poly(amidoxime)-modified MOF macroporous membrane for high-efficient uranium extraction from seawater. *E-Polymers* **2022**, *22*, 399–410. [CrossRef]
33. Chen, Y.; Li, S.; Li, X.; Mei, C.; Zheng, J.; Shiju, E.; Duan, G.; Liu, K.; Jiang, S. Liquid Transport and Real-Time Dye Purification via Lotus Petiole-Inspired Long-Range-Ordered Anisotropic Cellulose Nanofibril Aerogels. *ACS Nano* **2021**, *15*, 20666–20677. [CrossRef] [PubMed]
34. Chen, Y.; Hanshe, M.; Sun, Z.; Zhou, Y.; Mei, C.; Duan, G.; Zheng, J.; Shiju, E.; Jiang, S. Lightweight and anisotropic cellulose nanofibril/rectorite composite sponges for efficient dye adsorption and selective separation. *Int. J. Biol. Macromol.* **2022**, *207*, 130–139. [CrossRef] [PubMed]
35. Gregorio-Jauregui, K.M.; Pineda, M.G.; Rivera-Salinas, J.E.; Hurtado, G.; Saade, H.; Martinez, J.L.; Ilyina, A.; López, R.G. One-step method for preparation of magnetic nanoparticles coated with chitosan. *J. Nanomater.* **2012**, *2012*, 4. [CrossRef]
36. Ngah, W.S.W.; Fatinathan, S. Pb(II) biosorption using chitosan and chitosan derivatives beads: Equilibrium, ion exchange and mechanism studies. *J. Environ. Sci.* **2010**, *22*, 338–346. [CrossRef]
37. Fan, L.; Zhang, Y.; Luo, C.; Lu, F.; Qiu, H.; Sun, M. Synthesis and characterization of magnetic β-cyclodextrin-chitosan nanoparticles as nano-adsorbents for removal of methyl blue. *Int. J. Biol. Macromol.* **2012**, *50*, 444–450. [CrossRef]
38. Chen, W.; Mo, J.; Du, X.; Zhang, Z.; Zhang, W. Biomimetic dynamic membrane for aquatic dye removal. *Water Res.* **2019**, *151*, 243–251. [CrossRef]
39. Pal, P.; Pal, A. Methylene blue removal: An approach towards sludge management after adsorption of cadmium onto surfactant modified chitosan beads. *J. Indian Chem. Soc.* **2018**, *95*, 357–364.
40. Fu, C.C.; Tran, H.N.; Chen, X.H.; Juang, R.S. Preparation of polyaminated Fe_3O_4@chitosan core-shell magnetic nanoparticles for efficient adsorption of phosphate in aqueous solutions. *J. Ind. Eng. Chem.* **2020**, *83*, 235–246. [CrossRef]
41. Pap, S.; Kirk, C.; Bremner, B.; Turk Sekulic, M.; Gibb, S.W.; Maletic, S.; Taggart, M.A. Synthesis optimisation and characterisation of chitosan-calcite adsorbent from fishery-food waste for phosphorus removal. *Environ. Sci. Pollut. Res.* **2020**, *27*, 9790–9802. [CrossRef]
42. Liang, L.; Xi, F.; Tan, W.; Meng, X.; Hu, B.; Wang, X. Review of organic and inorganic pollutants removal by biochar and biochar-based composites. *Biochar* **2021**, *3*, 255–281. [CrossRef]
43. Qiu, M.; Hu, B.; Chen, Z.; Yang, H.; Zhuang, L.; Wang, X. Challenges of organic pollutant photocatalysis by biochar-based catalysts. *Biochar* **2021**, *3*, 117–123. [CrossRef]
44. Yu, S.; Pang, H.; Huang, S.; Tang, H.; Wang, S.; Qiu, M.; Chen, Z.; Yang, H.; Song, G.; Fu, D.; et al. Recent advances in metal-organic framework membranes for water treatment: A review. *Sci. Total Environ.* **2021**, *800*, 149662. [CrossRef]
45. Saxena, R.; Saxena, M.; Lochab, A. Recent Progress in Nanomaterials for Adsorptive Removal of Organic Contaminants from Wastewater. *ChemistrySelect* **2020**, *5*, 335–353. [CrossRef]
46. Saad, E.M.; Elshaarawy, R.F.; Mahmoud, S.A.; El-Moselhy, K.M. New Ulva lactuca Algae Based Chitosan Bio-composites for Bioremediation of Cd(II) Ions. *J. Bioresour. Bioprod.* **2021**, *6*, 223–242. [CrossRef]
47. Moradi Dehaghi, S.; Rahmanifar, B.; Moradi, A.M.; Azar, P.A. Removal of permethrin pesticide from water by chitosan-zinc oxide nanoparticles composite as an adsorbent. *J. Saudi Chem. Soc.* **2014**, *18*, 348–355. [CrossRef]
48. Crini, G. Studies on adsorption of dyes on beta-cyclodextrin polymer. *Bioresour. Technol.* **2003**, *90*, 193–198. [CrossRef] [PubMed]
49. Crini, G. Recent developments in polysaccharide-based materials used as adsorbents in wastewater treatment. *Prog. Polym. Sci.* **2005**, *30*, 38–70. [CrossRef]
50. Crini, G.; Badot, P.M. Application of chitosan, a natural aminopolysaccharide, for dye removal from aqueous solutions by adsorption processes using batch studies: A review of recent literature. *Prog. Polym. Sci.* **2008**, *33*, 399–447. [CrossRef]
51. Madni, A.; Kousar, R.; Naeem, N.; Wahid, F. Recent advancements in applications of chitosan-based biomaterials for skin tissue engineering. *J. Bioresour. Bioprod.* **2021**, *6*, 11–25. [CrossRef]
52. Azzaza, S.; Kumar, R.T.; Vijaya, J.J.; Bououdina, M. CHAPTER 7: Nanomaterials for Heavy Metal Removal. In *Biological Fluid-Surface Interactions in Detection and Medical Devices*; Royal Society of Chemistry: London, UK, 2017; ISBN 9781782620976.
53. Jeevanandam, J.; Barhoum, A.; Chan, Y.S.; Dufresne, A.; Danquah, M.K. Review on nanoparticles and nanostructured materials: History, sources, toxicity and regulations. *Beilstein J. Nanotechnol.* **2018**, *9*, 1050–1074. [CrossRef]

54. Ambashta, R.D.; Sillanpää, M. Water purification using magnetic assistance: A review. *J. Hazard. Mater.* **2010**, *180*, 38–49. [CrossRef] [PubMed]
55. Tuutijärvi, T.; Lu, J.; Sillanpää, M.; Chen, G. As(V) adsorption on maghemite nanoparticles. *J. Hazard. Mater.* **2009**, *166*, 1415–1420. [CrossRef] [PubMed]
56. Zhou, L.; Liu, Z.; Liu, J.; Huang, Q. Adsorption of Hg(II) from aqueous solution by ethylenediamine-modified magnetic crosslinking chitosan microspheres. *Desalination* **2010**, *258*, 41–47. [CrossRef]
57. Chang, Y.C.; Chang, S.W.; Chen, D.H. Magnetic chitosan nanoparticles: Studies on chitosan binding and adsorption of Co(II) ions. *React. Funct. Polym.* **2006**, *66*, 335–341. [CrossRef]
58. Xu, P.; Zeng, G.M.; Huang, D.L.; Feng, C.L.; Hu, S.; Zhao, M.H.; Lai, C.; Wei, Z.; Huang, C.; Xie, G.X.; et al. Use of iron oxide nanomaterials in wastewater treatment: A review. *Sci. Total Environ.* **2012**, *424*, 1–10. [CrossRef]
59. Lu, A.H.; Salabas, E.L.; Schüth, F. Magnetic nanoparticles: Synthesis, protection, functionalization, and application. *Angew. Chemie-Int. Ed.* **2007**, *46*, 1222–1244. [CrossRef]
60. Ohya, Y.; Shiratani, M.; Kobayashi, H.; Ouchi, T. Release behavior of 5-fluorouracil from chitosan-gel nanospheres immobilizing 5-fluorouracil coated with polysaccharides and their cell specific cytotoxicity. *J. Macromol. Sci. Part A* **1994**, *31*, 629–642. [CrossRef]
61. Banerjee, T.; Mitra, S.; Kumar Singh, A.; Kumar Sharma, R.; Maitra, A. Preparation, characterization and biodistribution of ultrafine chitosan nanoparticles. *Int. J. Pharm.* **2002**, *243*, 93–105. [CrossRef]
62. Alonso Fernandez, M.J.; Calvo Salve, P.; Remunan Lopes, C.; Vila Jato, J.L. B05s Application of nanoparticles based on hydrophilic polymers as pharmaceutical forms. Patent Cooperation Treaty No. EP0860166A1, 22 October 2016.
63. Hu, B.; Pan, C.; Sun, Y.; Hou, Z.; Ye, H.; Hu, B.; Zeng, X. Optimization of fabrication parameters to produce chitosan-tripolyphosphate nanoparticles for delivery of tea catechins. *J. Agric. Food Chem.* **2008**, *56*, 7451–7458. [CrossRef]
64. Qi, L.; Xu, Z. Lead sorption from aqueous solutions on chitosan nanoparticles. *Colloids Surfaces A Physicochem. Eng. Asp.* **2004**, *251*, 183–190. [CrossRef]
65. Ali, M.E.A.; Aboelfadl, M.M.S.; Selim, A.M.; Khalil, H.F.; Elkady, G.M. Chitosan nanoparticles extracted from shrimp shells, application for removal of Fe(II) and Mn(II) from aqueous phases. *Sep. Sci. Technol.* **2018**, *53*, 2870–2881. [CrossRef]
66. Du, W.L.; Xu, Z.R.; Han, X.Y.; Xu, Y.L.; Miao, Z.G. Preparation, characterization and adsorption properties of chitosan nanoparticles for eosin Y as a model anionic dye. *J. Hazard. Mater.* **2008**, *153*, 152–156. [CrossRef] [PubMed]
67. Malik, M.A.; Wani, M.Y.; Hashim, M.A. Microemulsion method: A novel route to synthesize organic and inorganic nanomaterials. 1st Nano Update. *Arab. J. Chem.* **2012**, *5*, 397–417. [CrossRef]
68. Brunel, F.; Véron, L.; David, L.; Domard, A.; Delair, T. A novel synthesis of chitosan nanoparticles in reverse emulsion. *Langmuir* **2008**, *24*, 11370–11377. [CrossRef] [PubMed]
69. Prakash, A.; Praveen, R.P.; Daisy, P.A.; Abeena, P.B. A Review on Nanoparticles. *Int. J. Pharm. Sci. Rev. Res.* **2020**, *64*, 64–68. [CrossRef]
70. Quiñones, J.P.; Gothelf, K.V.; Kjems, J.; Caballero, Á.M.H.; Schmidt, C.; Covas, C.P. N,O6-partially acetylated chitosan nanoparticles hydrophobically-modified for controlled release of steroids and vitamin e. *Carbohydr. Polym.* **2013**, *91*, 143–151. [CrossRef] [PubMed]
71. Liu, C.G.; Desai, K.G.H.; Chen, X.G.; Park, H.J. Linolenic acid-modified chitosan for formation of self-assembled nanoparticles. *J. Agric. Food Chem.* **2005**, *53*, 437–441. [CrossRef]
72. Morales, M.A.; De Souza Rodrigues, E.C.; De Amorim, A.S.C.M.; Soares, J.M.; Galembeck, F. Size selected synthesis of magnetite nanoparticles in chitosan matrix. *Appl. Surf. Sci.* **2013**, *275*, 71–74. [CrossRef]
73. Galhoum, A.A.; Mahfouz, M.G.; Gomaa, N.A.M.; Vincent, T.; Guibal, E.; Atia, A.A.; Mahfouz, M.G.; Abdel-Rehem, S.T.; Gomaa, N.A.M.; Vincent, T.; et al. Amino Acid Functionalized Chitosan Magnetic Nanobased Particles for Uranyl Sorption. *Ind. Eng. Chem. Res.* **2015**, *54*, 12374–12385. [CrossRef]
74. Barakat, M.A. New trends in removing heavy metals from industrial wastewater. *Arab. J. Chem.* **2011**, *4*, 361–377. [CrossRef]
75. Yang, C.H.; Wang, C.Y.; Huang, K.S.; Yeh, C.S.; Wang, A.H.J.; Wang, W.T.; Lin, M.Y. Facile Synthesis of Radial-Like Macroporous Superparamagnetic Chitosan Spheres with In-Situ Co-Precipitation and Gelation of Ferro-Gels. *PLoS ONE* **2012**, *7*, 49329. [CrossRef] [PubMed]
76. Shaumbwa, V.R.; Liu, D.; Archer, B.; Li, J.; Su, F. Preparation and application of magnetic chitosan in environmental remediation and other fields: A review. *J. Appl. Polym. Sci.* **2021**, *138*, 1–25. [CrossRef]
77. Namdeo, M.; Bajpai, S.K. Chitosan-magnetite nanocomposites (CMNs) as magnetic carrier particles for removal of Fe(III) from aqueous solutions. *Colloids Surfaces A Physicochem. Eng. Asp.* **2008**, *320*, 161–168. [CrossRef]
78. Brião, G.d.V.; de Andrade, J.R.; da Silva, M.G.C.; Vieira, M.G.A. Removal of toxic metals from water using chitosan-based magnetic adsorbents. A review. *Environ. Chem. Lett.* **2020**, *18*, 1145–1168. [CrossRef]
79. Yuwei, C.; Jianlong, W. Preparation and characterization of magnetic chitosan nanoparticles and its application for Cu(II) removal. *Chem. Eng. J.* **2011**, *168*, 286–292. [CrossRef]
80. Huang, G.; Yang, C.; Zhang, K.; Jeffrey, S. Adsorptive Removal of Copper Ions from Aqueous Solution Using Cross-linked Magnetic Chitosan Beads. *Chin. J. Chem. Eng.* **2009**, *17*, 960–966. [CrossRef]
81. Galhoum, A.A.; Mahfouz, M.G.; Abdel-Rehem, S.T.; Gomaa, N.A.; Atia, A.A.; Vincent, T.; Guibal, E. Diethylenetriamine-functionalized chitosan magnetic nano-based particles for the sorption of rare earth metal ions [Nd(III), Dy(III) and Yb(III)]. *Cellulose* **2015**, *22*, 2589–2605. [CrossRef]

82. Galhoum, A.A.; Mahfouz, M.G.; Gomaa, N.M.; Vincent, T.; Guibal, E. Chemical modifications of chitosan nano-based magnetic particles for enhanced uranyl sorption. *Hydrometallurgy* **2017**, *168*, 127–134. [CrossRef]
83. Galhoum, A.A.; Atia, A.A.; Mahfouz, M.G.; Abdel-Rehem, S.T.; Gomaa, N.A.; Vincent, T.; Guibal, E. Dy(III) recovery from dilute solutions using magnetic-chitosan nano-based particles grafted with amino acids. *J. Mater. Sci.* **2015**, *50*, 2832–2848. [CrossRef]
84. Galhoum, A.A.; Mafhouz, M.G.; Abdel-Rehem, S.T.; Gomaa, N.A.; Atia, A.A.; Vincent, T.; Guibal, E. Cysteine-Functionalized chitosan magnetic nano-based particles for the recovery of light and heavy rare earth metals: Uptake kinetics and sorption isotherms. *Nanomaterials* **2015**, *5*, 154–179. [CrossRef]
85. Liu, C.; Wang, B.; Deng, Y.; Cui, B.; Wang, J.; Chen, W.; He, S.Y. Performance of a new magnetic chitosan nanoparticle to remove arsenic and its separation from water. *J. Nanomater.* **2015**, *2015*, 964–967. [CrossRef]
86. Zhou, Z.; Lin, S.; Yue, T.; Lee, T.C. Adsorption of food dyes from aqueous solution by glutaraldehyde cross-linked magnetic chitosan nanoparticles. *J. Food Eng.* **2014**, *126*, 133–141. [CrossRef]
87. Ayati, A.; Ranjbari, S.; Tanhaei, B.; Sillanpää, M. Ionic liquid-modified composites for the adsorptive removal of emerging water contaminants: A review. *J. Mol. Liq.* **2019**, *275*, 71–83. [CrossRef]
88. Ranjbari, S.; Tanhaei, B.; Ayati, A.; Khadempir, S.; Sillanpää, M. Efficient tetracycline adsorptive removal using tricaprylmethylammonium chloride conjugated chitosan hydrogel beads: Mechanism, kinetic, isotherms and thermodynamic study. *Int. J. Biol. Macromol.* **2020**, *155*, 421–426. [CrossRef]
89. Ranjbari, S.; Tanhaei, B.; Ayati, A.; Sillanpää, M. Novel Aliquat-336 impregnated chitosan beads for the adsorptive removal of anionic azo dyes. *Int. J. Biol. Macromol.* **2019**, *125*, 989–998. [CrossRef]
90. Dasgupta, N.; Ranjan, S.; Mundekkad, D.; Ramalingam, C.; Shanker, R.; Kumar, A. Nanotechnology in agro-food: From field to plate. *Food Res. Int.* **2015**, *69*, 381–400. [CrossRef]
91. Fu, F.; Wang, Q. Removal of heavy metal ions from wastewaters: A review. *J. Environ. Manag.* **2011**, *92*, 407–418. [CrossRef]
92. Kuang, S.P.; Wang, Z.Z.; Liu, J.; Wu, Z.C. Preparation of triethylene-tetramine grafted magnetic chitosan for adsorption of Pb(II) ion from aqueous solutions. *J. Hazard. Mater.* **2013**, *260*, 210–219. [CrossRef]
93. Salehi, N.; Moghimi, A.; Shahbazi, H. Preparation of cross-linked magnetic chitosan with methionine-glutaraldehyde for removal of heavy metals from aqueous solutions. *Int. J. Environ. Anal. Chem.* **2020**, *102*, 2305–2321. [CrossRef]
94. Hritcu, D.; Dodi, G.; Popa, M.I. Heavy Metal Ions Adsorption on Chitosan-Magnetite Microspheres. *Int. J. Chem. Eng.* **2012**, *4*, 364–368.
95. Tang, Z.X.; Qian, J.Q.; Shi, L.E. Preparation of chitosan nanoparticles as carrier for immobilized enzyme. *Appl. Biochem. Biotechnol.* **2007**, *136*, 77–96. [CrossRef] [PubMed]
96. Sivakami, M.S.; Gomathi, T.; Venkatesan, J.; Jeong, H.S.; Kim, S.K.; Sudha, P.N. Preparation and characterization of nano chitosan for treatment wastewaters. *Int. J. Biol. Macromol.* **2013**, *57*, 204–212. [CrossRef] [PubMed]
97. Mohamed, A. Abd-Elhakeem, Maha, M. Ramadan, Faisal, S. Basaad, Removing of heavymetals from water by chitosan nanoparticles. *J. Adv. Chem.* **2016**, *11*, 3765–3771. [CrossRef]
98. Fan, L.; Luo, C.; Lv, Z.; Lu, F.; Qiu, H. Preparation of magnetic modified chitosan and adsorption of Zn^{2+} from aqueous solutions. *Colloids Surf. B. Biointerfaces* **2011**, *88*, 574–581. [CrossRef]
99. Olivera, S.; Muralidhara, H.B.; Venkatesh, K.; Guna, V.K.; Gopalakrishna, K.; Kumar, K.Y. Potential applications of cellulose and chitosan nanoparticles/composites in wastewater treatment: A review. *Carbohydr. Polym.* **2016**, *153*, 600–618. [CrossRef]
100. Rorrer, G.L.; Hsien, T.Y.; Way, J.D. Synthesis of Porous-Magnetic Chitosan Beads for Removal of Cadmium Ions from Waste Water. *Ind. Eng. Chem. Res.* **1993**, *32*, 2170–2178. [CrossRef]
101. Zhou, L.; Wang, Y.; Liu, Z.; Huang, Q. Characteristics of equilibrium, kinetics studies for adsorption of Hg(II), Cu(II), and Ni(II) ions by thiourea-modified magnetic chitosan microspheres. *J. Hazard. Mater.* **2009**, *161*, 995–1002. [CrossRef]
102. Sreeram, A.; Hadi, P.; Hui, C.W.; Al Ansari, T.; McKay, G. Optimisation of the removal of arsenate from water using nanochitosan. *Desalin. Water Treat.* **2017**, *70*, 235–243. [CrossRef]
103. Seyedi, S.M.; Anvaripour, B.; Motavassel, M.; Jadidi, N. Comparative Cadmium Adsorption from water by nanochitosan and chitosan. *Int. J. Eng. Innov. Technol.* **2013**, *5*, 145–148.
104. Amin, M.M.; Khodabakhshi, A.; Mozafari, M.; Bina, B.; Kheiri, S. Removal of Cr(VI) from simulated electroplating wastewater by magnetite nanoparticles. *Environ. Eng. Manag. J.* **2010**, *9*, 921–927. [CrossRef]
105. Varma, A.J.; Deshpande, S.V.; Kennedy, J.F. Metal complexation by chitosan and its derivatives: A review. *Carbohydr. Polym.* **2004**, *55*, 77–93. [CrossRef]
106. Ravi Kumar, M.N.V. A review of chitin and chitosan applications. *React. Funct. Polym.* **2000**, *46*, 1–27. [CrossRef]
107. Fan, L.; Luo, C.; Sun, M.; Qiu, H.; Li, X. Synthesis of magnetic β-cyclodextrin-chitosan/graphene oxide as nanoadsorbent and its application in dye adsorption and removal. *Colloids Surf. B Biointerfaces* **2013**, *103*, 601–607. [CrossRef] [PubMed]
108. Massoudinejad, M.; Rasoulzadeh, H.; Ghaderpoori, M. Magnetic chitosan nanocomposite: Fabrication, properties, and optimization for adsorptive removal of crystal violet from aqueous solutions. *Carbohydr. Polym.* **2019**, *206*, 844–853. [CrossRef] [PubMed]
109. Kadam, A.A.; Lee, D.S. Glutaraldehyde cross-linked magnetic chitosan nanocomposites: Reduction precipitation synthesis, characterization, and application for removal of hazardous textile dyes. *Bioresour. Technol.* **2015**, *193*, 563–567. [CrossRef]
110. Li, Y.; Zhou, Y.; Nie, W.; Song, L.; Chen, P. Highly efficient methylene blue dyes removal from aqueous systems by chitosan coated magnetic mesoporous silica nanoparticles. *J. Porous Mater.* **2015**, *22*, 1383–1392. [CrossRef]

111. Jaafari, J.; Barzanouni, H.; Mazloomi, S.; Amir Abadi Farahani, N.; Sharafi, K.; Soleimani, P.; Haghighat, G.A. Effective adsorptive removal of reactive dyes by magnetic chitosan nanoparticles: Kinetic, isothermal studies and response surface methodology. *Int. J. Biol. Macromol.* **2020**, *164*, 344–355. [CrossRef]
112. Zhou, Y.; Zhang, M.; Wang, X.; Huang, Q.; Min, Y.; Ma, T.; Niu, J. Removal of crystal violet by a novel cellulose-based adsorbent: Comparison with native cellulose. *Ind. Eng. Chem. Res.* **2014**, *53*, 5498–5506. [CrossRef]
113. Mahfouz, M.G.; Galhoum, A.A.; Gomaa, N.A.; Abdel-Rehem, S.S.; Atia, A.A.; Vincent, T.; Guibal, E. Uranium extraction using magnetic nano-based particles of diethylenetriamine-functionalized chitosan: Equilibrium and kinetic studies. *Chem. Eng. J.* **2015**, *262*, 198–209. [CrossRef]
114. Elwakeel, K.Z.; Atia, A.A.; Guibal, E. Fast removal of uranium from aqueous solutions using tetraethylenepentamine modified magnetic chitosan resin. *Bioresour. Technol.* **2014**, *160*, 107–114. [CrossRef]
115. Galhoum, A.A.; Mahfouz, M.G.; Atia, A.A.; Gomaa, N.A.; Abdel-Rehem, S.T.; Vincent, T.; Guibal, E. Alanine and serine functionalized magnetic nano-based particles for sorption of Nd(III) and Yb(III). *Adv. Environ. Res.* **2016**, *5*, 1–18. [CrossRef]
116. Cai, W.; Xue, W.; Jiang, Y. Facile Preparation of Magnetic Chitosan Coprecipitated by Ethanol/$NH_3 \cdot H_2O$ for Highly Efficient Removal toward Cr(VI). *ACS Omega* **2018**, *3*, 5725–5734. [CrossRef] [PubMed]
117. Özacar, M.; Şengil, I.A. Adsorption of metal complex dyes from aqueous solutions by pine sawdust. *Bioresour. Technol.* **2005**, *96*, 791–795. [CrossRef] [PubMed]
118. Liu, R.; Zhang, Y.; Hu, B.; Wang, H. Improved Pb(II) removal in aqueous solution by sulfide@biochar and polysaccharose-FeS@ biochar composites: Efficiencies and mechanisms. *Chemosphere* **2022**, *287*, 232087. [CrossRef]
119. Liu, F.; Hua, S.; Wang, C.; Qiu, M.; Jin, L.; Hu, B. Adsorption and reduction of Cr(VI) from aqueous solution using cost-effective caffeic acid functionalized corn starch. *Chemosphere* **2021**, *279*, 130539. [CrossRef]
120. Liu, F.; Hua, S.; Wang, C.; Hu, B. Insight into the performance and mechanism of persimmon tannin functionalized waste paper for U(VI) and Cr(VI) removal. *Chemosphere* **2022**, *287*, 132199. [CrossRef]
121. Crini, G.; Badot, P.-M. (coordonnateurs) Procédés membranaires, bioadsorption et oxydation chimique. In *Traitement et Epuration des Eaux Industrielles Polluées;* Presses Universitaires de Franche-Comté: Besançon, France, 2007; p. 352; ISBN 2848671971.
122. Obeid, L.; Bée, A.; Talbot, D.; Jaafar, S.B.; Dupuis, V.; Abramson, S.; Cabuil, V.; Welschbillig, M. Chitosan/maghemite composite: A magsorbent for the adsorption of methyl orange. *J. Colloid Interface Sci.* **2013**, *410*, 52–58. [CrossRef]
123. Chang, Y.C.; Chen, D.H. Preparation and adsorption properties of monodisperse chitosan-bound Fe_3O_4 magnetic nanoparticles for removal of Cu(II) ions. *J. Colloid Interface Sci.* **2005**, *283*, 446–451. [CrossRef]
124. Roberts, G.A.F. Preparation of Chitin and Chitosan BT–Chitin Chemistry. In *Chitin Chemistry;* Palgrave: London, UK, 1992; pp. 54–84; ISBN 978-1-349-11545-7.
125. Jin, L.; Bai, R. Mechanisms of lead adsorption on chitosan/PVA hydrogel beads. *Langmuir* **2002**, *18*, 9765–9770. [CrossRef]

nanomaterials

Review

Research Progress on the Preparation of Manganese Dioxide Nanomaterials and Their Electrochemical Applications

Chunsheng Xie [1,2], Zesheng Xu [1], Yujian Zheng [1], Shuo Wang [3,4], Min Dai [1,2] and Chun Xiao [1,2,*]

1 College of Environmental and Chemical Engineering, Zhaoqing University, Zhaoqing 526061, China; xiechsh@126.com (C.X.); xuzesheng0712@163.com (Z.X.); zhengyujian10032@163.com (Y.Z.); daimin1007@163.com (M.D.)
2 Guangdong Provincial Key Laboratory of Environmental Health and Land Resource, Zhaoqing University, Zhaoqing 526061, China
3 School of Environmental and Chemical Engineering, Xi'an Polytechnic University, Xi'an 710048, China; 220621098@stu.xpu.edu.cn
4 State Environmental Protection Key Laboratory of Water Environmental Simulation and Pollution Control, South China Institute of Environmental Sciences, Ministry of Ecology and Environment, Guangzhou 510655, China
* Correspondence: 13554301718@163.com

Abstract: Manganese dioxide (MnO_2) nanomaterials have shown excellent performance in catalytic degradation and other fields because of their low density and great specific surface area, as well as their tunable chemical characteristics. However, the methods used to synthesize MnO_2 nanomaterials greatly affect their structures and properties. Therefore, the present work systematically illustrates common synthetic routes and their advantages and disadvantages, as well as examining research progress relating to electrochemical applications. In contrast to previous reviews, this review summarizes approaches for preparing MnO_2 nanoparticles and describes their respective merits, demerits, and limitations. The aim is to help readers better select appropriate preparation methods for MnO_2 nanomaterials and translate research results into practical applications. Finally, we also point out that despite the significant progress that has been made in the development of MnO_2 nanomaterials for electrochemical applications, the related research remains in the early stages, and the focus of future research should be placed on the development of green synthesis methods, as well as the composition and modification of MnO_2 nanoparticles with other materials.

Keywords: nano-MnO_2; preparation method; structure; electrochemical applications

Citation: Xie, C.; Xu, Z.; Zheng, Y.; Wang, S.; Dai, M.; Xiao, C. Research Progress on the Preparation of Manganese Dioxide Nanomaterials and Their Electrochemical Applications. *Nanomaterials* **2024**, *14*, 1283. https://doi.org/10.3390/nano14151283

Academic Editor: Gian Andrea Rizzi

Received: 25 June 2024
Revised: 26 July 2024
Accepted: 27 July 2024
Published: 30 July 2024

1. Introduction

MnO_2 nanomaterials stand out among other nanomaterials owing to their good environmental compatibility, low cost, and strong oxidative and adsorptive properties. Owing to their good biocompatibility, optical physical properties, and chemical properties [1,2], these nanomaterials are used as catalysts [3] and in electrochemistry [4], biomedicine [5], and materials sciences [6], among other fields. MnO_2 nanomaterials have different spatial structures and therefore have different crystalline forms, mainly α-MnO_2, β-MnO_2, γ-MnO_2, δ-MnO_2, and λ-MnO_2. The surface physicochemical characteristics of MnO_2 vary considerably based on the crystal structure. Based on their spatial structure, MnO_2 nanomaterials can be categorized into having a one-dimensional (1D) tunnel structure, a two-dimensional (2D) layered structure, or a three-dimensional (3D) network structure [7].

The synthesis method of MnO_2 nanomaterials crucially impacts their electrochemical performance. It has been shown that chemical synthesis methods yield MnO_2 nanomaterials with poor electrochemical performance, such as low capacity attenuation and low cycling efficiency, due to side reactions and defects in synthesis. Contrarily, hydrothermal methods can yield MnO_2 nanomaterials with crystal structures that are favorable for charge transfer and ion diffusion. Therefore, they usually exhibit enhanced electrochemical performance. The electrochemical deposition method enables MnO_2 to be directly deposited on the electrode; therefore, it has high controllability and a long cycle life, as well as conferring enhanced electrochemical performance. By studying the synthesis of MnO_2 nanomaterials, their applications in electrochemistry have been substantially improved. However, there is an enormous gap between theory and practice. Therefore, this review summarizes the research progress on MnO_2 nanomaterials in recent years, with the aim of helping readers to better select preparation methods for MnO_2 nanomaterials and translate research results into practical applications, as shown in Figure 1. Compared with previous reviews [8,9], this review contains the latest research results in this field in recent years, and these advances have not only improved the performance and stability of MnO_2 nanomaterials but can also be applied across several fields. Recent research has placed much focus on preparing MnO_2 nanomaterials using the green synthesis method, a synthetic method that has been studied more and more in recent years, and which indicates that greening preparation methods will be an important direction for the future.

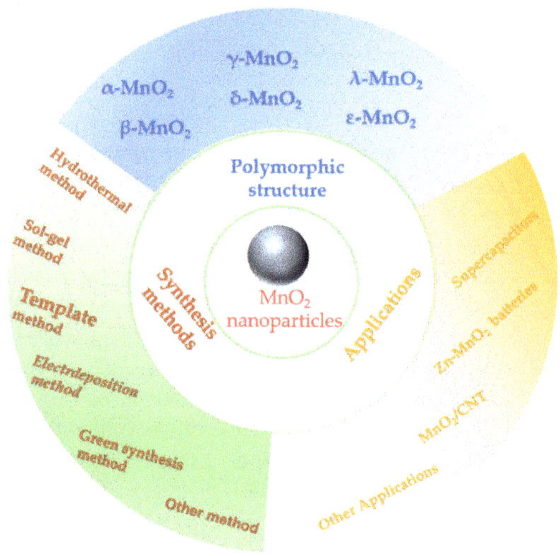

Figure 1. Preparation methods and applications of MnO_2 nanoparticles and polymorphic structures.

2. Structure of MnO_2 Nanoparticles

The polymorphic phases of MnO_2 usually comprise crystalline and amorphous phases. The crystalline phase comprises octahedral units; they can form either layered or chain/tunnel structures when different joining methods are used [10–12]. The interconnection of MnO_6 octahedra forms *c*-axis paralleling chains within the crystal structure, along with tunnels between these chains. The different polymorphs can be associated with Mn^{4+} arrangement, as each polymorph contains a hexagonal close-packed lattice structure composed of O^{2-} and Mn^{4+} [13].

The structures of different manganese dioxide materials are shown in Table 1. α-MnO_2, β-MnO_2, and γ-MnO_2 have 1D $(1 \times 1)/(2 \times 2)$, $(1 \times 1)/(1 \times 1)$, and $(1 \times 1)/(1 \times 2)$ tunneling structures, respectively. However, β-MnO_2 has a smaller tunneling structure, which is unfavorable for rapid ion transport, while α-MnO_2 has a larger tunneling structure, which is favorable for ion embedding and detachment. ε-MnO_2 has an alike structure to γ-MnO_2; however, the manganese lattice sites are arranged in a disorderly manner, with irregular tunneling. δ-MnO_2 has a 2D laminar structure formed on the MnO_6 octahedra side; this structure facilitates rapid ion transport with low preparation cost and high specific surface area. And λ-MnO_2 has the representative spinel structure with the 3D (1×1) tunnel structure; this structure excels in electrochemical performance. The varying atomic configurations within these different crystalline phases result in a diverse array of pores, which have implications on the electrolyte ion migration or electron transfer processes within the charge storage mechanisms.

Table 1. Structures of MnO_2 materials.

Crystalline Morphology	Structure Type	Tunnels (n × m)	Dimension	Reference
α-MnO_2	Hollandite	(2 × 2)	1D	[14]
β-MnO_2	Pyrolusite	(1 × 1)	1D	[15]
γ-MnO_2	Nsutite	(1 × 1)/(1 × 2)	1D	[16]
δ-MnO_2	Birnessite	(1 × ∞)	2D	[17]
λ-MnO_2	Spinel	(1 × 1)	3D	[18]
ε-MnO_2	-	(1 × 1)/(1 × 2)	3D	[19]

3. Synthesis of MnO_2 Nanomaterials

MnO_2 nanomaterials have been extensively studied as environmentally friendly catalysts. Their preparation methods include the hydrothermal method [20], sol–gel [21], template [22], electrochemical method [23], and coprecipitation [24] methods. Each of these methods has different degrees of effects on the particle size distribution, grain size, and crystal transformation of MnO_2 nanomaterials. Moreover, the properties, structure, and morphology of MnO_2 nanomaterials are considerably influenced by the synthesis conditions. To synthesize MnO_2 nanomaterials with specific structures, morphologies, and sizes for practical production or experiments, studying the synthesis methods and conditions is vital. Nine commonly used methods for synthesizing manganese dioxide are described below, all of which have unique advantages, potential drawbacks, and a wide range of applications.

3.1. Hydrothermal Method

The hydrothermal method involves synthesizing materials via chemical reactions in water under high temperature and pressure using the water solubility of inorganic compounds. MnO_2 nanomaterials with different morphologies can be obtained by changing temperature and pressure [25].

Chen et al. [26] prepared β-MnO_2, γ-MnO_2, and δ-MnO_2 using the hydrothermal approach and α-MnO_2 via solid-phase synthesis, and investigated their catalytic properties for the oxidation of benzene and formaldehyde. The results showed that α-MnO_2 and γ-MnO_2 outperformed δ-MnO_2 and β-MnO_2 in benzene oxidation, while δ-MnO_2 was more active in formaldehyde oxidation. Oxygen was found to exert the catalytic effect on oxidizing formaldehyde and benzene, as elucidated through the quantitative correlation between specific oxygen content and reaction rate. Yang et al. [27] prepared α-MnO_2 solid and hollow sea urchins via hydrothermal synthesis. The 3D α-MnO_2 hollow sea urchin was analyzed for the post-plasma toluene catalytic decomposition. The carbon dioxide selectivity, toluene decomposition, and carbon balance of α-MnO_2 hollow sea urchin were ~59%, ~100%, and ~81%, respectively, which were 96%, 43%, and 44% higher than the non-thermal plasma process. These values were also higher than those for the α-MnO_2 solid sea urchin. Aljafari et al. [28] used α-MnO_2 and Cu-MnO_2 nanoparticles as candidate materials for counter electrode materials (CEs) and synthesized them with the simple hydrothermal approach under 140 °C and 14 h. Among those prepared Dye-Sensitized Solar Cell (DSSCs), the 10 wt% Cu-doped MnO_2 cathode showed the highest energy conversion efficiency of 1.7%, whereas the Power Conversion Efficiency (PCE) of pristine MnO_2 was only 1.21%. The results indicated that Cu-MnO_2 nanoparticles exhibited superior electrocatalytic ability for DSSCs than α-MnO_2. Table 2 summarizes the environmental applications of MnO_2 prepared by the hydrothermal method. Clearly, MnO_2 has good applications in heavy metal adsorption, organic pollutant adsorption, and catalysts. Especially, Figure 2 illustrates the preparation process of porous ε-MnO_2 with the assistance of the solvent, MnO_2 showed high porosity and the best performance of the catalyst preparation at the 6-2-6 (ε-MnO_2 of

Mn-6-2-6) manganese glucose-urea ratio. Therefore, it is necessary to pay attention to the molar ratio of solvent in the hydrothermal synthesis of MnO_2.

Table 2. Hydrothermal preparation of MnO_2 and applications.

Structure of MnO_2	Targets	Synthesis Conditions	Results	Applications	Reference
δ-MnO_2	Pb (II) and U (VI)	-	The adsorption capacities were 41.32 and 492.61 mg g^{-1}, respectively	Adsorbent	[29]
Pristine ε-MnO_2 and ε-MnO_2 of Mn-6-2-6	Toluene	Manganese (II) nitrate hexahydrate, urea, glucose 180 °C	The conversion 41% and 85%, respectively	Catalysts	[30]
MnO_2	Tl (I)	$KMnO_4$, $MnSO_4 \cdot H_2O$, 240 °C	Adsorption capacity was 450 mg g^{-1}	For removing thallium (Tl) from wastewater	[31]
MnO_2 nanoparticles	MB (Methylene Blue)	$KMnO_4$, CH_3CH_2OH, HCl	The adsorption capacities 22.2 mg g^{-1} after 60 min.	Removal of MB	[32]
α-MnO_2, β-MnO_2, and δ-MnO_2	MG (Methyl Glucoside)	-	The removal efficiency of MG 96.42%, 46.58%, 99.75%, respectively	For typical organic pollutant removal	[33]
MnO_2 nanostructures	-	$KMnO_4$, Mn $(CH_3COO)_2$	The capacitance was 348.2 F g^{-1} and rate capability of 89% for 2000 cycles.	Electrode materials	[34]
δ-MnO_2	-	Mn-MOF, $KMnO_4$, 120 °C	The capacitance was 416 F g^{-1}	Capacitors	[35]

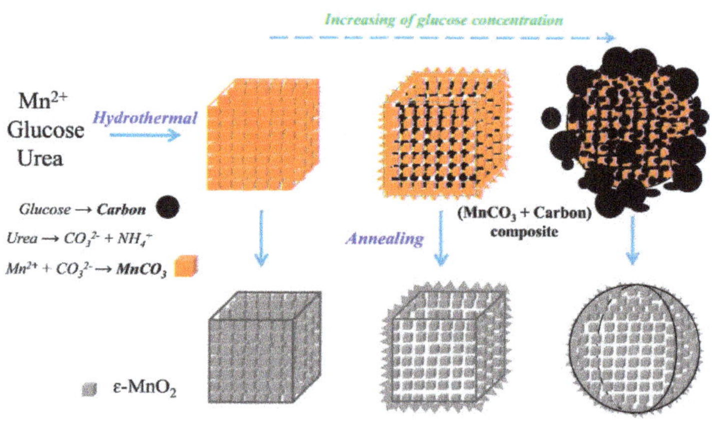

Figure 2. Schematic diagram of porous ε-MnO_2 microcubes [30].

In summary, hydrothermal synthesis is an economical and excellent method, and it has the following advantages. (1) It can yield high-purity products and (2) the as-synthesized products have excellent properties. (3) It enables us to precisely adjust nanoparticle size and morphology, and the final nanoparticle size and morphology are influenced by changing reaction system pH value, and (4) no organic solvent is needed by the synthesis reagents. However, its primary drawbacks lie in the requirement for costly equipment and stringent reaction conditions during synthesis, coupled with relatively lengthy reaction periods [36].

3.2. Sol–Gel Approach

In the sol–gel approach, homogeneous solutions are formed within the solvent using metal-alcohol salts or inorganic salts as precursors. The solute interacts with the solvent or other solutes via hydrolysis and condensation. The solution is condensed into a gel, which is then heated and subjected to later treatments to obtain the eventual target product. Different temperatures, pH, solution concentrations, and reaction duration may affect the reaction system and yield products with different phases [37]. Although the sol–gel method is less studied for synthesizing MnO_2, it has numerous possible merits compared with traditional synthetic approaches. For instance, it is often used to synthesize optical and photovoltaic hybrid materials [38].

Reddy et al. [39] synthesized MnO_2 with a sol–gel approach and by reducing aqueous $NaMnO_4$ using an organic reducing agent, i.e., fumaric acid. The difference in pore size distribution between the two materials resulted in two forms of manifold. Compared with the dry gel, the manifold exhibited higher capacitance. MnO_2 has a higher capacitance in two moles of sodium chloride electrolyte than other electrolytes like potassium chloride, sodium sulfate, and lithium chloride. According to previous research [40], MnO_2 nanoparticles and Ag-doped MnO_2 nanoparticles were prepared using the sol–gel approach. The decomposition of methyl orange (MO) and phenol via MnO_2 was evaluated upon visible light irradiation. The results showed that the MnO_2 catalyst doped with a Ag volume fraction of 10 exhibited higher photocatalytic efficiency for MO than phenol. In addition, Ag-doped MnO_2 catalysts can be used for wastewater treatment and for removing environmental contaminants. Kusworo et al. [41] prepared a photocatalyst composite (ZnO-MnO_2@SiO_2) using the sol–gel approach, and later prepared the polysulfone/ZnO-MnO_2@SiO_2 (PSf/ZnO-MnO_2@SiO_2) membrane through the non-solvent-induced phase separation technique. Incorporating the ZnO-MnO_2@SiO_2 photocatalyst could enhance membrane hydrophilicity, porosity, mechanical strength and water absorption capacity. Moreover, the recyclability, flux stability, and antifouling performances of the membrane improved under UV light irradiation, thereby preventing scale formation and prolonging the membrane life span. Thus, the PSf/ZnO-MnO_2@SiO_2 membrane was used for natural rubber-containing wastewater treatment. Table 3 summarizes the electrochemical applications of MnO_2 nanomaterials prepared by the sol–gel method, from which it is known that MnO_2 nanomaterials has good applications in supercapacitors. Figure 3 illustrates the preparation process of carbon fiber @cobaltferrite@manganese dioxide (CF@$CoFe_2O_4$@MnO_2) composites by sol–gel method and hydrothermal reaction. Notably, the CF@$CoFe_2O_4$@MnO_2 nanomaterials can also have good magnetic behavior in microwave absorbers.

Table 3. Structure and application of MnO_2 nanomaterials prepared using the sol–gel method.

MnO_2 Structure	Synthesis Conditions	Result	Applications	Reference
γ-MnO_2	$MnAc_2{\cdot}4H_2O$, $C_6H_8O_7{\cdot}H_2O$	Capacitance was 317 F g^{-1}	Supercapacitors	[42]
Mesoporous Silica/MnO_2 composite (MS/MnO_2)	Tetraethyl Orthosilicate, $KMnO_4$	Capacitance was 1158.50 F g^{-1}	Supercapacitors	[43]
Nanostructured MnO_2	-	The capacitance was 627.9 F g^{-1}	Supercapacitors	[44]
Nickel-doped layered MnO_2	$KMnO_4$, Ni $(NO_3)_2{\cdot}6H_2O$	The capacitance was 140 mAh g^{-1}	Sodium-ion batteries	[45]
CF@$CoFe_2O_4$@MnO_2	$FeCl_3{\cdot}6H_2O$, $CoCl_2{\cdot}6H_2O$, CF (Carbon Fiber), $KMnO_4$	The microwave absorbing capacity can reach up −41 dB	Microwave absorbers	[46]
α-MnO_2 and Cu-α-MnO_2	$CuSO_4{\cdot}5H_2O$, $KMnO_4$	The maximum degradation of Methylene Blue (MB) by α-MnO_2, 1% Cu-α-MnO_2, 5% Cu-α-MnO_2, and 10% Cu-αMnO_2 were 97.9%, 98.3%, 98.7%, and 99.5%, respectively	Degradable MB	[47]

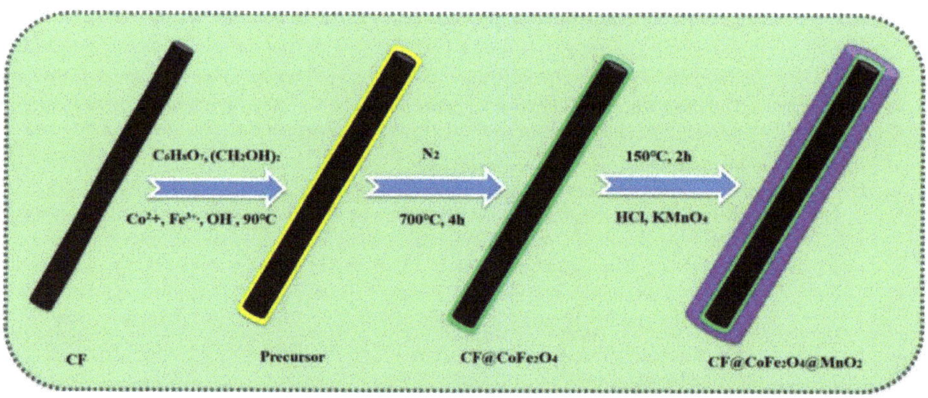

Figure 3. Process for the preparation of CF@$CoFe_2O_4$@MnO_2 composites [46].

In summary, the sol–gel approach can be a simple technique for controlling the reaction at a molecular level, which yields products with fine, high-purity, homogeneous morphologies and crystal structure [48]. Using the sol–gel method, a thin protective coating can be fabricated to ensure good adhesion between the substrate and the top layer. However, the method has drawbacks like long synthesis time and complex operation steps.

3.3. Template Method

In recent years, the template method is commonly used for nanomaterial synthesis using various structure-directing agents or templates. Using organic molecules as template agents, guest species and surfactant molecules are co-assembled to the regular microstructure via template guidance to generate self-assembled nanomaterials with ordered structures. Template methods can be categorized into soft and hard template methods according to the used template type. The entire process is broadly divided into three steps: (1) template synthesis, (2) MnO_2 synthesis according to the template, and (3) template deletion or retention in line with the requirements [49].

3.3.1. Soft Template Approach

The soft template approach typically utilizes nonrigid nanostructures as the templates, which are generated through intermolecular interactions. Subsequently, inorganic source deposition onto nonrigid soft template interior and surface yields mesostructures with well-defined dimensions and pore structures. Surfactants, flexible organic molecules, and block copolymers are generally used as soft templates for interacting with metal ions and merging to liquid crystal phases using the sol–gel method. The mesostructures with open pores were acquired when the soft template was removed via calcination. Depending on concentration of surfactants, micelles of different morphologies are formed. These micellar structures allowed inorganic materials to exhibit specific distribution trends driven by electrostatic interactions between surfactant molecules and nanomaterials, hydrogen bonding, and van der Waals forces [50]. Hou et al. [51] found that micelles are an important factor in controlling shape synthesis. However, soft templates can be adjusted to produce various MnO_2 nanomaterials by adjusting precursor concentrations and reaction conditions [52].

Yuan et al. [53] used polymers as soft templates for directing MnO_2 nanowire growth and stabilizing their structure to form the special graphene-loaded MnO_2 nanowires. The nanostructures exhibited excellent catalytic activity for oxidizing organic pollutants in neutral and alkali environments. They demonstrated that the morphology of MnO_2 considerably influenced the catalytic performance of MnO_2. Tran et al. [54] synthesized mesoporous MnO_2 nanoparticles by olefinic oxidation using permanganate within a soft template solution. Asymmetric capacitors, with activated carbon and MnO_2 as the cathode and anode separately, were assembled and investigated in aqueous potassium sulfate solution. Experimental results showed that mesoporous MnO_2 nanoparticles were the candidate electrode material used in electrochemical energy storage because of their superb low-power capacitive performance. Yang et al. [55] developed the in situ soft template reduction method for the deposition of exposed and well-dispersed MnO_2 nanoparticles in mesoscopic channels within the regular ordered mesoporous Ce-based metal-organic framework (OMUiO-66(Ce)). The substrate channel promoted hydrogen peroxide decomposition with MnO_2 as the catalyst; it also exhibited great efficiency, persistent intracellular antioxidant effects and low-dose activity. The developed MnO_2@OMUiO-66(Ce) had considerable potential for application and could efficiently reduce the oxidative stress.

The soft template approach has numerous merits. For instance, soft templates are available in various forms, and can be prepared by an easy and economical way, with no need of complex instrument. Indeed, soft templates also possess certain drawbacks such as imprecise control over size and shape, difficulties in template removal, challenges in achieving high product purity, and potential contamination from byproducts. These limitations need to be carefully considered when utilizing soft templates for various applications.

3.3.2. Hard Template Approach

The above-mentioned soft template approach has limitations such as uneasy control of product size, morphology or uniformity. Moreover, the remaining macromolecules, organic compounds, and surfactants probably enhance ionic resistivity [56]. On the contrary, the hard template method effectively decreases interference as no surfactant is involved. Compared with the soft template approach, the hard template approach shows a promising application in synthesizing MnO_2 nanostructures.

Bai et al. [57] used KIT-6 to be the hard template to synthesize a 3D regular mesoporous MnO_2 (3D-MnO_2). 3D-MnO_2 had large specific surface area, templated mesoporous properties, and cubic symmetry. 3D-MnO_2 made it possible for formaldehyde to completely convert into water and carbon dioxide. The excellent catalytic activity of 3D-MnO_2 might be associated with the great specific surface area, special mesoporous structure, and numerous surfaces Mn^{4+} ions. Zhang et al. [58] synthesized mesoporous MnO_2 (M-MnO_2) via nano-casting by using porous silica SBA-15 as a hard template. M-MnO_2 exhibited an 8-fold increased adsorption capacity for phenol compared with control MnO_2 (C-MnO_2). Hydroxyl radicals were identified as major reactive oxygen species, while the concentration of hydroxyl radical from M-MnO_2 was increased by about two times compared with that from C-MnO_2. Zhang et al. [59] also synthesized M-MnO_2 catalysts with increased pore size, pore volume, and specific surface area The oxidation reactivity of M-MnO_2 for oxalic acid (OA) and MO was evaluated. The results showed that the M-MnO_2 catalysts were most potent for catalyzing MO and OA degradation, with degradation efficiencies of 98.37% and 92.96%, respectively. Figure 4 displays the above MnO_2 synthetic process. Table 4 summarizes the environmental applications of MnO_2 synthesized using the hard template approach, MnO_2 has good application properties in supercapacitors, batteries, catalysts, etc.

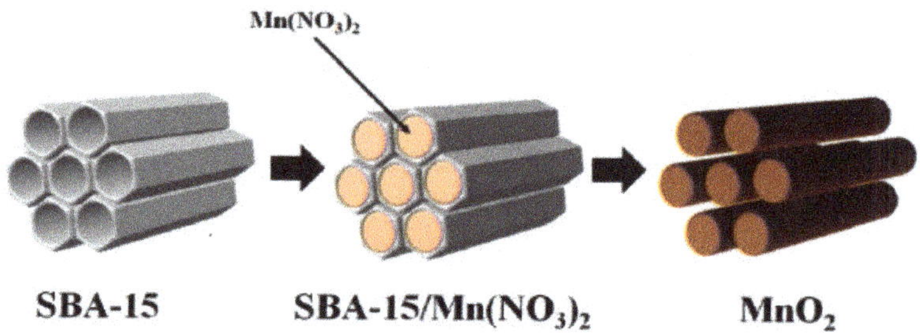

Figure 4. Schematic showing the mesoporous MnO_2 preparation through nano-casting with the ordered mesoporous SBA-15 material being the hard template [59].

Table 4. Structure and application of MnO_2 prepared using the hard template method.

Preparation Products	Formwork	Experimental Data	Applications	Reference
MnO_2@polypyrrole	Polystyrene	The specific capacitance, energy density, and power density were 63 F g^{-1}, 42 Wh kg^{-1} and 1100 W kg^{-1}, separately.	Supercapacitors	[60]
S/MnO_2-280H	S	The capacitances of 1053 and 551 mAh g^{-1} following 400 cycles	Cathodes with Li-S batteries	[61]
MnO_2 (KIT-6)	KIT-6	The bifunctional activity measurable value of 1.28 V	Electrocatalysts	[62]
Flower-like MnO_2	$MnCO_3$ microspheres	90% removal of 1000-ppm toluene	Catalyst	[63]

Nonetheless, template utilization leads to the higher cost of synthesis. Wang et al. [64] prepared graded MnO_2, in which cotton and potassium permanganate were the template and precursor, respectively. Compared to additional templates, the biomaterial is environmentally-friendly and easily available, and cotton fibers have homogeneous morphology compared with other plant fibers. From an economic point of view, the environmentally friendly, cost-effective, and sustainable bio-template approach is applicable to synthesizing MnO_2 nanomaterials.

Generally, the hard template method has the following advantages compared with other synthesis methods: (1) the template can be used as a carrier for synthesizing nanomaterials of various shapes, (2) it solves the problem of the dispersion stability of nanomaterials, with the realization of the synthesis and assembly of the integration, and (3) the operation process is simple and suitable for mass production [65]. However, there are drawbacks like the high cost of the templates and the contamination from byproducts.

3.4. Electrodeposition Method

Electrodeposition is commonly used to prepare thin films and nanoparticles [66]. The deposit morphology, physicochemical properties and crystal structure are adjusted by changing the electrodeposition conditions, such as voltage, current [67], deposition time [68], and electrolyte concentrations [69]. Therefore, the electrochemical method is advantageous relative to others, and its properties include that it is (1) controllable, (2) simple and easy to operate, (3) has a relatively low processing temperature, and (4) has mild reaction conditions. MnO_2 electrodeposition proceeds as follows:

$$Mn^{2+} + 2H_2O \rightarrow MnO_2 + 2e^- + 4H^+.$$

The deposition potential and conditions considerably affect the oxidation state, structure, surface area, and properties of MnO_2 [70]. Ren et al. [71] used the easy electrodeposition approach to prepare Na^+ pre-intercalated δ-MnO_2 nanosheets ($Na_{0.11}MnO_2$) onto 3D graphene (3DG). The specific capacitance of $Na_{0.11}MnO_2$/3DG electrodes was 1240 F g^{-1} at the 0.2 A g^{-1} current density. Moreover, $Na_{0.11}MnO_2$/3DG showed high cycle stability, and the capacitance retention of the electrolyte was 90% following 9000 cycles within 2 mol $ZnSO_4$/0.2 mol or $MnSO_4$ aqueous solution. The above study provided a new perspective for δ-MnO_2 to be a cathode with excellent energy and power density for energy-storage devices. The $Na_{0.11}MnO_2$/3DG material preparation process is shown in Figure 5. Shi et al. [72] deposited reduced nickel (rNi) bases via secondary construction on nanocore nickel foam materials. These bases had a great specific surface area and improved active substance mass utilization. The electrodeposition of MnO_2 on reduced nickel bases could be achieved via pre-intercalation treatment using Na^+, K^+, and NH_4^+ three cations. Moreover, the mechanism of diverse monovalent cations guiding MnO_2 material growth was analyzed. The rNi/MnO_2 composite with the unique nano-sintered structure could be acquired via electrodeposition on reduced nickel bases. Supercapacitors assembled using this electrode exhibited extremely high special capacitance as well as energy densities of 80.22 and 24.90 W kg^{-1} at the 599.99 and 11,997.98 W kg^{-1} power densities, respectively. Zhao et al. [73] used ultrathin nanosheets to prepare MnS_2/MnO_2-Carbon Cloth (MnS_2/MnO_2-CC) heterostructure bifunctional catalysts via the two-step electrodeposition approach for MB degradation in organic wastewater. These catalysts required overpotentials as low as 66 and 116 mV for achieving 10 and 100 mA cm^{-2} current densities within the MB/H_2SO_4 medium. They also had superb stability (with performance retention during 24-h testing) and a low Tafel slope (26.72 mV dec^{-1}). The MB degradation rate reached 97.76%, which is considerably increased relative to the 72.10% rate of the MnO_X-CC catalyst. The study provided a novel idea for synthesizing stable and high-efficiency nonprecious metal bifunctional electrocatalysts to conduct out HER and degradation of organic wastewater. Table 5 summarizes the environmental applications of MnO_2 synthesized through electrodeposition, MnO_2 can be used in supercapacitors, catalysts, and batteries.

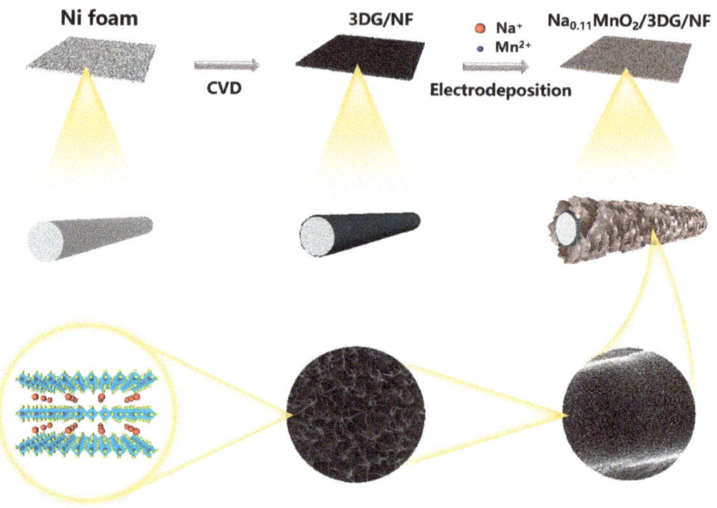

Figure 5. Schematic of the synthesis process of $Na_{0.11}MnO_2$/3DG composites [71].

Table 5. Structure and application of MnO_2 prepared by the electrodeposition method.

Synthetic Structure	Measurement Conditions	Performance	Applications	Reference
MnO_2/poly (3,4-ethylenedioxythiophene) (PEDOT)	10 mV s^{-1}	Capacitance was 89.7 mF cm^{-2}	Supercapacitors	[74]
γ-MnO_2	0.025 V s^{-1}	The capacitance was 43.1 F g^{-1}	Capacitor electrodes	[75]
ε-MnO_2	100 mAh g^{-1}	The discharge capacity delivered by the cell was 5700 mAh g^{-1}	Li-O_2 Catalysts	[76]
MnO_2 nanostructures	1 A g^{-1}	Capacitance and stability were 369 F g^{-1} and 97% following 1000 cycles	Supercapacitors	[77]
MnO_2 nanowires	1 mA cm^{-2}	The stability was 92.6% after 10,000 cycles	Supercapacitors	[78]
MnO_2@Mn	0.86 V	The catalyst showed good stability after a 30h timed current test with little or no decay	Catalysts	[79]
MnO_2-NiFe/Ni	50 mA cm^{-2}	The power density was 93.95 mW cm^{-2}	Oxygen electrocatalysts	[80]
α-MnO_2/γ-MnO_2	193 μW cm^{-2}	The energy density was 93.8 μWh cm^{-2}	Supercapacitors	[81]

In summary, the performance of MnO_2 nanomaterials prepared via electrochemical deposition does not have high electrical conductivity, resistance, and specific capacitance compared with those synthesized using other methods [82]. However, nanocomposites such as carbon nanorods synthesized using this method have promising applications.

3.5. Reflux Approach

The reflux method is the wet chemical preparation approach, which requires no high-temperature calcination. The reflux method can synthesize nanomaterials with the same particle size and excellent catalytic performance directly. This approach is advantageous in the simple operation, mild reaction conditions, and excellent purity of synthesized materials [83]. Moreover, it can be used in large-scale MnO_2 nanoparticle synthesis.

Zhang et al. [84] used the simple microwave-assisted reflux method without using templates and surfactants to synthesize γ-MnO_2 and α-MnO_2 nanoparticles via 5 min refluxing in neutral and acidic environments separately. Similarly, single-crystal β-MnO_2 nanorods (length, 0.5~2 μm; diameter, 20~50 nm) were prepared via reflux treatment with potassium permanganate and manganese (II) sulfate within a nitric acid solution. In addition, the reflux method is applicable for synthesizing doped nanoparticles. Said et al. [85] converted γ-MnO_2 into ε-MnO_2 morphology by controlling the reflux reaction temperature and time. The TGA/DTA results showed γ-MnO_2 had higher stability compared with ε-MnO_2. Moreover, reaction temperature considerably impacted the product phase and surface properties according to surface area analysis. The thermal behavior and magnetic properties of MnO_2 were also investigated. May et al. [86] synthesized α-MnO_2 via refluxing using nitric oxide and investigated how two synthesis methods affected the catalytic activities of CuO/α-MnO_2 catalysts. The relations of catalytic CO oxidation capacity with structural properties were explored. The results showed that CO and abundant surface oxygen could be found at the catalyst's interfacial sites, inferring that the catalytic performance of the CuO/MnO_2 catalyst depended on CO adsorption onto the reduced copper oxide.

However, the reflux method has some drawbacks, and the quality of the as-obtained product is influenced by several factors. Kijima et al. [87] prepared α-, β-, and γ-MnO_2 with three phase structures by acid digestion using MnO_2 trioxide under repetitive conditions. MnO_2 products had a polymorphic type, which was tightly associated with reaction temperature as well as the acid type and concentration. α-MnO_2 was formed by reaction at high sulfuric acid concentrations and low temperatures. On the contrary, β-MnO_2 could be acquired by reaction under low sulfuric acid concentrations and high temperatures. γ-MnO_2 was obtained under intermediate conditions between β-MnO_2 and α-MnO_2. Only β-MnO_2 and γ-MnO_2 were synthesized using nitric acid, whereas β-MnO_2 was formed under harsher conditions compared to γ-MnO_2, with higher temperatures and higher nitric acid concentrations.

In summary, the reflux method for preparing MnO_2 nanomaterials boasts advantages such as simplicity of operation, mild reaction conditions, and high purity of the synthesized material. Furthermore, it is suitable for large-scale synthesis of MnO_2 nanoparticles. However, the quality of the obtained MnO_2 nanomaterials was influenced by a multitude of factors.

3.6. Microemulsion Approach

Microemulsions are clear liquid phases (monophases) with high thermodynamic stability formed from water, oil, surfactants, and co-surfactants. Water and oil are immiscible, and surfactants are amphiphilic. Different from common emulsions, microemulsions can be generated after blending water, oil and surfactants without the requirement of high-shear conditions. Direct (oil dispersed within water), reverse (water dispersed in oil), and bi-continuous and supercritical carbon dioxide are four microemulsion types. The microemulsion method is used to synthesize well-controlled, narrow, monodispersed nanoparticles [88]. It is mainly used to homogeneously synthesize metal nanoparticles (diameters, 5–50 nm) [89]. This method demonstrates high practicability and efficiency in synthesizing and processing inorganic nanomaterials, which is beneficial for uniform volume heating, energy saving and higher reaction rate than conventional heating methods.

Xu et al. [90] synthesized MnO_2 with a particle diameter of ~4 nm by the microemulsion method. Compared to chemical coprecipitation, the particle size of MnO_2 considerably decreased. The capacitance value of MnO_2 was 246.2 F g^{-1}, which considerably increased relative to chemically coprecipitated MnO_2 (146.5 F g^{-1}). The specific capacitance was reduced by just 6% following 600 cycles due to the high material cycling performance. Zefirov et al. [91] used an organometallic compound dissolved in supercritical carbon dioxide in an organometallic precursor to prepare MnO_2 nanoparticles with small grains and a low polydispersity index.

In summary, the advantages of the microemulsion method had a simple experimental set-up and low energy consumption, enabled easy handling, and had potential for commercial production. However, the microemulsion process requires excessive solvent [92].

3.7. Chemical Coprecipitation

Chemical coprecipitation is used for synthesizing composites containing two or more metallic elements. In this method, nanoparticle precipitates can be generated through a controlled reaction of cations with anions. This reaction may be impacted by temperature, pH, and reactant concentration [93].

Sivakumar et al. [94] synthesized α-MnO_2 nanoparticles via chemical coprecipitation. The results of cyclic voltammetry analysis showed α-MnO_2 nanoparticles had good capacitive behavior. Yaday et al. [95] synthesized MnO_2 nanoparticles using simple chemical coprecipitation and reflux-assisted coprecipitation methods at different reflux durations and annealing temperatures separately. XRD, FTIR spectroscopy, UV-vis spectroscopy, BET surface area analyzer, and thermogravimetric analysis were utilized to examine sample optical, structural and thermal performances. The Scherrer equation was utilized to evaluate the mean sample grain size, which was determined to be 6~8 nm (6~7 nm) and 15~30 nm (20~46 nm) for reflux-assisted and coprecipitation approaches, separately. The peaks correspond to Mn-O bonds on the FTIR spectra, verifying that MnO_2 nanoparticles were formed. According to FESEM analysis, the samples had nanorod-type morphology. MnO_2 nanoparticles exhibited pseudo-capacitive behavior and excellent photocatalytic performance for the degradation of bright green dyes. Figure 6 displays the MnO_2 nanoparticles preparation route. Pan et al. [96] prepared five crystalline forms of MnO_2 with manganese sulfate being the manganese source and investigated differences in physicochemical properties based on specific surface area, phase morphology, pore volume, pore size, surface structure and particle size. The performance tests and electrode reaction kinetics for the five crystal batteries and capacitors showed that δ-MnO_2 and γ-MnO_2 are more suitable for capacitors and batteries, respectively.

The chemical co-precipitation method requires low reaction temperature and simple equipment and has low energy consumption, safe operation, simplicity, and low cost. However, chemical co-precipitation also has drawbacks: (1) the prepared manganese dioxide material is relatively low in purity and (2) poor homogeneity and being prone to agglomeration problems, which affects material properties.

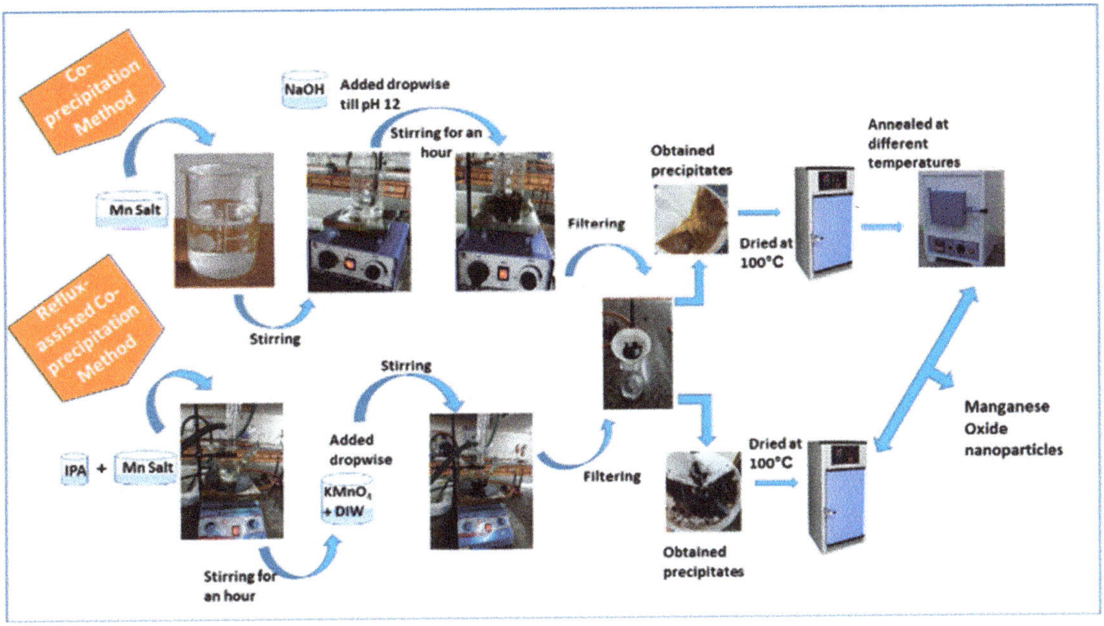

Figure 6. Schematic representation of MnO_2 nanoparticle synthesis process [95].

3.8. Chemical Reduction Method

Chemical reduction has been developed as the high-efficiency wet chemical approach used to synthesize zero-valent nanoparticles. It is commonly used for synthesizing magnetic metal nanoparticles such as iron, cobalt, and nickel [97]. The average particle size and distribution can be controlled by adjusting the preparation conditions such as the concentrations of solvents, surfactants, and reducing agents [98].

Li et al. [99] provided an easy and high-efficiency solid-solution reaction pathway at a low temperature (60 °C) without using templates or surfactants for the large-scale synthesis of α-MnO_2. α-MnO_2 is a new candidate material for lithium battery applications. Khan et al. [100] prepared MnO_2 nanoparticles and MnO_2 nanoparticle/activated carbon (MnO_2/AC) composites via chemical reduction. The results showed that the MnO_2/AC composite contributed to degrading CR (Congo Red) dye by ~98.53%, whereas MnO_2 nanoparticles degraded CR dye by 66.57% under the identical irradiation time. Moreover, the MnO_2/AC composite was highly sustainable and could be used for repeated degradation of CR dye after rinsing and thermal treatment. Cremonezzi et al. [101] synthesized highly capacitive δ-MnO_2 using a new easy route by reducing potassium permanganate. The capacitance of δ-MnO_2 was 190 F g^{-1} at 0.25 A g^{-1}.

In summary, the chemical reduction method for the preparation of MnO_2 nanoparticles has advantages such as the low cost and ready availability of raw materials, straightforward operational procedures, and ease of control. However, the chemical reduction method has some limitations in the reducing agent such as high toxicity, low purity, and high synthesis costs.

3.9. Green Synthesis Method

Recently, more and more studies have been conducted to prepare manganese dioxide nanoparticles by the green synthesis methods. This phenomenon indicates that the greening of preparation methods will be a significant direction in the future. Green synthesis is an ecofriendly, cleaner, and cheaper method for nanoparticle synthesis. This method is viable for synthesizing biocompatible nanoparticles, thereby bridging materials science and biotechnology. Moreover, nanoparticles with controllable shapes and sizes can be prepared via green synthesis [102]. Fruits, vegetables, plant extracts, fungi and microorganisms have been used as raw materials to prepare manganese and manganese-oxide nanoparticles via green synthesis [103].

3.9.1. Plant Extraction Method

The environmentally friendly preparation of MnO_2 nanoparticles with plant extracts can be an economical and effective method [104]. In this method, plant extracts are added to a metal salt solution at room temperature, and this reaction can be completed within several minutes. Metal reduction can be achieved by diverse compounds including terpenoids, polysaccharides, phenolics and flavonoids in plant extracts [105]. At present, some plant extracts are adopted for synthesizing MnO_2 nanoparticles.

Hashem et al. [106] prepared MnO_2 nanomaterials via the green synthesis of lemon peel (P) or juice (J). The crystalline and electrochemical properties of P-MnO_2 and J-MnO_2 were improved since lemon peel possesses 3 reducing reagents, and lemon juice contains citric acid and ascorbic acid. P-MnO_2 have the same electrochemical properties as conventional reducing reagents, but P-MnO_2 was expensive. The novel preparation method is simple, cost-effective, environmentally friendly, and scalable for large-scale α-MnO_2 nanoparticle synthesis. The MnO_2 nanoparticles can be applied to electrochemical energy storage. Shehroz et al. [107] prepared the three MnO_2 three phases (α-, β-, and γ-MnO_2) in a single individual. For this purpose, natural surfactants were synthesized using bitter apple extract as a green solvent. MnO_2 nanoparticles were synthesized under the same conditions with/without plant extracts. Experimental results showed that the average size of products was 20~50 nm by the green synthesis method, while that was 20~25 nm for nanoparticles prepared by chemical methods. Dye and nitroaromatic reduction was investigated by using MnO_2 nanoparticles as the catalysts. Moreover, the apparent rate constants, reduction rates, reduction concentrations, and reduction time were analyzed. The nanoparticles prepared by the environmentally friendly method showed superior catalytic performance to those prepared by the chemical method. Ramesh et al. [108] synthesized green MnO_2 nanoparticles using medicinal plant extracts. The results of XRD analysis proved the crystal structure of MnO_2 nanoparticles. The results of SEM illustrated that MnO_2 nanoparticles prepared by the environmentally friendly method showed a spherical shape. Moreover, 72% of methylene blue (MB) dye was degraded after 150 min under UV light irradiation. Table 6 summarizes the applications of MnO_2 synthesized using the plant extraction method, the prepared MnO_2 nanoparticles have different sizes with the different plant sources. MnO_2 has good applications in heavy metal adsorption, organic pollutant adsorption and so on. The degradation mechanism of toxic dyes by green synthesized manganese dioxide nanoparticles is shown in Figure 7.

Table 6. Preparation of MnO_2 nanoparticles with plant extracts and their applications.

Plant Organism	Nanoparticle Structures of MnO_2	Particle Size	Effect	Appliance	Reference
Flower extract	MnO_2 nanorods	100 nm	Decolorization of the target dye was 91.3%. TOC and COD were reduced by 90.6% and 92.1% separately.	Removal of crystalline violet dye	[109]
Saraca asoca leaves extract	MnO_2 nanoparticles	18 nm	The semi-inhibitory concentration values of 20 μg/mL for both MCF-7 and MDAMB-231 cells	Considerable cytotoxic effects on cancer cells	[110]
Yucca gloriosa leaf extract	MnO_2 nanoparticle	80 nm	The photocatalytic efficiency for 20 min was 33%	Photocatalytic activity and good degradation of organic dyes	[111]

Table 6. *Cont.*

Plant Organism	Nanoparticle Structures of MnO_2	Particle Size	Effect	Appliance	Reference
Potato leaf extract	MnO_2 nanoparticle	26 nm	Significant increases of 67.1% in plant growth activity, 52.8% in photosynthetic pigments, and 56.25% in non-enzymatic antioxidant activity in soil, respectively	Multi-aspect enhancer	[112]
Extract of viola betonicifolia	Green synthesized MnO_2 nanoparticles and Chemically Synthesized MnO_2 Nanoparticles	10.5 ± 0.85 nm	Cell survival (79.33 ± 0.75%), (73.54 ± 0.82%), respectively	Used to provide antimicrobial coatings	[113]
Extract of ficus retusa plant	α-MnO_2 nanoparticles	30~50 nm	The adsorption capacities for Mo and Mr dyes were 116.1 and 74.02 mg g^{-1}, separately	Adsorbent	[114]
Papaya leaf extract	MnO_2 nano-conjugate	30~40 nm	The urea and cholesterol reduced to 94 ± 2.16	For the treatment of hyperbilirubi-naemia	[115]
Chamomile flower extract	MnO_2 nanoparticles	16.5 nm	The percentage of apoptotic cells in RS-2 ranged from 0.97% to 99.94%	Strong inhibitory effect on rice strain RS-2	[116]
Plant extracts	α-MnO_2	2.8~4.5 nm	The capacitance and stability were 500 F g^{-1} and 71%, separately, after 7000 cycles	Supercapacitors	[117]
Mango lead extract	δ-MnO_2 nanoparticles	1.5~2.5 nm	The efficiency with >96% removal of cationic pollutants	Cation adsorbent	[118]

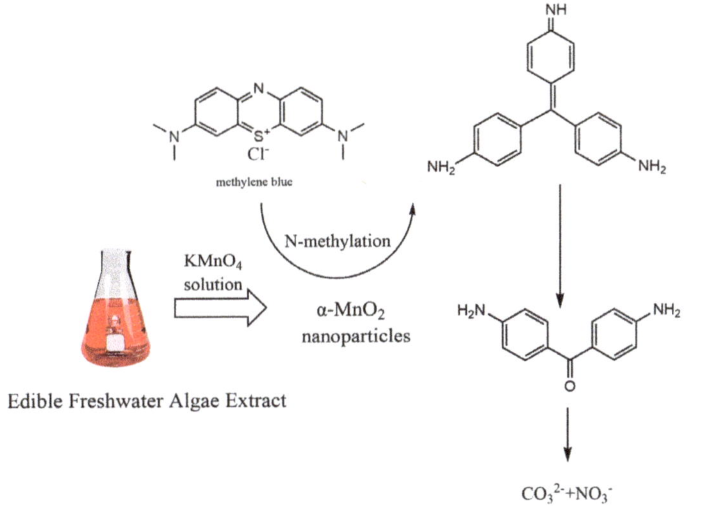

Figure 7. Degradation of toxic dyes by green synthetic manganese dioxide nanoparticles.

In summary, the yield of MnO_2 nanoparticles prepared by the plant extraction method is low compared with those prepared by other methods, and it can hardly control the generation conditions of nanoparticles precisely, and the products prepared by the plant extraction method still have certain toxins [119]. In contrast to other preparation methods, the plant extraction method possesses the advantages of being cleaner and more conducive to sustainable development.

3.9.2. Environmentally Friendly Synthetic Methods Based on Microorganisms

Microorganisms are promising for nanoparticle synthesis. Metal salts are reduced into metal nanoparticles via the domestication of enzymes. Fungi exhibit superior bioaccumulation and resistance, which contribute to synthesizing metal nanoparticles. The interaction of microorganisms with metals is also extensively investigated [120]. Microorganisms can be adopted for accumulating or extracting metals via bioleaching, bioremediation, and heavy metal elimination [121].

Sinha et al. [122] synthesized intracellular MnO_2 nanoparticles by the simultaneous manganese remediation from the highly mono-disperse medium using Bacillus sphaericus. Those prepared nanoparticles were orthorhombic crystalline MnO_2. When cells were challenged with manganese, MnO_2 nanoparticles (mean size, 4.62 ± 0.14 nm) were prepared. The above study offered the merits of synthesizing relevant oxide nanoparticles to prevent manganese pollution. Borah et al. [123] achieved a compositionally controllable, room-temperature, and simple environmentally friendly preparation route of high-purity α-MnO_2 nanoparticles by reducing $KMnO_4$ aqueous solution with an edible freshwater red algae aqueous extract. The synthesized MnO_2 nanoparticles showed excellent photocatalytic performance for rhodamine B (RhB), methylene blue (MB) and methyl Orange (MO), with degradation rate constants of 0.06781, 0.03831 and 0.04323 min^{-1}, separately. The photocatalysts were easily recycled and highly stable. In total, 3 mg of MnO_2 nanoparticles exhibited nearly total degradation efficiency (92%) within 30 min. Alvares et al. [124] used *Haloarchaea alexandrina* GUSF-1 cell lysates to obtain Mn_3O_4-MnO_2 nanocomposites. The antimicrobial activity of these nanocomposites satisfied *Pseudomonas aeruginosa* > *Salmonella typhimurium* > *Escherichia coli* > *Amoebacterium* commonly known as *Proteus mirabilis* > *Candida albicans* > *Staphylococcus aureus*.

In summary, the environmentally friendly synthetic methods based on microorganisms exhibit the advantages of being environmentally friendly and possessing good biocompatibility. However, the synthesis method is easily influenced by different factors [125,126], like strain type and environmental conditions such as temperature, pH, salt concentration and growth medium, all of which have direct or indirect influence on nanoparticle composition, size and morphology. As a result, it also faces challenges such as high technical difficulty, low stability, issues with purity, and concerns over biosafety.

4. Electrochemical Applications

MnO_2 nanomaterials can be used in supercapacitors and batteries due to them being inexpensive, widely available, and malleable [127]. Table 7 summarizes other electrochemical applications of manganese dioxide, which can be used in zinc-ion batteries, pneumatic actuators, and so on.

Table 7. Electrochemical applications of MnO_2.

Synthetic Structure	Measurement Conditions	Performance	Applications	Reference
β-MnO_2/Polypyrrole	0.2 A g^{-1}	Specific discharge capacity of 361.7 mAh g^{-1}	Zinc-ion batteries	[128]

Table 7. *Cont.*

Synthetic Structure	Measurement Conditions	Performance	Applications	Reference
Manganese dioxide/gelatin-glycerol	±2 V	High bending actuation (20-mm deflection, >360° scan angle, and 2.5-mm radius of curvature) and different shape change	Air-working actuator	[129]
α-MnO_2	0.1 A g $^{-1}$	Capacity was 190 mAh g^{-1} and the stability was after 50,000 cycles in $(NH_4)_2SO_4$	Ammonium-ion energy storage	[130]
MnO_2/graphitic carbon nitride (g-CN)	5 mV/s	The optimal composite system achieved a current density of 10 mA/cm^2 with an overpotential of 430 mV and exhibited a Tafel slope of approximately 70 mV/dec	Electrocatalysts	[131]

4.1. Supercapacitors

Supercapacitors are energy-storage technologies widely researched recently. Unlike batteries, supercapacitors can be rapidly recharged, operate at a wider temperature range, are environmentally friendly, and offer better safety, higher reliability, and maintenance-free operation [132]. Moreover, the electrochemical properties are largely determined by active substances contained within the electrodes. MnO_2 nanomaterials can be used to prepare high-performance electrode materials for supercapacitors because of their large specific capacity and good electrochemical performance. MnO_2 materials can store and release electrical energy quickly and display excellent cycle stability. Additionally, the energy-storage properties of supercapacitors are further enhanced by controlling MnO_2 material characteristics, such as morphology, crystal structure, and pore structure. Supercapacitors can be used in printed electronics [133], electric vehicles [134], smart devices [135], and energy-storage systems [136].

Conventional supercapacitors use activated carbon-based materials as electrodes. This material had typical carbon-based material advantages, including abundant material sources, environmentally friendly properties, excellent electroconductivity, high specific surface area, and broad operating temperatures [137]. Electrochemical capacitors containing carbon-based materials were electrochemical double-layer capacitor types. The capacitance depends on the accessible electrolyte ion surface area rather than the capacitor material body. The carbon-based materials provide a high specific surface area, their pore size distribution and pore structure affect the energy storage rate of EDLC supercapacitors [138].

Electrochemical double-layer capacitors use materials with limited capacitance and supercapacitor materials with pseudo-capacitance may be 10~100 times more capacitive. The store charge is similar to conventional capacitor electrodes and exhibits a Faraday reaction between the electrode material and ions. Such pseudo-capacitive supercapacitor materials are divided into two types: excessive metal oxides or conducting polymers [139–141]. Excessive metal oxides include ruthenium oxides, manganese oxides, and nickel oxides [142–144]. To be specific, metal oxides offer increased energy density compared with traditional carbon-based materials. The pseudo-capacitance of metal oxides is affected by physical properties and chemical factors [145,146]; however, they can yield higher performances by modifying or using composite materials as well as the adjustment of electrode structure.

Yao et al. [147] obtained an excellent capacitance of MnO_2 electrode material by printing pseudo-capacitor electrodes. The MnO_2 electrode was loaded with 182.2 mg cm^{-2} and its capacitance was 44.13 F cm^{-2}. The specific capacitance of the 2D MnO_2/pSiNW electrode prepared by Bagal et al. [148] was 311.89 F g^{-1} at 2 A g^{-1}. Using it as the anode, the density and power density of this capacitor were the highest (93.31 mWh cm^{-2} and 1.51 mW cm^{-2}, separately), while its capacitance retention was 89.5% over 10,000 cycles.

Tynan et al. [149] uniformly deposited MnO_2 nanoparticles with pseudo-capacitance on carbon nanotubes using the chemical method, and the capacitance of MnO_2 nanoparticle electrodes could be enhanced by a factor of 9 relative to the benchmark material at a loading of 95 wt% of MnO_2. Moreover, MnO_2 nanoparticles enhanced the structure of hybrid electrodes, such as a 110% and 430% increase in tensile strength and stiffness compared to the benchmark material. Table 8 summarizes the different synthesis methods and forms of MnO_2 used as supercapacitors. It elucidates the specific capacitance, energy density, scan rate, and cycling stability, with the results indicating that the manganese dioxide prepared via the hydrothermal method and doped with Ag exhibits a maximum specific capacitance of 1027 F g^{-1}, at a scan rate of 1 A g^{-1}. Although the two-dimensional layered δ-MnO_2 prepared by the chemical reduction method possesses a relatively low energy density, it exhibits exceptional cycling stability, retaining 98.7% of its initial performance after 10,000 cycles. In contrast, the cycling stability of manganese dioxide prepared by electrodeposition is relatively poor, achieving only 56.81% after 1000 cycles, as compared to other methods. δ-MnO_2 materials successfully prepared by a chemical reduction method, and thoroughly evaluated the electrochemical properties of these materials, as well as their composites with carbon (C, labeled as C/MnO_2 with varying reaction times of 0.5 h, 1 h, and 2 h), using cyclic voltammetry (CV) and galvanostatic charge–discharge (GCD) tests in a standard three-electrode system with 1.0 M sodium sulfate electrolyte. At a scan rate of 200 mV/s, the CV curve of pure carbon (C) exhibited a near-rectangular shape, clearly indicating its excellent electric double-layer capacitance behavior during both anodic and cathodic scans. Similarly, the CV curves of the C/MnO_2 nanocomposites also displayed a quasi-rectangular shape with no pronounced redox peaks, revealing a synergistic effect between the electric double-layer capacitance and the rapid, reversible Faradaic redox reactions occurring on the MnO_2 surface, operating at a pseudo-constant rate across the entire potential range. Notably, the C/MnO_2 sample prepared for 1 h exhibited the largest CV area, signifying its possession of the highest specific capacitance. During the charge–discharge tests, the GCD curves of all samples maintained an almost perfect triangular shape, which not only attested to the materials' extended charge–discharge durations but also highlighted the substantial positive contribution of pseudocapacitive mechanisms to the overall specific capacitance. Across a wide range of current densities from 0.25 to 10 A g^{-1}, the GCD curves of all samples remained close to triangular, demonstrating ideal capacitive behavior and high Coulombic efficiency. It is noteworthy that while the pure carbon material (C) displayed good rate capability, its specific capacitance fell below 50 F g^{-1}. In contrast, the C/MnO_2 sample prepared for 1 h achieved the highest specific capacitance of 116.61 F g^{-1} at a current density of 1 A g^{-1}, significantly surpassing that of C/MnO_2 prepared for 0.5 h (84.65 F g^{-1}) and 2 h (58.37 F g^{-1}), likely due to the optimized nanosheet structure and appropriate composition. It was also observed that as the current density increased, the specific capacitance of all electrode materials decreased gradually. This phenomenon can be attributed to the fact that electrolyte ions can diffuse sufficiently and uniformly into the internal pores of the electrode materials at low current densities, enabling a higher specific capacitance. However, the electrolyte ions are time-constrained and fail to adequately access all active sites within the electrode under the high current densities, leading to insufficient Faradaic redox reactions and, consequently, a lower specific capacitance.

Table 8. Comparison of energy storage performance of different synthesized and formed manganese dioxide in supercapacitors.

Material	Preparation Method	Specific Capacitance	Cycling Life	Energy Density	Reference
α-MnO_2	Plant extraction method	90 F g^{-1} at 1 A g^{-1}	98% after 1000 cycles	37 Wh kg^{-1}	[150]
δ-MnO_2	Chemical reduction method	116.61 F g^{-1} at 1 A g^{-1}	98.7% after 10,000 cycles	22.7 Wh kg^{-1}	[151]

Table 8. *Cont.*

Material	Preparation Method	Specific Capacitance	Cycling Life	Energy Density	Reference
MnO_2/Ag	Chemical reduction method	115 F g^{-1} at 0.2 A g^{-1}	75% after 1000 cycles	45 Wh kg^{-1}	[152]
MnO_2-NiO	Electrodeposition method	375 F g^{-1} at 0.5 A g^{-1}	56.81% after 1000 cycles	-	[153]
$Ag_{0.05}$ MnO_2	Hydrothermal method	1027 F g^{-1} at 1 A g^{-1}	93.16% after 10,000 cycles	-	[154]
ZnO@MnO_2	Hydrothermal method	839.9 F g^{-1} at 0.3 A g^{-1}	92% after 10,000 cycles	74.6 Wh kg^{-1}	[155]
α-MnO_2	Hydrothermal method	47 F g^{-1} at 0.5 A g^{-1}	94% after 5000 cycles	21 Wh kg^{-1}	[156]
λ-MnO_2/polyaniline	Hydrothermal method	232.1 F g^{-1} at 0.2 A g^{-1}	78.65% after 3000 cycles	66.4 Wh kg^{-1}	[157]
β-MnO_2	Hydrothermal method	212.85 F g^{-1} at 0.2 A g^{-1}	97.5% after 5000 cycles	-	[158]
γ-MnO_2	Hydrothermal method	103 F g^{-1} at 1 A g^{-1}	-	-	[159]
Polyaniline-MnO_2	Templates method	765 F g^{-1} at 0.25 A g^{-1}	80% after 14,000 cycles	-	[160]
Polyaniline-MnO_2	chemical co-precipitation method	417 F g^{-1} at 5 mV s^{-1}	-	7.2 Wh kg^{-1}	[161]

In summary, supercapacitors can be used in many applications, but their performance is considerably affected by electrode materials. MnO_2 and its composites can improve cycle life, power density, and energy density compared with traditional carbon-based materials and may be potentially applied in large-scale energy storage.

4.2. *Zn-MnO_2 Batteries*

Zn-MnO_2 batteries are a common type of disposable batteries and typically comprise Zn and MnO_2 as anode and cathode, separately, and an electrolyte. Compared with other batteries, Zn-MnO_2 batteries are inexpensive, have better stability and longer storage life, and are environmentally friendly and recyclable [162]. Zn-MnO_2 batteries are mainly used in electronic devices [163].

In neutral and weakly acidic electrolytes, MnO_2 in Zn-MnO_2 batteries are first reduced to MnOOH. As the acidic solubility increases, MnOOH is reduced to Mn^{2+} and Zn metal is oxidized to Zn ions. This redox reaction generates an electric current in the battery, thus realizing electrical energy conversion and storage. The Zn-MnO_2 battery performance is affected by physical conditions and chemical factors [164,165]. However, the performance is enhanced after adding electrolytes. Shen et al. [166] found that redox conversion of MnO_2 with Mn^{2+} could be achieved by maintaining critical range conditions. Zn-MnO_2 batteries based on this electrochemical property can withstand 16,000 cycles without significant capacity degradation, and the stored energy density was 602 Wh kg^{-1}. Liu et al. [167] modulated the electrolyte composition by adding acetic acid and chromium chloride (Cr^{3+}) and using combined strategies such as pre-cycling and sonication. MnO_2 suspension was mitigated, and a more stable and reversible cycling reaction was achieved after combining pre-cycling and sonication. The modified zinc-MnO_2 batteries showed higher Coulombic efficiency at 1.4 V and maintained 7500 stable cycles, and the capacity and current density were 0.5 mAh cm^{-2} and 10 mA cm^{-2} separately. Ma et al. [168] added an aqueous organic electrolyte of tetra-ethylene glycol dimethyl ether to inhibit water molecule activity, thus avoiding the generation of by-products. The specific capacity of Zn-MnO_2 batteries was as high as 132 mAh g^{-1}. The capacity retention reached >98% following 1000 cycles at the 1.25 V operating voltage and the 200 mA g^{-1} current density. Each of these studies demonstrates that Zn-MnO_2 cell performance may be improved by adjusting the electrolyte composition and employing specific strategies. Table 9 summarizes the comparison of different crystals of manganese dioxide in zinc–manganese batteries. The results indicate that δ-MnO_{2-x} exhibits higher capacitance compared to several other electrode materials, primarily due to the importance of layered structure in enhancing capacitive performance. This unique structure favorably facilitates the surface adsorption and intercalation of metal cations such as Na^+, K^+, and H^+. Consequently, it enables the reversible transition between Mn (IV) and Mn (III) valence states, which is vital for charge storage. Essentially, the layered structure of δ-MnO_2 promotes efficient ion transport and electron transfer, thereby elevating its overall capacitive performance. However, β-MnO_2 electrodes prepared via the electrodeposition method exhibit the longest cycle life. This is attributed to the ability of the electrodeposition process to precisely control the thickness and structure of the β-MnO_2 deposit, resulting in a uniform and dense layer. Additionally, the tunnel structure and chemical stability of β-MnO_2 facilitate rapid ion transport and charge storage, while minimizing material degradation during cycling. Consequently, β-MnO_2 electrodes produced through the electrodeposition method are able to demonstrate extended cycle life. The electrochemical properties of β-MnO_2 material were comprehensively evaluated using cyclic voltammetry (CV). Within the discharge potential range of approximately 1.8–2 V, paired with the Zn^{2+}/Zn system, the material exhibited stable areal capacity performance over the initial 20 cycle periods. Notably, under a 2.2 V charging condition, a uniform layer of manganese dioxide was observed to cover the C-cloth CC substrate, clearly indicating the achievement of homogeneous and dense deposition of manganese dioxide nanoflowers on the C-cloth surface. This phenomenon was robustly supported by the stability of the discharge platform over the first 20 cycles, further attesting to the remarkable thermodynamic stability of β-MnO_2 within the Zn–manganese dioxide battery system and its ability to maintain a more regular morphological structure.

Table 9. Comparison of different crystalline manganese dioxide in zinc–manganese batteries.

Cathode	Preparation Method	Electrolyte	Plateau (V)	Capacity (mAh g^{-1})	Cycling Life	Reference
α-MnO_2	Hydrothermal method	2 M $ZnSO_4$ + 0.1 M $MnSO_4$	0.8–2.0	302	78.4% after 2000 cycles	[169]
β-MnO_2	Hydrothermal method	2 M $ZnSO_4$ + 0.1 M $MnSO_4$ + 0.1 M Na_2SO_4	1.0–1.9	325	94% after 1000 cycles	[170]
δ-MnO_{2-x}	Hydrothermal method	2 M $ZnSO_4$ + 0.1 M $MnSO_4$	0.9–1.9	551.8	83% after 1500 cycles	[171]
ε-MnO_2	Hydrothermal method	3 M $MnSO_4$ + 0.3 M H_2SO_4 + 0.06 M $NiSO_4$	1.16–3.4	270	99% after 450 cycles	[172]
β-MnO_2	Electrodeposition method	1 M $ZnSO_4$ + 1 M $MnSO_4$	1.8–2.2	-	≈100% after 400 cycles	[173]
γ-MnO_2	Electrodeposition method	0.5 M Mn $(CH_3COO)_2$ + 0.5 M Na_2SO_4	-	391.2	92.17% after 3000 cycles	[174]

In general, Zn-MnO_2 batteries, as a kind of low-cost battery, offer a reasonable capacity and energy density. Therefore, Zn-MnO_2 is widely adopted for electronic devices, and its performance, although affected by many factors, can still be improved by optimizing the electrolyte composition. Zn-MnO_2 batteries have a broad application prospect in the energy-storage field

4.3. MnO_2/Carbon Nanomaterial Composites

MnO_2/carbon nanomaterials composites (MnO_2/CNTs) are nanocomposites integrated with the unique properties of MnO_2 and carbon nanotubes. Since carbon nanotubes possess superb mechanical stability, increased surface area, and great electrical conductivity, the integration of CNTs with manganese dioxide efficiently enhances the specific capacity, conductivity, as well as other electrochemical properties of the composite [175]. In addition, CNTs contribute to improving cycle stability and charge/discharge rate capability of the MnO_2/CNTs composites [176]. Therefore, MnO_2/CNT composites exhibit superior electrochemical properties. MnO_2/CNTs are mainly used in supercapacitors [177], biosensors [178], catalysts [179] and other applications.

Li et al. [180] successfully recovered MnO_2/CNT cathodes from MnO_2 electrodes by simple calcination at mild temperatures and used then to be electrodes in supercapacitors. The specific capacity was 253.86 F g^{-1} within the 0.5 M Na_2SO_4 at 0.5 A g^{-1}. The sustainability of carbon-based materials for high-performance electrochemical applications was demonstrated through recycling. Rosaiah et al. [181] prepared and investigated electrochemical performances of pure manganese dioxide and MnO_2/CNTs composites by a hydrothermal synthesis method, and discharge capacities were 1225 and 1589 mAh g^{-1}, separately. The MnO_2/CNTs composites exhibited high stability, and the capacitance was 957 mAh g^{-1} following 60 cycles. The superb specific capacity and cycling performance are associated with the synergistic effect of carbon matrix materials with MnO_2, and this synergistic effect also indicates that carbon matrix materials are important for MnO_2/CNT composites. Zhou et al. [182] prepared MnO_2@CNTs composite electrodes that possessed the 3D nanostructure. The capacitance of the prepared MnO_2@CNTs composite electrodes reached 256 mAh g^{-1} at 0.1 A g^{-1} and remained stable following 700 cycles. This work explains the mechanisms by which carbon nanotubes enhance MnO_2 cathode performances, providing a new perspective on designing efficient electrochemical energy storage devices.

To sum up, the carbon matrix material plays a central role and significantly improves the composite electrochemical performances in the MnO_2/CNTs composites. Meanwhile, the sustainable characteristics of carbon matrix materials contribute to their broad uses in energy storage.

5. Summary and Outlook

In conclusion, MnO_2 is being widely used due to its unique properties. In contrast to the previous review, this review summarizes approaches for preparing MnO_2 nanoparticles and describes the corresponding respective merits, demerits or limitations, which is believed to help researchers to better select the synthesis methods. In addition, the multifunctional extension applications of MnO_2 nanomaterials are also presented. Although great progress is achieved in MnO_2 nanoparticle studies, the preparation methods and the electrochemical applications of MnO_2 nanoparticles need to be further investigated. Here, the future research prospects of MnO_2 nanoparticles are briefly discussed, as shown in Figure 8.

Up to now, the preparation method of MnO_2 nanoparticles has been continuously improved, and the performance of MnO_2 nanoparticles has been greatly improved. However, the research on MnO_2 nanoparticles is still in its infancy and has not been fully applied to practical applications. Addressing this challenge will require focused efforts in several areas in the future.

i. Although there are several ways to prepare MnO_2 nanoparticles, realizing large-scale, cost-effective and high-quality synthesis remains challenging. The high production cost makes the commercialization of manganese dioxide nanoparticles difficult, especially in cost-sensitive industries.
ii. Although MnO_2 nanoparticles have demonstrated excellent performance on the laboratory scale, a series of technical challenges need to be addressed in practical applications, for example, how to improve the stability and electrochemical properties of MnO_2 nanoparticles. These issues need to be addressed by continuous research and technological innovations.

iii. Despite the excellent MnO_2 nanoparticle performances, the process of MnO_2 preparation may generate some hazardous substances and wastes. This requires manufacturers to take environmental protection measures during the production process. In addition, the environmental impacts of nanomaterials need to be further studied and evaluated.

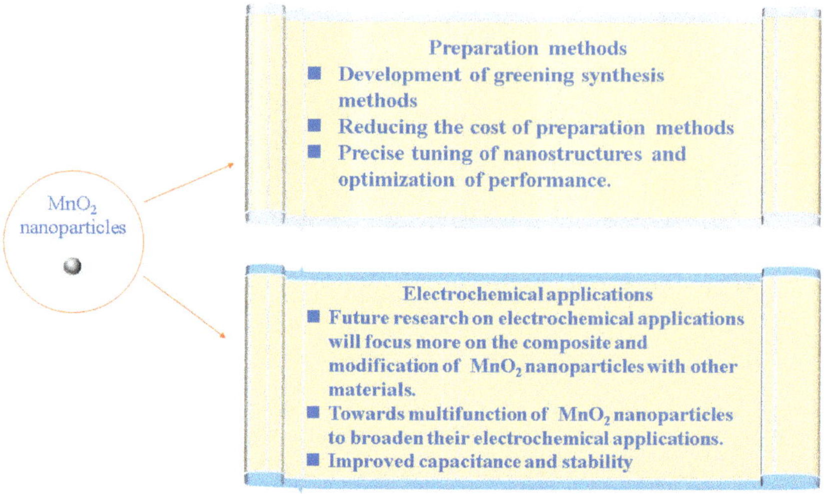

Figure 8. Future aspects of MnO_2 nanoparticles.

Author Contributions: Conceptualization, C.X. (Chunsheng Xie) and Z.X.; writing—original draft preparation, Z.X.; writing—review and editing, Z.X., Y.Z. and S.W.; supervision, M.D. and C.X. (Chun Xiao); funding acquisition, C.X. (Chunsheng Xie) and C.X. (Chun Xiao). All authors have read and agreed to the published version of the manuscript.

Funding: This work was jointly supported by Special Innovation Projects in Key Areas of Guangdong Provincial Department of Education (2023ZDZX4061), Zhaoqing University Innovation Research Team Project ([2021]31), Youth Foundation of SCIES (PM-zx097-202304-147), College Student Innovation and Entrepreneurship Training Program Project of Guangdong Province (X202210580133).

Acknowledgments: The authors would like to thank Guangdong Provincial Key Laboratory of Environmental Health and Land Resource for providing scientific research conditions.

Conflicts of Interest: The authors declare no conflicts of interest.

References

1. Liu, J.; Meng, L.; Fei, Z.; Dyson, P.J.; Jing, X.; Liu, X. MnO_2 nanosheets as an artificial enzyme to mimic oxidase for rapid and sensitive detection of glutathione. *Biosens. Bioelectron.* **2017**, *90*, 69–74. [CrossRef] [PubMed]
2. Prasad, K.S.; Patra, A. Green synthesis of MnO_2 nanorods using Phyllanthus amarus plant extract and their fluorescence studies. *Green Process. Synth.* **2017**, *6*, 549–554. [CrossRef]
3. Jin, S.B.; Jeong, J.-M.; Son, S.G.; Park, S.H.; Lee, K.G.; Choi, B.G. Synthesis of two-dimensional holey MnO_2/graphene oxide nanosheets with high catalytic performance for the glycolysis of poly(ethylene terephthalate). *Mater. Today Commun.* **2021**, *26*, 101857. [CrossRef]
4. Gaire, M.; Liang, K.; Luo, S.; Subedi, B.; Adireddy, S.; Schroder, K.; Farnsworth, S.; Chrisey, D.B. Nanostructured manganese oxides electrode with ultra-long lifetime for electrochemical capacitors. *RSC Adv.* **2020**, *10*, 16817–16825. [CrossRef] [PubMed]
5. Zhang, Z.; Ji, Y. Nanostructured manganese dioxide for anticancer applications: Preparation, diagnosis, and therapy. *Nanoscale* **2020**, *12*, 17982–18003. [CrossRef] [PubMed]
6. Yang, R.; Fan, Y.; Ye, R.; Tang, Y.; Cao, X.; Yin, Z.; Zeng, Z. MnO_2-Based Materials for Environmental Applications. *Adv. Mater.* **2021**, *33*, 2004862. [CrossRef] [PubMed]
7. Huang, S.-D.; Shang, C.; Zhang, X.-J.; Liu, Z.-P. Material discovery by combining stochastic surface walking global optimization with a neural network. *Chem. Sci.* **2017**, *8*, 6327–6337. [CrossRef] [PubMed]

8. Liu, X.; Chen, C.; Zhao, Y.; Jia, B. A Review on the Synthesis of Manganese Oxide Nanomaterials and Their Applications on Lithium-Ion Batteries. *J. Nanomater.* **2013**, *2013*, 736375. [CrossRef]
9. Yadav, P.; Bhaduri, A.; Thakur, A. Manganese Oxide Nanoparticles: An Insight into Structure, Synthesis and Applications. *ChemBioEng Rev.* **2023**, *10*, 510–528. [CrossRef]
10. Hayashi, E.; Yamaguchi, Y.; Kamata, K.; Tsunoda, N.; Kumagai, Y.; Oba, F.; Hara, M. Effect of MnO_2 Crystal Structure on Aerobic Oxidation of 5-Hydroxymethylfurfural to 2,5-Furandicarboxylic Acid. *J. Am. Chem. Soc.* **2019**, *141*, 890–900. [CrossRef]
11. Robinson, D.M.; Go, Y.B.; Mui, M.; Gardner, G.; Zhang, Z.; Mastrogiovanni, D.; Garfunkel, E.; Li, J.; Greenblatt, M.; Dismukes, G.C. Photochemical water oxidation by crystalline polymorphs of manganese oxides: Structural requirements for catalysis. *J. Am. Chem. Soc.* **2013**, *135*, 3494–3501. [CrossRef] [PubMed]
12. Chabre, Y.; Pannetier, J. Structural and electrochemical properties of the proton/γ-MnO_2 system. *Prog. Solid State Chem.* **1995**, *23*, 1–130. [CrossRef]
13. Biswal, A.; Chandra Tripathy, B.; Sanjay, K.; Subbaiah, T.; Minakshi, M. Electrolytic manganese dioxide (EMD): A perspective on worldwide production, reserves and its role in electrochemistry. *RSC Adv.* **2015**, *5*, 58255–58283. [CrossRef]
14. Salvador, G.M.S.; Silva, A.L.; Silva, L.P.C.; Passos, F.B.; Carvalho, N.M.F. Enhanced activity of Pd/α-MnO_2 for electrocatalytic oxygen evolution reaction. *Int. J. Hydrogen Energy* **2021**, *46*, 26976–26988. [CrossRef]
15. Gowrisankar, A.; Thangavelu, S. Effect of β-MnO_2 on Controlled Polymorphism of VO_2(x) (x = A, B, M Polymorphs) Microstructures Anchored on Two-Dimensional Reduced Graphene Oxide Nanosheets for Overall Water Splitting. *J. Phys. Chem. C* **2022**, *126*, 3419–3431. [CrossRef]
16. Rojas, S. Durable MnO_2 electrocatalysts by stronger Mn–O bonds. *Nat. Catal.* **2024**, *7*, 227–228. [CrossRef]
17. Guo, C.; Liu, H.; Li, J.; Hou, Z.; Liang, J.; Zhou, J.; Zhu, Y.; Qian, Y. Ultrathin δ-MnO_2 nanosheets as cathode for aqueous rechargeable zinc ion battery. *Electrochim. Acta* **2019**, *304*, 370–377. [CrossRef]
18. Li, L.; Yang, Q.; Wang, D.; Peng, Y.; Yan, J.; Li, J.; Crittenden, J. Facile synthesis λ-MnO_2 spinel for highly effective catalytic oxidation of benzene. *Chem. Eng. J.* **2021**, *421*, 127828. [CrossRef]
19. Kim, C.-H.; Akase, Z.; Zhang, L.; Heuer, A.H.; Newman, A.E.; Hughes, P.J. The structure and ordering of ε-MnO_2. *J. Solid State Chem.* **2006**, *179*, 753–774. [CrossRef]
20. Zeng, L.; Zhang, G.; Huang, X.; Wang, H.; Zhou, T.; Xie, H. Tuning crystal structure of MnO_2 during different hydrothermal synthesis temperature and its electrochemical performance as cathode material for zinc ion battery. *Vacuum* **2021**, *192*, 110398. [CrossRef]
21. Hasegawa, Y.; Fukumoto, K.; Ishima, T.; Yamamoto, H.; Sano, M.; Miyake, T. Preparation of copper-containing mesoporous manganese oxides and their catalytic performance for CO oxidation. *Appl. Catal. B Environ.* **2009**, *89*, 420–424. [CrossRef]
22. Peng, J.; Deng, F.; Shi, H.; Wang, Z.; Li, X.; Zou, J.; Luo, X. Target recognition and preferential degradation of toxic chemical groups by innovative group-imprinted photocatalyst with footprint cavity. *Appl. Catal. B Environ.* **2024**, *340*, 123179. [CrossRef]
23. Mishra, R.K.; Prajapati, C.S.; Shahi, R.R.; Kushwaha, A.K.; Sahay, P.P. Influence of electrodeposition modes on the electrochemical performance of MnO_2 films prepared using anionic MnO_4^- (Mn^{7+}) precursor. *Ceram. Int.* **2018**, *44*, 5710–5718. [CrossRef]
24. Cherian, E.; Rajan, A.; Gurunathan, D.B. Synthesis of manganese dioxide nanoparticles using co-precipitation method and its antimicrobial activity. *Int. J. Mod. Sci. Technol.* **2016**, *01*, 17–22.
25. Burda, C.; Chen, X.; Narayanan, R.; El-Sayed, M.A. Chemistry and Properties of Nanocrystals of Different Shapes. *Chem. Rev.* **2005**, *105*, 1025–1102. [CrossRef]
26. Chen, B.; Wu, B.; Yu, L.; Crocker, M.; Shi, C. Investigation into the Catalytic Roles of Various Oxygen Species over Different Crystal Phases of MnO_2 for C_6H_6 and HCHO Oxidation. *ACS Catal.* **2020**, *10*, 6176–6187. [CrossRef]
27. Yang, S.; Yang, H.; Yang, J.; Qi, H.; Kong, J.; Bo, Z.; Li, X.; Yan, J.; Cen, K.; Tu, X. Three-dimensional hollow urchin α-MnO_2 for enhanced catalytic activity towards toluene decomposition in post-plasma catalysis. *Chem. Eng. J.* **2020**, *402*, 126154. [CrossRef]
28. Aljafari, B.; Vijaya, S.; Takshi, A.; Anandan, S. Copper doped manganese dioxide as counter electrode for dye-sensitized solar cells. *Arab. J. Chem.* **2022**, *15*, 104068. [CrossRef]
29. Yang, C.; Zhong, Y.; Li, L.; Ren, X.; Sun, Y.; Niu, D.; Liu, Y.; Yin, M.; Zhang, D. Lead and uranium sorption characteristics on hydrothermal synthesized delta manganese dioxide. *J. Radioanal. Nucl. Chem.* **2018**, *317*, 1399–1408. [CrossRef]
30. Nguyen Dinh, M.T.; Nguyen, C.C.; Truong Vu, T.L.; Ho, V.T.; Truong, Q.D. Tailoring porous structure, reducibility and Mn^{4+} fraction of ε-MnO_2 microcubes for the complete oxidation of toluene. *Appl. Catal. A Gen.* **2020**, *595*, 117473. [CrossRef]
31. Li, K.; Li, H.; Xiao, T.; Long, J.; Zhang, G.; Li, Y.; Liu, X.; Liang, Z.; Zheng, F.; Zhang, P. Synthesis of manganese dioxide with different morphologies for thallium removal from wastewater. *J. Environ. Manag.* **2019**, *251*, 109563. [CrossRef] [PubMed]
32. Abdullah, T.A.; Rasheed, R.T.; Juzsakova, T.; Al-Jammal, N.; Mallah, M.A.; Cuong, L.P.; Salman, A.D.; Domokos, E.; Ali, Z.; Cretescu, I. Preparation and characterization of MnO_2-based nanoparticles at different annealing temperatures and their application in dye removal from water. *Int. J. Environ. Sci. Technol.* **2021**, *18*, 1499–1512. [CrossRef]
33. Zhong, M.; Li, M.; Fan, Z.; Huang, W.; Hao, H.; Xia, Z.; Zhang, Q.; Peng, H.; Zhang, Y. Tuning the crystallinity of MnO_2 oxidant to achieve highly efficient pollutant degradation. *Chin. Chem. Lett.* **2023**, *34*, 107189. [CrossRef]
34. Kumar, Y.; Chopra, S.; Gupta, A.; Kumar, Y.; Uke, S.J.; Mardikar, S.P. Low temperature synthesis of MnO_2 nanostructures for supercapacitor application. *Mater. Sci. Energy Technol.* **2020**, *3*, 566–574. [CrossRef]
35. Yuan, Y.; Zhu, J.; Wang, Y.; Li, S.; Jin, P.; Chen, Y. Facile synthesis of manganese oxide nanostructures with different crystallographic phase and morphology for supercapacitors. *J. Alloys Compd.* **2020**, *830*, 154524. [CrossRef]

36. Yadav, S.; Sharma, A. Importance and challenges of hydrothermal technique for synthesis of transition metal oxides and composites as supercapacitor electrode materials. *J. Energy Storage* **2021**, *44*, 103295. [CrossRef]
37. Patra, T.; Mohanty, A.; Singh, L.; Muduli, S.; Parhi, P.K.; Sahoo, T.R. Effect of calcination temperature on morphology and phase transformation of MnO_2 nanoparticles: A step towards green synthesis for reactive dye adsorption. *Chemosphere* **2022**, *288*, 132472. [CrossRef] [PubMed]
38. Zayat, M.; Pardo, R.; Castellón, E.; Torres, L.; Almendro, D.; Parejo, P.G.; Álvarez, A.; Belenguer, T.; García-Revilla, S.; Balda, R.; et al. Optical and Electro-optical Materials Prepared by the Sol-Gel Method. *Adv. Mater.* **2011**, *23*, 5318–5323. [CrossRef] [PubMed]
39. Reddy, R.N.; Reddy, R.G. Sol–gel MnO_2 as an electrode material for electrochemical capacitors. *J. Power Sources* **2003**, *124*, 330–337. [CrossRef]
40. Panimalar, S.; Logambal, S.; Thambidurai, R.; Inmozhi, C.; Uthrakumar, R.; Muthukumaran, A.; Rasheed, R.A.; Gatasheh, M.K.; Raja, A.; Kennedy, J.; et al. Effect of Ag doped MnO_2 nanostructures suitable for wastewater treatment and other environmental pollutant applications. *Environ. Res.* **2022**, *205*, 112560. [CrossRef]
41. Kusworo, T.D.; Kumoro, A.C.; Aryanti, N.; Kurniawan, T.A.; Dalanta, F.; Alias, N.H. Photocatalytic polysulfone membrane incorporated by ZnO-MnO_2@SiO_2 composite under UV light irradiation for the reliable treatment of natural rubber-laden wastewater. *Chem. Eng. J.* **2023**, *451*, 138593. [CrossRef]
42. Wang, X.; Yuan, A.; Wang, Y. Supercapacitive behaviors and their temperature dependence of sol–gel synthesized nanostructured manganese dioxide in lithium hydroxide electrolyte. *J. Power Sources* **2007**, *172*, 1007–1011. [CrossRef]
43. Pal, A.; Das, T.; Ghosh, S.; Nandi, M. Supercapacitor behaviour of manganese dioxide decorated mesoporous silica synthesized by a rapid sol–gel inverse micelle method. *Dalton Trans.* **2020**, *49*, 12716–12730. [CrossRef] [PubMed]
44. Siddique, M.A.B.; Bithi, U.H.; Ahmed, A.N.; Gafur, M.A.; Reaz, A.H.; Roy, C.K.; Islam, M.M.; Firoz, S.H. Preparation of Manganese Oxide Nanoparticles with Enhanced Capacitive Properties Utilizing Gel Formation Method. *ACS Omega* **2022**, *7*, 48007–48017. [CrossRef] [PubMed]
45. Nguyen, V.H.; Huynh, L.T.N.; Nguyen, T.H.; Vu, T.P.; Le, M.L.P.; Grag, A.; Tran, V.M. Promising electrode material using Ni-doped layered manganese dioxide for sodium-ion batteries. *J. Appl. Electrochem.* **2018**, *48*, 793–800. [CrossRef]
46. Feng, A.; Hou, T.; Jia, Z.; Wu, G. Synthesis of a hierarchical carbon fiber@cobalt ferrite@manganese dioxide composite and its application as a microwave absorber. *RSC Adv.* **2020**, *10*, 10510–10518. [CrossRef] [PubMed]
47. Anggraini, R.; Siregar, S.S.; Awaluddin, A.; Linggawati, A. The Preliminary Studies on the Tremendous Degradation Rate of Methylene Blue with Cu-doped α-MnO_2 Photocatalyst Under UV Light Irradation. *J. Phys. Conf. Ser.* **2021**, *2049*, 012060. [CrossRef]
48. Fu, L.J.; Liu, H.; Li, C.; Wu, Y.P.; Rahm, E.; Holze, R.; Wu, H.Q. Electrode materials for lithium secondary batteries prepared by sol–gel methods. *Prog. Mater. Sci.* **2005**, *50*, 881–928. [CrossRef]
49. Petkovich, N.D.; Stein, A. Controlling macro- and mesostructures with hierarchical porosity through combined hard and soft templating. *Chem. Soc. Rev.* **2013**, *42*, 3721–3739. [CrossRef]
50. Zhang, L.; Jin, L.; Liu, B.; He, J. Templated Growth of Crystalline Mesoporous Materials: From Soft/Hard Templates to Colloidal Templates. *Front. Chem.* **2019**, *7*, 22. [CrossRef]
51. Hou, B.; Liu, Y.; Li, Y.; Yuan, B.; Jia, M.; Jiang, F. Evolvement of soft templates in surfactant/cosurfactant system for shape control of ZnSe nanocrystals. *Mater. Sci. Eng. B* **2012**, *177*, 411–415. [CrossRef]
52. Chen, J.; Meng, H.; Tian, Y.; Yang, R.; Du, D.; Li, Z.; Qu, L.; Lin, Y. Recent advances in functionalized MnO_2 nanosheets for biosensing and biomedicine applications. *Nanoscale Horiz.* **2019**, *4*, 321–338. [CrossRef] [PubMed]
53. Yuan, W.; Shen, P.K.; Jiang, S.P. Controllable synthesis of graphene supported MnO_2 nanowires via self-assembly for enhanced water oxidation in both alkaline and neutral solutions. *J. Mater. Chem. A* **2014**, *2*, 123–129. [CrossRef]
54. Tran, C.C.H.; Santos-Peña, J.; Damas, C. Theoretical and Practical Approach of Soft Template Synthesis for the Preparation of MnO_2 Supercapacitor Electrode. *J. Phys. Chem. C* **2018**, *122*, 16–29. [CrossRef]
55. Yang, J.; Li, K.; Li, C.; Gu, J. In Situ Coupling of Catalytic Centers into Artificial Substrate Mesochannels as Super-Active Metalloenzyme Mimics. *Small* **2021**, *17*, 2101455. [CrossRef] [PubMed]
56. Bai, B.; Li, J.; Hao, J. 1D-MnO_2, 2D-MnO_2 and 3D-MnO_2 for low-temperature oxidation of ethanol. *Appl. Catal. B Environ.* **2015**, *164*, 241–250. [CrossRef]
57. Bai, B.; Qiao, Q.; Li, J.; Hao, J. Synthesis of three-dimensional ordered mesoporous MnO_2 and its catalytic performance in formaldehyde oxidation. *Chin. J. Catal.* **2016**, *37*, 27–31. [CrossRef]
58. Zhang, J.; Zhuang, T.; Liu, S.; Zhang, G.C.; Huo, K. Catalytic ozonation of phenol enhanced by mesoporous MnO_2 prepared through nanocasting method with SBA-15 as template. *J. Environ. Chem. Eng.* **2020**, *8*, 103967. [CrossRef]
59. Zhang, X.; Zhou, B.; Yin, S.; Wang, Y.; Zhang, X.; Meng, Q.; Meng, F.; Wei, C.; Wen, G. Mesoporous manganese dioxide prepared by nano-casting: An efficient catalyst for of methyl orange and oxalic acid degradation in aqueous solution. *Vacuum* **2022**, *206*, 111495. [CrossRef]
60. Wang, C.; Liu, Z.; Wang, Q.; Guo, J.; Zhao, Q.; Lu, Y. MnO_2@polypyrrole composite with hollow microsphere structure for electrode material of supercapacitors. *J. Electroanal. Chem.* **2021**, *901*, 115780. [CrossRef]
61. Wu, J.; Ma, Q.; Lian, C.; Yuan, Y.; Long, D. Promoting polythionate intermediates formation by oxygen-deficient manganese oxide hollow nanospheres for high performance lithium-sulfur batteries. *Chem. Eng. J.* **2019**, *370*, 556–564. [CrossRef]

62. Selvakumar, K.; Kumar, S.M.S.; Thangamuthu, R.; Rajput, P.; Bhattacharyya, D.; Jha, S.N. 2D and 3D Silica-Template-Derived MnO_2 Electrocatalysts towards Enhanced Oxygen Evolution and Oxygen Reduction Activity. *ChemElectroChem* **2018**, *5*, 3980–3990. [CrossRef]
63. Gu, W.; Li, C.; Qiu, J.; Yao, J. Facile fabrication of flower-like MnO_2 hollow microspheres as high-performance catalysts for toluene oxidation. *J. Hazard. Mater.* **2021**, *408*, 124458. [CrossRef]
64. Wang, H.-q.; Zheng, M.-b.; Chen, J.-h.; Ji, G.-b.; Cao, J.-m. Synthesis of MnO_2 Microfiber with Secondary Nanostructure by Cotton Template. *J. Nanotechnol.* **2010**, *2010*, 479172. [CrossRef]
65. Hsueh, H.-Y.; Yao, C.-T.; Ho, R.-M. Well-ordered nanohybrids and nanoporous materials from gyroid block copolymer templates. *Chem. Soc. Rev.* **2015**, *44*, 1974–2018. [CrossRef] [PubMed]
66. Eric, M.G.; Vanessa, F.C.L.; Tulio, M. Metallic and Oxide Electrodeposition. In *Modern Surface Engineering Treatments*; Mahmood, A., Ed.; IntechOpen: Rijeka, Croatia, 2013; p. Ch. 5.
67. Shirole, D.; Volpatti, G.; Guerini, A.; Zampini, D.; Cusatis, G.; Rotta Loria, A.F. Effects of electrodeposition in concrete mediated by electric currents of variable polarity. *Cem. Concr. Res.* **2023**, *172*, 107254. [CrossRef]
68. Dai, X.; Zhang, M.; Li, J.; Yang, D. Effects of electrodeposition time on a manganese dioxide supercapacitor. *RSC Adv.* **2020**, *10*, 15860–15869. [CrossRef] [PubMed]
69. Surendranath, Y.; Dincă, M.; Nocera, D.G. Electrolyte-Dependent Electrosynthesis and Activity of Cobalt-Based Water Oxidation Catalysts. *J. Am. Chem. Soc.* **2009**, *131*, 2615–2620. [CrossRef] [PubMed]
70. Ye, Z.; Li, T.; Ma, G.; Peng, X.; Zhao, J. Morphology controlled MnO_2 electrodeposited on carbon fiber paper for high-performance supercapacitors. *J. Power Sources* **2017**, *351*, 51–57. [CrossRef]
71. Ren, H.; Zhang, L.; Zhang, J.; Miao, T.; Yuan, R.; Chen, W.; Wang, Z.; Yang, J.; Zhao, B. Na^+ pre-intercalated $Na_{0.11}MnO_2$ on three-dimensional graphene as cathode for aqueous zinc ion hybrid supercapacitor with high energy density. *Carbon* **2022**, *198*, 46–56. [CrossRef]
72. Shi, Y.; Zhang, M.; Zhao, J.; Zhang, L.; Cui, X.M.; Zhu, X.; Jin, D.; Gong, J.; Yang, D.; Li, J. Preparation of Manganese Dioxide Supercapacitors by Secondary Construction of Three-Dimensional Substrates and Ion Embedding. *Electron. Mater. Lett.* **2022**, *18*, 475–488. [CrossRef]
73. Zhao, W.K.; Ma, Z.Q.; Zheng, J.Y.; Han, C.B.; Zhou, K.L.; Hao, M.Y.; Fang, D.C.; Xia, Y.; Yan, H. Dual-functional MnS_2/MnO_2 heterostructure catalyst for efficient acidic hydrogen evolution reaction and assisted degradation of organic wastewater. *J. Energy Chem.* **2023**, *87*, 215–224. [CrossRef]
74. Zhou, H.; Zhi, X.; Zhai, H.-J. Promoted supercapacitive performances of electrochemically synthesized poly(3,4-ethylenedioxythiophene) incorporated with manganese dioxide. *J. Mater. Sci. Mater. Electron.* **2018**, *29*, 3935–3942. [CrossRef]
75. Forghani, M.; McCarthy, J.; Cameron, A.P.; Davey, S.B.; Donne, S.W. Semiconductor Properties of Electrodeposited Manganese Dioxide for Electrochemical Capacitors: Mott-Schottky Analysis. *J. Electrochem. Soc.* **2021**, *168*, 020508. [CrossRef]
76. Hu, X.; Cheng, F.; Han, X.; Zhang, T.; Chen, J. Oxygen Bubble-Templated Hierarchical Porous ε-MnO_2 as a Superior Catalyst for Rechargeable Li–O_2 Batteries. *Small* **2015**, *11*, 809–813. [CrossRef] [PubMed]
77. Liu, J.; Yang, L.; Song, Z.; Xu, C. Microstructures and capacitance performance of MnO_2 films fabricated by ultrasonic-assisted electrodeposition. *Appl. Surf. Sci.* **2019**, *478*, 94–102. [CrossRef]
78. Zhao, P.; Wang, N.; Yao, M.; Ren, H.; Hu, W. Hydrothermal electrodeposition incorporated with CVD-polymerisation to tune PPy@MnO_2 interlinked core-shell nanowires on carbon fabric for flexible solid-state asymmetric supercapacitors. *Chem. Eng. J.* **2020**, *380*, 122488. [CrossRef]
79. Meng, X.; Liu, T.; Qin, M.; Liu, Z.; Wang, W. Carbon-Free, Binder-Free MnO_2@Mn Catalyst for Oxygen Reduction Reaction. *ACS Appl. Mater. Interfaces* **2023**, *15*, 20110–20119. [CrossRef]
80. Wang, P.; Lin, Y.; Wan, L.; Wang, B. Construction of a Janus MnO_2-NiFe Electrode via Selective Electrodeposition Strategy as a High-Performance Bifunctional Electrocatalyst for Rechargeable Zinc–Air Batteries. *ACS Appl. Mater. Interfaces* **2019**, *11*, 37701–37707. [CrossRef]
81. Jeong, J.-H.; Park, J.W.; Lee, D.W.; Baughman, R.H.; Kim, S.J. Electrodeposition of α-MnO_2/γ-MnO_2 on Carbon Nanotube for Yarn Supercapacitor. *Sci. Rep.* **2019**, *9*, 11271. [CrossRef]
82. Li, X.-b.; Xu, G.-r. Hydrothermal vs electrodeposition: How does deposition method affect the electrochemical capacitor performance of manganese dioxide? *Ceram. Int.* **2017**, *43*, 8963–8969. [CrossRef]
83. Ali, H.; Mahto, B.; Barhoi, A.; Hussain, S. Visible light-driven photocatalytic thiol–ene/yne reactions using anisotropic 1D Bi_2S_3 nanorods: A green synthetic approach. *Nanoscale* **2023**, *15*, 14551–14563. [CrossRef] [PubMed]
84. Zhang, X.; Sun, X.; Zhang, H.; Zhang, D.; Ma, Y. Microwave-assisted reflux rapid synthesis of MnO_2 nanostructures and their application in supercapacitors. *Electrochim. Acta* **2013**, *87*, 637–644. [CrossRef]
85. Said, M.I. Akhtenskite-nsutite phases: Polymorphic transformation, thermal behavior and magnetic properties. *J. Alloys Compd.* **2020**, *819*, 152976. [CrossRef]
86. May, Y.A.; Wei, S.; Yu, W.Z.; Wang, W.W.; Jia, C.J. Highly Efficient CuO/α-MnO_2 Catalyst for Low-Temperature CO Oxidation. *Langmuir ACS J. Surf. Colloids* **2020**, *36*, 11196–11206. [CrossRef] [PubMed]
87. Kijima, N.; Sakata, Y.; Takahashi, Y.; Akimoto, J.; Kumagai, T.; Igarashi, K.; Shimizu, T. Synthesis and lithium ion insertion/extraction properties of hollandite-type MnO_2 prepared by acid digestion of Mn_2O_3. *Solid State Ion.* **2009**, *180*, 616–620. [CrossRef]
88. Antonio, C. Synthesis of NPs by Microemulsion Method. In *Microemulsion*; Juan, C.M., Ed.; IntechOpen: Rijeka, Croatia, 2018; p. Ch. 5.

89. Tavakoli, A.; Sohrabi, M.; Kargari, A. A review of methods for synthesis of nanostructured metals with emphasis on iron compounds. *Chem. Pap.* **2007**, *61*, 151–170. [CrossRef]
90. Xu, C.; Li, B.; Du, H.; Kang, F.; Zeng, Y. Electrochemical properties of nanosized hydrous manganese dioxide synthesized by a self-reacting microemulsion method. *J. Power Sources* **2008**, *180*, 664–670. [CrossRef]
91. Zefirov, V.V.; Elmanovich, I.V.; Levin, E.E.; Abramchuk, S.S.; Kharitonova, E.P.; Khokhlov, A.A.; Kondratenko, M.S.; Gallyamov, M.O. Synthesis of manganese oxide electrocatalysts in supercritical carbon dioxide. *J. Mater. Sci.* **2018**, *53*, 9449–9462. [CrossRef]
92. Tatarchuk, T.; Bououdina, M.; Judith Vijaya, J.; John Kennedy, L. Spinel Ferrite Nanoparticles: Synthesis, Crystal Structure, Properties, and Perspective Applications. In *Nanophysics, Nanomaterials, Interface Studies, and Applications*; Springer: Berlin/Heidelberg, Germany, 2017; pp. 305–325.
93. Liu, Z.; Ji, H.; Wang, S.; Zhao, W.; Huang, Y.; Feng, H.; Wei, J.; Li, M. Enhanced Electrical and Mechanical Properties of a Printed Bimodal Silver Nanoparticle Ink for Flexible Electronics. *Phys. Status Solidi A* **2018**, *215*, 1800007. [CrossRef]
94. Sivakumar, S.; Nelson Prabu, L. Synthesis and Characterization of α-MnO_2 nanoparticles for Supercapacitor application. *Mater. Today Proc.* **2021**, *47*, 52–55. [CrossRef]
95. Yadav, P.; Bhaduri, A. Chemically synthesized manganese oxide nanorods for effectual organic dye removal and energy storage application. *Mater. Chem. Phys.* **2023**, *299*, 127495. [CrossRef]
96. Yang, P.; Wang, J.; Wang, H.; Wang, S.; Yang, C.; He, Y. Physicochemical properties of different crystal forms of manganese dioxide prepared by a liquid phase method and their quantitative evaluation in capacitor and battery materials. *Nanoscale Adv.* **2023**, *5*, 3396–3413. [CrossRef]
97. Quintero-Quiroz, C.; Acevedo, N.; Zapata-Giraldo, J.; Botero, L.E.; Quintero, J.; Zárate-Triviño, D.; Saldarriaga, J.; Pérez, V.Z. Optimization of silver nanoparticle synthesis by chemical reduction and evaluation of its antimicrobial and toxic activity. *Biomater. Res.* **2019**, *23*, 27. [CrossRef] [PubMed]
98. Ferrando, R.; Jellinek, J.; Johnston, R.L. Nanoalloys: From Theory to Applications of Alloy Clusters and Nanoparticles. *Chem. Rev.* **2008**, *108*, 845–910. [CrossRef] [PubMed]
99. Li, B.; Rong, G.; Xie, Y.; Huang, L.; Feng, C. Low-Temperature Synthesis of α-MnO_2 Hollow Urchins and Their Application in Rechargeable Li^+ Batteries. *Inorg. Chem.* **2006**, *45*, 6404–6410. [CrossRef] [PubMed]
100. Khan, I.; Sadiq, M.; Khan, I.; Saeed, K. Manganese dioxide nanoparticles/activated carbon composite as efficient UV and visible-light photocatalyst. *Environ. Sci. Pollut. Res.* **2019**, *26*, 5140–5154. [CrossRef] [PubMed]
101. Cremonezzi, J.M.d.O.; Tiba, D.Y.; Domingues, S.H. Fast synthesis of δ-MnO_2 for a high-performance supercapacitor electrode. *SN Appl. Sci.* **2020**, *2*, 1689. [CrossRef]
102. Narayanan, K.B.; Sakthivel, N. Green synthesis of biogenic metal nanoparticles by terrestrial and aquatic phototrophic and heterotrophic eukaryotes and biocompatible agents. *Adv. Colloid Interface Sci.* **2011**, *169*, 59–79. [CrossRef]
103. Ahmed, S.; Annu; Chaudhry, S.A.; Ikram, S. A review on biogenic synthesis of ZnO nanoparticles using plant extracts and microbes: A prospect towards green chemistry. *J. Photochem. Photobiol. B Biol.* **2017**, *166*, 272–284. [CrossRef]
104. Joshi, N.C.; Siddqui, F.; Salman, M.; Singh, A. Antibacterial Activity, Characterizations, and Biological Synthesis of Manganese Oxide Nanoparticles using the Extract of Aloe vera. *Asian Pac. J. Health Sci.* **2020**, *7*, 27–29. [CrossRef]
105. Souri, M.; Shakeri, A. Optimization of Total Phenol and Tannin Content and Biological Activity of *Dittrichia graveolens* (L.) GREUTER. *Curr. Bioact. Compd.* **2020**, *16*, 124–132. [CrossRef]
106. Hashem, A.M.; Abuzeid, H.; Kaus, M.; Indris, S.; Ehrenberg, H.; Mauger, A.; Julien, C.M. Green synthesis of nanosized manganese dioxide as positive electrode for lithium-ion batteries using lemon juice and citrus peel. *Electrochim. Acta* **2018**, *262*, 74–81. [CrossRef]
107. Shehroz, H.; Ali, S.; Bibi, G.; Khan, T.; Jamil, S.; Khan, S.R.; Hashaam, M.; Naz, S. Comparative investigation of the catalytic application of $\alpha/\beta/\gamma$-MnO_2 nanoparticles synthesized by green and chemical approaches. *Environ. Technol.* **2024**, *45*, 1081–1091. [CrossRef] [PubMed]
108. Ramesh, P.; Rajendran, A. Green synthesis of manganese dioxide nanoparticles photocatalytic and antimicrobial investigations. *Int. J. Environ. Anal. Chem.* **2023**, 1–13. [CrossRef]
109. Rahmat, M.; Kiran, S.; Gulzar, T.; Yusuf, M.; Nawaz, R.; Khalid, J.; Fatima, N.; Ullah, A.; Azam, M. Plant-assisted synthesis and characterization of MnO_2 nanoparticles for removal of crystal violet dye: An environmental remedial approach. *Environ. Sci. Pollut. Res.* **2023**, *30*, 57587–57598. [CrossRef] [PubMed]
110. Majani, S.S.; Sathyan, S.; Manoj, M.V.; Vinod, N.; Pradeep, S.; Shivamallu, C.; Venkatachalaiah, K.N.; Kollur, S.P. Eco-friendly synthesis of MnO_2 nanoparticles using Saraca asoca leaf extract and evaluation of in vitro anticancer activity. *Curr. Res. Green Sustain. Chem.* **2023**, *6*, 100367. [CrossRef]
111. Hoseinpour, V.; Souri, M.; Ghaemi, N. Green synthesis, characterisation, and photocatalytic activity of manganese dioxide nanoparticles. *Micro Nano Lett.* **2018**, *13*, 1560–1563. [CrossRef]
112. Samsoon, S.; Azam, M.; Khan, A.; Ashraf, M.; Bhatti, H.N.; Alshawwa, S.Z.; Iqbal, M. Green-synthesized MnO_2 nanofertilizer impact on growth, photosynthetic pigment, and non-enzymatic antioxidant of Vigna unguiculata cultivar. *Biomass Convers. Biorefinery* **2022**, 1–10. [CrossRef]
113. Lu, H.; Zhang, X.; Khan, S.A.; Li, W.; Wan, L. Biogenic Synthesis of MnO_2 Nanoparticles with Leaf Extract of Viola betonicifolia for Enhanced Antioxidant, Antimicrobial, Cytotoxic, and Biocompatible Applications. *Front. Microbiol.* **2021**, *12*, 761084. [CrossRef]
114. Srivastava, V.; Choubey, A.K. Study of adsorption of anionic dyes over biofabricated crystalline α-MnO_2 nanoparticles. *Environ. Sci. Pollut. Res.* **2021**, *28*, 15504–15518. [CrossRef]

115. Bhattacharjee, S.; Bardhan, M.; Ghosh, S.; Banerjee, A.; Pal, K.; Guha, A.; Mondal, D.; Basu, R.; Das, S.; Sinha, S.K. An in-vivo interpretation for validating the ameliorative efficacy of green synthesized MnO_2 nano-conjugate using Carica Papaya (Papaya) leaf extract against acute hepatic damage. *J. Drug Deliv. Sci. Technol.* **2021**, *66*, 102774. [CrossRef]
116. Ogunyemi, S.O.; Zhang, F.; Abdallah, Y.; Zhang, M.; Wang, Y.; Sun, G.; Qiu, W.; Li, B. Biosynthesis and characterization of magnesium oxide and manganese dioxide nanoparticles using Matricaria chamomilla L. extract and its inhibitory effect on Acidovorax oryzae strain RS-2. *Artif. Cells Nanomed. Biotechnol.* **2019**, *47*, 2230–2239. [CrossRef]
117. Elsherif, S.A.; Abuzeid, H.M.; Hashem, A.M.; Abdel Ghany, N.A. Green synthesis of MnO_2 via plant extracts and its composite with exfoliated graphene for high-performance asymmetric supercapacitors. *J. Energy Storage* **2023**, *74*, 109341. [CrossRef]
118. Ghosh, A.; Hegde, R.V.; Limaye, A.S.; Thrilokraj, R.; Patil, S.A.; Dateer, R.B. Biogenic synthesis of δ-MnO_2 nanoparticles: A sustainable approach for C-alkylation and quinoline synthesis via acceptorless dehydrogenation and borrowing hydrogen reactions. *Appl. Organomet. Chem.* **2023**, *37*, e7119. [CrossRef]
119. Affrald, J. A comprehensive review of manganese dioxide nanoparticles and strategy to overcome toxicity. *Nanomedicine* **2022**, *10*, 1–15. [CrossRef]
120. Beveridge, T.J.; Hughes, M.N.; Lee, H.; Leung, K.T.; Poole, R.K.; Savvaidis, I.; Silver, S.; Trevors, J.T. Metal-Microbe Interactions: Contemporary Approaches. *Adv. Microb. Physiol.* **1996**, *38*, 177–243.
121. Bharde, A.; Rautaray, D.; Bansal, V.; Ahmad, A.; Sarkar, I.; Yusuf, S.M.; Sanyal, M.; Sastry, M. Extracellular Biosynthesis of Magnetite using Fungi. *Small* **2006**, *2*, 135–141. [CrossRef] [PubMed]
122. Sinha, A.; Singh, V.N.; Mehta, B.R.; Khare, S.K. Synthesis and characterization of monodispersed orthorhombic manganese oxide nanoparticles produced by Bacillus sp. cells simultaneous to its bioremediation. *J. Hazard. Mater.* **2011**, *192*, 620–627. [CrossRef]
123. Borah, D.; Rout, J.; Gogoi, D.; Nath Ghosh, N.; Bhattacharjee, C.R. Composition controllable green synthesis of manganese dioxide nanoparticles using an edible freshwater red alga and its photocatalytic activity towards water soluble toxic dyes. *Inorg. Chem. Commun.* **2022**, *138*, 109312. [CrossRef]
124. Alvares, J.; Gaonkar, S.; Naik, C.; Asogekar, P.; Furtado, I. Characterization of Mn_3O_4-MnO_2 nanocomposites biosynthesized by cell lysate of Haloferax alexandrinus GUSF-1. *J. Basic Microbiol.* **2023**, *63*, 996–1006. [CrossRef]
125. Iravani, S.; Varma, R.S. Bacteria in Heavy Metal Remediation and Nanoparticle Biosynthesis. *ACS Sustain. Chem. Eng.* **2020**, *8*, 5395–5409. [CrossRef]
126. Gahlawat, G.; Choudhury, A.R. A review on the biosynthesis of metal and metal salt nanoparticles by microbes. *RSC Adv.* **2019**, *9*, 12944–12967. [CrossRef] [PubMed]
127. Zhang, K.; Han, X.; Hu, Z.; Zhang, X.; Tao, Z.; Chen, J. Nanostructured Mn-based oxides for electrochemical energy storage and conversion. *Chem. Soc. Rev.* **2015**, *44*, 699–728. [CrossRef] [PubMed]
128. Liao, X.; Pan, C.; Pan, Y.; Yin, C. Synthesis of three-dimensional β-MnO_2/PPy composite for high-performance cathode in zinc-ion batteries. *J. Alloys Compd.* **2021**, *888*, 161619. [CrossRef]
129. Wu, R.; Kwan, K.W.; Wang, Y.; Ngan, A.H.W. Air-Working Electrochemical Actuator and Ionic Sensor Based on Manganese Dioxide/Gelatin-Glycerol Composites. *Adv. Mater. Technol.* **2023**, *8*, 2202062. [CrossRef]
130. Wu, Y.; Xu, Z.; Ren, R.; Lv, N.; Yang, J.; Zhang, J.; Ren, H.; Dong, S.; Dong, X. Flexible Ammonium-Ion Pouch Cells Based on a Tunneled Manganese Dioxide Cathode. *ACS Appl. Mater. Interfaces* **2023**, *15*, 12434–12442. [CrossRef]
131. Benedet, M.; Gallo, A.; Maccato, C.; Rizzi, G.A.; Barreca, D.; Lebedev, O.I.; Modin, E.; McGlynn, R.; Mariotti, D.; Gasparotto, A. Controllable Anchoring of Graphitic Carbon Nitride on MnO_2 Nanoarchitectures for Oxygen Evolution Electrocatalysis. *ACS Appl. Mater. Interfaces* **2023**, *15*, 47368–47380. [CrossRef] [PubMed]
132. Miller, J.R.; Simon, P. Electrochemical Capacitors for Energy Management. *Science* **2008**, *321*, 651–652. [CrossRef]
133. Zhang, Y.-Z.; Wang, Y.; Cheng, T.; Yao, L.-Q.; Li, X.; Lai, W.-Y.; Huang, W. Printed supercapacitors: Materials, printing and applications. *Chem. Soc. Rev.* **2019**, *48*, 3229–3264. [CrossRef]
134. Krishnamoorthy, K.; Pazhamalai, P.; Mariappan, V.K.; Manoharan, S.; Kesavan, D.; Kim, S.-J. Two-Dimensional Siloxene–Graphene Heterostructure-Based High-Performance Supercapacitor for Capturing Regenerative Braking Energy in Electric Vehicles. *Adv. Funct. Mater.* **2021**, *31*, 2008422. [CrossRef]
135. Jiao, X.; Wang, J.; Yuan, Z.; Zhang, C. Smart current collector for high-energy-density and high-contrast electrochromic supercapacitors toward intelligent and wearable power application. *Energy Storage Mater.* **2023**, *54*, 254–265. [CrossRef]
136. Jin, W.-Y.; Ovhal, M.M.; Lee, H.B.; Tyagi, B.; Kang, J.-W. Scalable, All-Printed Photocapacitor Fibers and Modules based on Metal-Embedded Flexible Transparent Conductive Electrodes for Self-Charging Wearable Applications. *Adv. Energy Mater.* **2021**, *11*, 2003509. [CrossRef]
137. Fic, K.; Platek, A.; Piwek, J.; Frackowiak, E. Sustainable materials for electrochemical capacitors. *Mater. Today* **2018**, *21*, 437–454. [CrossRef]
138. Ideta, K.; Kim, D.-W.; Kim, T.; Nakabayashi, K.; Miyawaki, J.; Park, J.-I.; Yoon, S.-H. Effect of pore size in activated carbon on the response characteristic of electric double layer capacitor. *J. Ind. Eng. Chem.* **2021**, *102*, 321–326. [CrossRef]
139. Teng, W.; Zhou, Q.; Wang, X.; Che, H.; Hu, P.; Li, H.; Wang, J. Hierarchically interconnected conducting polymer hybrid fiber with high specific capacitance for flexible fiber-shaped supercapacitor. *Chem. Eng. J.* **2020**, *390*, 124569. [CrossRef]
140. Wang, H.; Diao, Y.; Lu, Y.; Yang, H.; Zhou, Q.; Chrulski, K.; D'Arcy, J.M. Energy storing bricks for stationary PEDOT supercapacitors. *Nat. Commun.* **2020**, *11*, 3882. [CrossRef] [PubMed]

141. Chang, X.; El-Kady, M.F.; Huang, A.; Lin, C.-W.; Aguilar, S.; Anderson, M.; Zhu, J.Z.J.; Kaner, R.B. 3D Graphene Network with Covalently Grafted Aniline Tetramer for Ultralong-Life Supercapacitors. *Adv. Funct. Mater.* **2021**, *31*, 2102397. [CrossRef]
142. Zhang, L.; Shi, D.; Liu, T.; Jaroniec, M.; Yu, J. Nickel-based materials for supercapacitors. *Mater. Today* **2019**, *25*, 35–65. [CrossRef]
143. Zhang, A.; Zhao, R.; Hu, L.; Yang, R.; Yao, S.; Wang, S.; Yang, Z.; Yan, Y.-M. Adjusting the Coordination Environment of Mn Enhances Supercapacitor Performance of MnO_2. *Adv. Energy Mater.* **2021**, *11*, 2101412. [CrossRef]
144. Brousse, K.; Pinaud, S.; Nguyen, S.; Fazzini, P.-F.; Makarem, R.; Josse, C.; Thimont, Y.; Chaudret, B.; Taberna, P.-L.; Respaud, M.; et al. Facile and Scalable Preparation of Ruthenium Oxide-Based Flexible Micro-Supercapacitors. *Adv. Energy Mater.* **2020**, *10*, 1903136. [CrossRef]
145. Wang, L.; Xie, X.; Dinh, K.N.; Yan, Q.; Ma, J. Synthesis, characterizations, and utilization of oxygen-deficient metal oxides for lithium/sodium-ion batteries and supercapacitors. *Coord. Chem. Rev.* **2019**, *397*, 138–167. [CrossRef]
146. Devi, N.; Goswami, M.; Saraf, M.; Singh, B.; Mobin, S.M.; Singh, R.K.; Srivastava, A.K.; Kumar, S. Physicochemical and electrochemical behaviours of manganese oxide electrodes for supercapacitor application. *J. Energy Storage* **2020**, *28*, 101228. [CrossRef]
147. Yao, B.; Chandrasekaran, S.; Zhang, J.; Xiao, W.; Qian, F.; Zhu, C.; Duoss, E.B.; Spadaccini, C.M.; Worsley, M.A.; Li, Y. Efficient 3D Printed Pseudocapacitive Electrodes with Ultrahigh MnO_2 Loading. *Joule* **2019**, *3*, 459–470. [CrossRef]
148. Bagal, I.V.; Chodankar, N.R.; Waseem, A.; Ali Johar, M.; Patil, S.J.; Abdullah, A.; Afifi Hassan, M.; Han, Y.-K.; Ryu, S.-W. CF_4 plasma-treated porous silicon nanowire arrays laminated with MnO_2 nanoflakes for asymmetric pseudocapacitors. *Chem. Eng. J.* **2021**, *419*, 129515. [CrossRef]
149. Tynan, B.; Zhou, Y.; Brown, S.A.; Dai, L.; Rider, A.N.; Wang, C.H. Structural supercapacitor electrodes for energy storage by electroless deposition of MnO_2 on carbon nanotube mats. *Compos. Sci. Technol.* **2023**, *238*, 110016. [CrossRef]
150. Mladenova, B.; Pashova, K.; Hinkov, I.; Dimitrova, M.; Stoyanova, A. Green synthesis of MnO_2 using Calendula officinalis and Tilia cordata extracts for application in supercapacitors. *Monatshefte Für Chem.-Chem. Mon.* **2024**, *155*, 341–348. [CrossRef]
151. Li, P.; Wu, J.; Tang, L.; Liu, H.; Xu, Y.; Zhang, D. Interconnected δ–MnO_2 nanosheets anchored on porous carbon derived from reed residue waste as high-performance electrode for supercapacitor. *Ionics* **2023**, *29*, 3629–3639. [CrossRef]
152. Mladenova, B.; Dimitrova, M.; Stoyanova, A. MnO_2/AgNPs Composite as Flexible Electrode Material for Solid-State Hybrid Supercapacitor. *Batteries* **2024**, *10*, 122. [CrossRef]
153. Sayah, A.; Boumaza, N.; Habelhames, F.; Bahloul, A.; Bencherif, H.; Tounsi, A.; Lamiri, L.; Nessark, B. Electrodeposition mode effects on the electrochemical performance of MnO_2–NiO eco-friendly material for supercapacitor electrode application. *J. Mater. Sci. Mater. Electron.* **2024**, *35*, 62. [CrossRef]
154. Khalid, M.U.; Zulfiqar, S.; Khan, M.N.; Shakir, I.; Warsi, M.F.; Cochran, E.W. Electrochemical performance enhancement of MnO_2 nanowires through silver incorporation for next-generation supercapacitors. *Mater. Adv.* **2024**, *5*, 6170–6184. [CrossRef]
155. Gupta, M.K.; Kumar, Y.; Shukla, V.K. Hydrothermal Synthesis of a Layered ZnO@MnO_2 Nanocomposite for High-Performance Supercapacitor Electrodes. *J. Electron. Mater.* **2024**, *53*, 2050–2061. [CrossRef]
156. Mofokeng, T.P.; Shabalala, S.; Haruna, A.B.; Mwonga, P.V.; Tetana, Z.N.; Ozoemena, K.I. Scalable synthesis of K^+/Na^+ pre-intercalated α-MnO_2 via Taylor fluid flow-assisted hydrothermal reaction for high-performance asymmetric supercapacitors. *J. Electroanal. Chem.* **2023**, *948*, 117809. [CrossRef]
157. Zhu, Y.; Xu, H.; Tang, J.; Jiang, X.; Bao, Y. Synthesis of γ-MnO_2/PANI Composites for Supercapacitor Application in Acidic Electrolyte. *J. Electrochem. Soc.* **2021**, *168*, 030542. [CrossRef]
158. Pundir, S.; Upadhyay, S.; Priya, R.; Kumar, N.; Chetana, S.; Hossain, I.; Joshi, N.C.; Pandey, O.P. Synthesis of 1D β-MnO_2 for high-performance supercapacitor application. *J. Solid State Electrochem.* **2023**, *27*, 531–538. [CrossRef]
159. Devi, R.; Kumar, V.; Kumar, S.; Bulla, M.; Sharma, S.; Sharma, A. Electrochemical Analysis of MnO_2 (α, β, and γ)-Based Electrode for High-Performance Supercapacitor Application. *Appl. Sci.* **2023**, *13*, 7907. [CrossRef]
160. Sinan-Tatli, N.; Unur-Yilmaz, E. PANI-grafted radially porous MnO_2 for supercapacitor applications. *J. Solid State Electrochem.* **2024**, *28*, 2593–2603. [CrossRef]
161. Jadhav, S.A.; Dhas, S.D.; Patil, K.T.; Moholkar, A.V.; Patil, P.S. Polyaniline (PANI)-manganese dioxide (MnO_2) nanocomposites as efficient electrode materials for supercapacitors. *Chem. Phys. Lett.* **2021**, *778*, 138764. [CrossRef]
162. Xie, C.; Li, T.; Deng, C.; Song, Y.; Zhang, H.; Li, X. A highly reversible neutral zinc/manganese battery for stationary energy storage. *Energy Environ. Sci.* **2020**, *13*, 135–143. [CrossRef]
163. Chen, J.; Liang, J.; Zhou, Y.; Sha, Z.; Lim, S.; Huang, F.; Han, Z.; Brown, S.A.; Cao, L.; Wang, D.-W.; et al. A vertical graphene enhanced Zn–MnO_2 flexible battery towards wearable electronic devices. *J. Mater. Chem. A* **2021**, *9*, 575–584. [CrossRef]
164. Sambandam, B.; Mathew, V.; Kim, S.; Lee, S.; Kim, S.; Hwang, J.Y.; Fan, H.J.; Kim, J. An analysis of the electrochemical mechanism of manganese oxides in aqueous zinc batteries. *Chem* **2022**, *8*, 924–946. [CrossRef]
165. Wu, D.; King, S.T.; Sadique, N.; Ma, L.; Ehrlich, S.N.; Ghose, S.; Bai, J.; Zhong, H.; Yan, S.; Bock, D.C.; et al. Operando investigation of aqueous zinc manganese oxide batteries: Multi-stage reaction mechanism revealed. *J. Mater. Chem. A* **2023**, *11*, 16279–16292. [CrossRef]
166. Shen, X.; Wang, X.; Zhou, Y.; Shi, Y.; Zhao, L.; Jin, H.; Di, J.; Li, Q. Highly Reversible Aqueous Zn-MnO_2 Battery by Supplementing Mn^{2+}-Mediated MnO_2 Deposition and Dissolution. *Adv. Funct. Mater.* **2021**, *31*, 2101579. [CrossRef]
167. Liu, Z.; Yang, Y.; Lu, B.; Liang, S.; Fan, H.J.; Zhou, J. Insights into complexing effects in acetate-based Zn-MnO_2 batteries and performance enhancement by all-round strategies. *Energy Storage Mater.* **2022**, *52*, 104–110. [CrossRef]

168. Ma, K.; Yang, G.; Wang, C. Towards storable and durable Zn-MnO_2 batteries with hydrous tetraglyme electrolyte. *J. Energy Chem.* **2023**, *80*, 432–441. [CrossRef]
169. Xie, J.; Liu, G.; Sun, J.; Zheng, R.; Zhao, W.; Chu, T.; Lin, H.; Xu, Y.; Gao, S.; Sui, Z. α-MnO_2/CNTs with cross-linked reticular structure: Towards ultralong life zinc-ion batteries. *Diam. Relat. Mater.* **2022**, *125*, 109024. [CrossRef]
170. Cai, X.; Li, H.; Li, J.; Yan, H.; Liu, Y.; Yu, H.; Yan, L.; Zhang, L.; Shu, J. Hydrothermal synthesis of β-MnO_2 nanorods for highly efficient zinc-ion storage. *Ionics* **2021**, *27*, 3943–3950. [CrossRef]
171. Wang, Y.; Zhang, Y.; Gao, G.; Fan, Y.; Wang, R.; Feng, J.; Yang, L.; Meng, A.; Zhao, J.; Li, Z. Effectively Modulating Oxygen Vacancies in Flower-Like δ-MnO_2 Nanostructures for Large Capacity and High-Rate Zinc-Ion Storage. *Nano-Micro Lett.* **2023**, *15*, 219. [CrossRef] [PubMed]
172. Chao, D.; Ye, C.; Xie, F.; Zhou, W.; Zhang, Q.; Gu, Q.; Davey, K.; Gu, L.; Qiao, S.-Z. Atomic Engineering Catalyzed MnO_2 Electrolysis Kinetics for a Hybrid Aqueous Battery with High Power and Energy Density. *Adv. Mater.* **2020**, *32*, 2001894. [CrossRef]
173. Panda, M.R.; El Meragawi, S.; Mirshekarloo, M.S.; Chen, W.; Shaibani, M.; Majumder, M. Acidity-Aided Surface Modification Strategy to Enhance In Situ MnO_2 Deposition for High Performance Zn-MnO_2 Battery Prototypes. *Small* **2024**, 2311933. [CrossRef]
174. Lv, W.; Shen, Z.; Li, X.; Meng, J.; Yang, W.; Ding, F.; Ju, X.; Ye, F.; Li, Y.; Lyu, X.; et al. Discovering Cathodic Biocompatibility for Aqueous Zn–MnO_2 Battery: An Integrating Biomass Carbon Strategy. *Nano-Micro Lett.* **2024**, *16*, 109. [CrossRef] [PubMed]
175. Shi, Y.-Y.; Liao, S.-Y.; Wang, Q.-F.; Xu, X.-Y.; Wang, X.-Y.; Gu, X.-Y.; Hu, Y.-G.; Zhu, P.-L.; Sun, R.; Wan, Y.-J. Enhancing the Interaction of Carbon Nanotubes by Metal–Organic Decomposition with Improved Mechanical Strength and Ultra-Broadband EMI Shielding Performance. *Nano-Micro Lett.* **2024**, *16*, 134. [CrossRef] [PubMed]
176. Ouda, E.; Yousf, N.; Magar, H.S.; Hassan, R.Y.A.; Duraia, E.-S.M. Electrochemical properties of MnO_2-based carbon nanomaterials for energy storage and electrochemical sensing. *J. Mater. Sci. Mater. Electron.* **2023**, *34*, 731. [CrossRef]
177. Tian, W.; Ren, P.; Hou, X.; Xue, R.; Chen, Z.; Guo, Z.; Jin, Y.; Ren, F. MnO_2 porous carbon composite from cellulose enabling high gravimetric/volumetric performance for supercapacitor. *Int. J. Biol. Macromol.* **2024**, *261*, 129977. [CrossRef] [PubMed]
178. Jereil, S.D.; Vijayalakshmi, K.; Monamary, A. Substantial effect of Pd incorporation in MnO_2 synthesized by spray pyrolysis on MWCNTs/Ta electrode for better H_2O_2 sensitivity. *Ceram. Int.* **2019**, *45*, 3782–3790. [CrossRef]
179. Xu, N.; Nie, Q.; Luo, L.; Yao, C.; Gong, Q.; Liu, Y.; Zhou, X.-D.; Qiao, J. Controllable Hortensia-like MnO_2 Synergized with Carbon Nanotubes as an Efficient Electrocatalyst for Long-Term Metal–Air Batteries. *ACS Appl. Mater. Interfaces* **2019**, *11*, 578–587. [CrossRef]
180. Li, Z.; Xiao, D.; Xu, C.; Li, Z.; Bi, S.; Xu, H.; Dou, H.; Zhang, X. MnO_2/carbon nanotube free-standing electrode recycled from spent manganese-oxygen battery as high-performance supercapacitor material. *J. Mater. Sci.* **2022**, *57*, 8818–8827. [CrossRef]
181. Rosaiah, P.; Divya, P.; Sambasivam, S.; Tighezza, A.M.; Kalaivani, V.; Muthukrishnaraj, A.; Ayyar, M.; Niyitanga, T.; Kim, H. Carbon based manganese oxide (MnO_2, MnO_2/MWCNT and MnO_2/rGO) composite electrodes for high-stability Li-ion batteries. *Carbon Lett.* **2024**, *34*, 215–225. [CrossRef]
182. Zhou, X.; Chen, S.; Zhang, Y.; Yu, B.; Chen, Y.; Liu, Y.; Li, S.; Liu, L.; Jin, H.; Deng, J.; et al. Three-Dimensional Conductive Interface and Tip Structure of MnO_2 Electrode Facilitate Superior Zinc Ion Batteries. *Small Struct* **2024**, 2400057. [CrossRef]

MDPI AG
Grosspeteranlage 5
4052 Basel
Switzerland
Tel.: +41 61 683 77 34

Nanomaterials Editorial Office
E-mail: nanomaterials@mdpi.com
www.mdpi.com/journal/nanomaterials

www.ingramcontent.com/pod-product-compliance
Lightning Source LLC
Chambersburg PA
CBHW061548210926
51850CB00009B/2041
* 9 7 8 3 7 2 5 8 2 2 1 9 5 *